McGraw-Hill Dictionary of Mathematics

McGraw-Hill

Dictionary of Mathematics

Sybil P. Parker
Editor in Chief

McGraw-Hill

New York San Francisco Washington, D.C. Auckland
Bogotá Caracas Lisbon London Madrid Mexico City
Milan Montreal New Delhi San Juan Singapore
Sydney Tokyo Toronto

Library of Congress Cataloging in Publication Data

McGraw-Hill dictionary of mathematics / Sybil P. Parker, editor in chief.
p. cm.
"All text in this dictionary was published previously in the McGraw-Hill dictionary of scientific and technical terms, fifth edition"—T.p. verso.
ISBN 0-07-052433-5 (alk. paper)
1. Mathematics—Dictionaries. I. Parker, Sybil P. II. McGraw-Hill dictionary of scientific and technical terms.
QA5.M425 1997
510'.3—dc21 97-5933

McGraw-Hill 
A Division of The ***McGraw-Hill*** *Companies*

1 2 3 4 5 6 7 8 9 0 DOC/DOC 9 0 2 1 0 9 8 7

ISBN 0-07-052433-5

This book is printed on recycled, acid-free paper containing a minimum of 50% recycled, de-inked fiber.

This book was set in Helvetica Bold *and* Century ITC Book *by* Progressive Information Technologies, Emigsville, Pennsylvania. *It was printed and bound by* R. R. Donnelley & Sons Company, The Lakeside Press.

McGraw-Hill *books are available at special quantity discounts to use as premiums and sales promotions, or for use in corporate training programs. For more information, please write to the* Director of Special Sales, McGraw-Hill, 11 West 19th Street, New York, NY 10011. *Or contact your local bookstore.*

Preface

The *McGraw-Hill Dictionary of Mathematics* concentrates on the vocabulary of mathematics and related fields. With 4000 terms, it serves as a major compendium of the specialized language that is essential to understanding mathematics and its many distinct branches. Students, teachers, librarians, writers, scientists, engineers, researchers, and the general public will appreciate the convenience of a single comprehensive reference.

The Dictionary covers all branches of mathematics, including those taught in high school, college, and university, such as algebra, geometry, analytic geometry, trigonometry, calculus, and vector analysis, as well as more advanced areas of mathematics, such as group theory and topology. Mathematics is the deductive study of shape, quantity, and dependence; the two main areas are applied mathematics and pure mathematics, the former arising from the study of physical phenomena, the latter involving the intrinsic study of mathematical structures.

The terms selected for this Dictionary are fundamental to understanding mathematics and related fields. All definitions were drawn from the *McGraw-Hill Dictionary of Scientific and Technical Terms* (5th ed., 1994). Along with definitions and pronunciations, defining terms include synonyms, acronyms, and abbreviations where appropriate. Such synonyms, acronyms, and abbreviations also appear in the alphabetical sequence as cross references to the defining terms.

The *McGraw-Hill Dictionary of Mathematics* is a reference that the editors hope will facilitate the communication of ideas and information, and thus serve the needs of readers with either professional or pedagogical interests in this field.

Sybil P. Parker
EDITOR IN CHIEF

Editorial Staff

Sybil P. Parker, Editor in Chief

Arthur Biderman, Senior editor
Jonathan Weil, Editor
Betty Richman, Editor
Patricia W. Albers, Editorial administrator

Dr. Henry F. Beechhold, Pronunciation Editor
Professor of English
Chairman, Linguistics Program
College of New Jersey
Trenton, New Jersey

Joe Faulk, Editing manager
Frank Kotowski, Jr., Senior editing supervisor
Ruth W. Mannino, Senior editing supervisor

Suzanne W. B. Rapcavage, Senior production supervisor

How to Use the Dictionary

ALPHABETIZATION. The terms in the *McGraw-Hill Dictionary of Mathematics* are alphabetized on a letter-by-letter basis; word spacing, hyphen, comma, solidus, and apostrophe in a term are ignored in the sequencing. For example, an ordering of terms would be:

Abelian group	**binary system**
Abel's problem	**binary-to-decimal conversion**
Abel theorem	**binomial**

FORMAT. The basic format for a defining entry provides the term in boldface and the single definition in lightface:

term Definition.

A term may be followed by multiple definitions, each introduced by a boldface number:

term **1.** Definition. **2.** Definition. **3.** Definition.

A simple cross-reference entry appears as:

term *See* another term.

A cross reference may also appear in combination with definitions:

term **1.** Definition. **2.** *See* another term.

CROSS REFERENCING. A cross-reference entry directs the user to the defining entry. For example, the user looking up "abac" finds:

abac *See* nomograph.

The user then turns to the "N" terms for the definition. Cross references are also made from variant spellings, acronyms, abbreviations, and symbols.

AD *See* average deviation.
cot *See* cotangent.
geodetic triangle *See* spheroidal triangle.

ALSO KNOWN AS . . ., etc. A definition may conclude with a mention of a synonym of the term, a variant spelling, an abbreviation for the term, or other such information, introduced by "Also known as . . .," "Also spelled . . .," "Abbreviated . . .," "Symbolized . . .," "Derived from . . .." When a term has more than one definition, the positioning of any of these phrases conveys the extent of applicability. For example:

term **1.** Definition. Also known as synonym. **2.** Definition. Symbolized T.

In the above arrangement, "Also known as . . ." applies only to the first definition; "Symbolized . . ." applies only to the second definition.

term Also known as synonym. **1.** Definition. **2.** Definition.

In the above arrangement, "Also known as . . ." applies to both definitions.

Pronunciation Key

Vowels

a as in b**a**t, th**a**t
ā as in b**ai**t, cr**a**te
ä as in b**o**ther, f**a**ther
e as in b**e**t, n**e**t
ē as in b**ee**t, tr**ea**t
i as in b**i**t, sk**i**t
ī as in b**i**te, l**igh**t
ō as in b**oa**t, n**o**te
ȯ as in b**ough**t, t**au**t
u̇ as in b**oo**k, p**u**ll
ü as in b**oo**t, p**oo**l
ə as in b**u**t, sof**a**
au̇ as in cr**ow**d, p**ow**er
ȯi as in b**oi**l, sp**oi**l
yə as in form**u**la, spectac**u**lar
yü as in f**ue**l, m**u**le

Semivowels/Semiconsonants

w as in **w**ind, t**w**in
y as in **y**et, on**ion**

Stress (Accent)

ˈ precedes syllable with primary stress

ˌ precedes syllable with secondary stress

¦ precedes syllable with variable or indeterminate primary/secondary stress

Consonants

b as in **b**i**b**, dri**bb**le
ch as in **ch**arge, stre**tch**
d as in **d**og, ba**d**
f as in **f**ix, sa**f**e
g as in **g**ood, si**g**nal
h as in **h**and, be**h**ind
j as in **j**oint, di**g**it
k as in **c**ast, bri**ck**
k̲ as in Ba**ch** (used rarely)
l as in **l**oud, be**ll**
m as in **m**ild, su**mm**er
n as in **n**ew, de**n**t
n̲ indicates nasalization of preceding vowel
ŋ as in ri**ng**, si**ng**le
p as in **p**ier, sli**p**
r as in **r**ed, sca**r**
s as in **s**ign, po**s**t
sh as in **su**gar, **sh**oe
t as in **t**imid, ca**t**
th as in **th**in, brea**th**
t̲h̲ as in **th**en, brea**the**
v as in **v**eil, wea**v**e
z as in **z**oo, crui**se**
zh as in bei**ge**, trea**s**ure

Syllabication

· Indicates syllable boundary when following syllable is unstressed

Contents

A

abac *See* nomograph. { ə'bak }

abacus An instrument for performing arithmetical calculations manually by sliding markers on rods or in grooves. { 'ab·ə‚kəs }

Abelian domain *See* Abelian field. { ə'bēl·yən dō'mān }

Abelian extension A Galois extension whose Galois group is Abelian. { ə'bēl·yən ik'sten·chən }

Abelian field A set of elements a, b, c, . . . forming Abelian groups with addition and multiplication as group operations where $a(b + c) = ab + ac$. Also known as Abelian domain; domain. { ə'bēl·yən 'fēld }

Abelian group A group whose binary operation is commutative; that is, $ab = ba$ for each a and b in the group. Also known as commutative group. { ə'bēl·yən 'grüp }

Abelian operation *See* commutative operation. { ə'bēl·yən ‚äp·ə'rā·shən }

Abelian ring *See* commutative ring. { ə'bēl·yən 'riŋ }

Abelian theorems A class of theorems which assert that if a sequence or function behaves regularly, then some average of the sequence or function behaves regularly; examples include the Abel theorem (second definition) and the statement that if a sequence converges to s, then its Cesaro summation exists and is equal to s. { ə'bēl·yən 'thir·əmz }

Abel's inequality An inequality which states that the absolute value of the sum of n terms, each in the form ab, where the bs are positive numbers, is not greater than the product of the largest b with the largest absolute value of a partial sum of the as. { 'ä·bəlz ‚in·ē'kwäl·i·dē }

Abel's integral equation The equation

$$f(x) = \int_a^x u(z)(x - z)^{-a} dz \quad (0 < a < 1, x \geq a)$$

where $f(x)$ is a known function and $u(z)$ is the function to be determined; when $a = 1/2$, this equation has application to Abel's problem. { 'ä·bəlz 'in·tə·grəl i'kwā·zhən }

Abel's problem The problem which asks what path a particle will follow if it moves under the influence of gravity alone and its altitude-time function is to follow a specific law. { 'ä·bəlz 'präb·ləm }

Abel's summation method A method of attributing a sum to an infinite series whose nth term is a_n by taking the limit on the left at $x = 1$ of the sum of the series whose nth term is $a_n x^n$ { 'ä·bəlz sə'mā·shən ˌmeth·əd }

Abel theorem **1.** A theorem stating that if a power series in z converges for $z = a$, it converges absolutely for $z < a$. **2.** A theorem stating that if a power series in z converges to $f(z)$ for $z < 1$ and to a for $z = 1$, then the limit of $f(z)$ as z approaches 1 equals a. **3.** A theorem stating that if the three series with nth term a_n, b_n, and $c_n = a_0 b_n + a_1 b_{n-1} + \cdots + a_n b_0$, respectively, converge, then the third series equals the product of the first two series. { 'ä·bəl 'thir·əm }

abscissa One of the coordinates of a two-dimensional coordinate system, usually the horizontal coordinate, denoted by x. { ab'sis·ə }

absolute convergence That property of an infinite series (or infinite product) of real or complex numbers if the series (product) of absolute values converges; absolute convergence implies convergence. { 'ab·səˌlüt kən'vərj·əns }

absolute coordinates Coordinates given with reference to a fixed point of origin. { 'ab·səˌlüt kō'ȯrd·ən·əts }

absolute error In an approximate number, the numerical difference between the number and a number considered exact. { 'ab·səˌlüt 'er·ər }

absolutely continuous function A function defined on a closed interval with the property that for any positive number ϵ there is another positive number η such that, for any finite set of nonoverlapping intervals, (a_1,b_1), (a_2,b_2), . . . , (a_n,b_n), whose lengths have a sum less than η, the sum over the intervals of the absolute values of the differences in the values of the function at the ends of the intervals is less than ϵ. { ¦ab·səˌlüt·lē kən¦tin·yə·wəs 'fəŋk·shən }

absolute inequality *See* unconditional inequality. { 'ab·səˌlüt ˌin·ē'kwäl·ə·dē }

absolute magnitude The absolute value of a number or quantity. { 'ab·səˌlüt 'mag·nə·tüd }

absolute number A number represented by numerals rather than by letters. { 'ab·səˌlüt 'nəm·bər }

absolute term *See* constant term. { 'ab·səˌlüt 'tərm }

absolute value Also known as magnitude. **1.** For a real number, the number if it is nonnegative, and the negative of the number if it is negative. Also known as numerical value. **2.** For a complex number, the square root of the sum of the squares of its real and imaginary parts. Also known as modulus. **3.** The length of a vector, disregarding its direction; the square root of the sum of the squares of its orthogonal components. { 'ab·səˌlüt 'val·yü }

absorbing state A special case of recurrent state in a Markov process in which the transition probability, P_{ii}, equals 1; a process will never leave an absorbing state once it enters. { əb'sȯrb·iŋ ˌstāt }

absorptive laws Either of two laws satisfied by the operations, usually denoted $\cup$ and $\cap$, on a Boolean algebra, namely $a \cup (a \cap b) = a$ and $a \cap (a \cup b) = a$, where a and b are any two elements of the algebra; if the elements of the algebra are sets, then $\cup$ and $\cap$ represent union and intersection of sets. { əb'sȯrp·tiv ˌlȯz }

abstract algebra The study of mathematical systems consisting of a set of elements, one or more binary operations by which two elements may be combined to yield a third, and several rules (axioms) for the interaction of the elements and the operations; includes group theory, ring theory, and number theory. { 'abz·trakt 'al·jə·brə }

abundant number A positive integer that is greater than the sum of all its divisors, including unity. { ə'bən·dənt 'nəm·bər }

accessibility condition The condition that any state of a finite Markov chain can be reached from any other state. { ak׳ses·ə'bil·əd·ē kən׳dish·ən }

accretive operator A linear operator T defined on a subspace D of a Hilbert space which satisfies the following condition: the real part of the inner product of Tu with u is nonnegative for all u belonging to D. { ə¦krēd·iv 'äp·ə׳rād·ər }

accumulation factor The quantity $(1 + r)$ in the formula for compound interest, where r is the rate of interest; measures the rate at which the principal grows. { ə·kyü·myə'lā·shən 'fak·tər }

accumulation point *See* cluster point. { ə·kyü·myə'lā·shən ׳point }

acnode *See* isolated point. { 'ak·nōd }

acute angle An angle of less than 90°. { ə'kyüt 'aŋ·gəl }

acute triangle A triangle each of whose angles is less than 90°. { ə'kyüt 'trī׳aŋ·gəl }

acyclic **1.** A transformation on a set to itself for which no nonzero power leaves an element fixed. **2.** A chain complex all of whose homology groups are trivial. { ā'sik·lik }

AD *See* average deviation.

Adams-Bashforth process A method of numerically integrating a differential equation of the form $(dy/dx) = f(x,y)$ that uses one of Gregory's interpolation formulas to expand f. { 'a·dəmz 'bash׳fórth ׳prä·səs }

adaptive integration A numerical technique for obtaining the definite integral of a function whose smoothness, or lack thereof, is unknown, to a desired degree of accuracy, while doing only as much work as necessary on each subinterval of the interval in question. { ə'dap·tiv ׳int·ə'grā·shən }

add To perform addition. { ad }

addend One of a collection of numbers to be added. { 'a׳dend }

addition **1.** An operation by which two elements of a set are combined to yield a third; denoted +; usually reserved for the operation in an Abelian group or the group operation in a ring or vector space. **2.** The combining of complex quantities in which the individual real parts and the individual imaginary parts are separately added. **3.** The combining of vectors in a prescribed way; for example, by algebraically adding corresponding components of vectors or by forming the third side of the triangle whose other sides each represent a vector. Also known as composition. { ə'di·shən }

addition sign The symbol +, used to indicate addition. Also known as plus sign. { ə'di·shən ׳sīn }

additive Pertaining to addition. { 'ad·əd·iv }

additive function Any function f that preserves addition; that is, $f(x + y) = f(x) + f(y)$. { 'ad·əd·iv 'fəŋ·shən }

additive identity In a mathematical system with an operation of addition denoted +, an element 0 such that $0 + e = e + 0 = e$ for any element e in the system. { 'ad·ə·div ī'den·ə·dē }

additive inverse In a mathematical system with an operation of addition denoted +, an additive inverse of an element e is an element $-e$ such that $e + (-e) = (-e) + e = 0$, where 0 is the additive identity. { 'ad·ə·div 'in,vərs }

additive set function A set function with the property that the value of the function at a finite union of disjoint sets is equal to the sum of the values at each set in the union. { ¦ad·əd·iv ¦set ,fəŋk·shən }

adjacency structure A listing, for each vertex of a graph, of all the other vertices adjacent to it. { ə'jās·ən·sē ,strək·chər }

adjacent angle One of a pair of angles with a common side formed by two intersecting straight lines. { ə'jās·ənt 'aŋ·gəl }

adjacent side For a given vertex of a polygon, one of the sides of the polygon that terminates at the vertex. { ə'jās·ənt 'sīd }

adjoined number A number z that is added to a number field F to form a new field consisting of all numbers that can be derived from z and the numbers in F by the operations of addition, subtraction, multiplication, and division. { ə¦jȯind 'nəm·bər }

adjoint of a matrix *See* adjugate; Hermitian conjugate. { 'aj,ȯint əv ə 'mā·triks }

adjoint operator An operator B such that the inner products (Ax,y) and (x,By) are equal for a given operator A and for all elements x and y of a Hilbert space. Also known as associate operator; Hermitian conjugate operator. { 'aj,ȯint 'äp·ə,rād·ər }

adjoint vector space The complete normed vector space constituted by a class of bounded, linear, homogeneous scalar functions defined on a normed vector space. { 'aj,ȯint 'vek·tər ,spās }

adjugate For a matrix A, the matrix obtained by replacing each element of A with the cofactor of the transposed element. Also known as adjoint of a matrix. { 'aj·ə,gāt }

affine connection A structure on an n-dimensional space that, for any pair of neighboring points P and Q, specifies a rule whereby a definite vector at Q is associated with each vector at P; the two vectors are said to be parallel. { ə'fīn kə'nek·shən }

affine geometry The study of geometry using the methods of linear algebra. { ə'fīn jē'äm·ə·trē }

affine Hjelmslev plane A generalization of an affine plane in which more than one line may pass through two distinct points. Also known as Hjelmslev plane. { ə¦fīn 'hyelm,slev ,plān }

affine plane In projective geometry, a plane in which (1) every two points lie on exactly one line, (2) if p and L are a given point and line such that p is not on L, then there exists exactly one line that passes through p and does not intersect L, and (3) there exist three noncollinear points. { ə'fīn ˌplān }

affine space An n-dimensional vector space which has an affine connection defined on it. { ə'fīn ˌspās }

affine transformation A function on a linear space to itself, which is the sum of a linear transformation and a fixed vector. { ə'fīn ˌtranz·fər'mā·shən }

Airy differential equation The differential equation $(d^2f/dz^2) - zf = 0$, where z is the independent variable and f is the value of the function; used in studying the diffraction of light near caustic surface. { ¦er·ē ˌdif·ə¦ren·chəl i'kwā·zhən }

Airy function Either of the solutions of the Airy differential equation. { ¦er·ē ¦fəŋk·shən }

aleph null The cardinal number of any set which can be put in one-to-one correspondence with the set of positive integers. { ¦äˌlef ¦nəl }

Alexandroff compactification *See* one-point compactification. { al·ik¦sanˌdrȯf kəm ˌpak·tə·fə'kā·shən }

algebra **1.** A method of solving practical problems by using symbols, usually letters, for unknown quantities. **2.** The study of the formal manipulations of equations involving symbols and numbers. **3.** An abstract mathematical system consisting of a vector space together with a multiplication by which two vectors may be combined to yield a third, and some axioms relating this multiplication to vector addition and scalar multiplication. Also known as hypercomplex system. { 'al·jə·brə }

algebraic addition The addition of algebraic quantities in the sense that adding a negative quantity is the same as subtracting a positive one. { ¦al·jə¦brā·ik ə'dish·ən }

algebraically closed field **1.** A field F such that every polynomial of degree equal to or greater than 1 with coefficients in F has a root in F. **2.** A field F is said to be algebraically closed in an extension field K if any root in K of a polynominal with coefficients in F also lies in F. { ¦al·jə¦brā·ik·lē ¦klōzd 'fēld }

algebraically independent A subset S of a commutative ring B is said to be algebraically independent over a subring A of B (or the elements of S are said to be algebraically independent over A) if, whenever a polynominal in elements of S, with coefficients in A, is equal to 0, then all the coefficients in the polynomial equal 0. { ¦al·jə¦brā·ik·lē ˌin·də'pen·dənt }

algebraic closure of a field An algebraic extension field which has no algebraic extensions but itself. { ¦al·jə¦brā·ik 'klō·zhər əv ə 'fēld }

algebraic curve **1.** The set of points in the plane satisfying a polynomial equation in two variables. **2.** More generally, the set of points in n-space satisfying a polynomial equation in n variables. { ¦al·jə¦brā·ik 'kərv }

algebraic equation An equation in which zero is set equal to an algebraic expression. { ¦al·jə¦brā·ik i'kwā·zhən }

algebraic expression An expression which is obtained by performing a finite number of the following operations on symbols representing numbers: addition, subtraction, multiplication, division, raising to a power. { ¦al·jə¦brā·ik ik'spresh·ən }

algebraic extension of a field A field which contains both the given field and all roots of polynomials with coefficients in the given field. { ¦al·jə¦brā·ik ik'sten·shən əv ə 'fēld }

algebraic function A function whose value is obtained by performing only the following operations to its argument: addition, subtraction, multiplication, division, raising to a rational power. { ¦al·jə¦brā·ik 'fəŋk·shən }

algebraic geometry The study of geometric properties of figures using methods of abstract algebra. { ¦al·jə¦brā·ik jē'äm·ə·trē }

algebraic identity A relation which holds true for all possible values of the literal symbols occurring in it; for example, $(x + y)(x - y) = x^2 - y^2$. { ¦al·jə¦brā·ik i'den·ə·tē }

algebraic integer The root of a polynomial whose coefficients are integers and whose leading coefficient is equal to 1. { ¦al·jə¦brā·ik 'in·tə·jər }

algebraic invariant A polynomial in coefficients of a quadratic or higher form in a collection of variables whose value is unchanged by a specified class of linear transformations of the variables. { ¦al·jə¦brā·ik in'ver·ē·ənt }

algebraic K theory The study of the mathematical structure resulting from associating with each ring A the group $K(A)$, the Grothendieck group of A. { ¦al·jə¦brā·ik 'kā ˌthē·ə·rē }

algebraic language The conventional method of writing the symbols, parentheses, and other signs of formulas and mathematical expressions. { ¦al·jə¦brā·ik 'laŋ·gwij }

algebraic number Any root of a polynomial with rational coefficients. { ¦al·jə¦brā·ik 'nəm·bər }

algebraic number field A finite extension field of the field of rational numbers. { ¦al·jə¦brā·ik 'nəm·bər ˌfēld }

algebraic number theory The study of properties of real numbers, especially integers, using the methods of abstract algebra. { ¦al·jə¦brā·ik 'nəm·bər ˌthē·ə·rē }

algebraic object Either an algebraic structure, such as a group, ring, or field, or an element of such an algebraic structure. { ¦al·jə¦brā·ik 'äbˌjekt }

algebraic operation Any of the operations of addition, subtraction, multiplication, division, raising to a power, or extraction of roots. { ¦al·jə·¦brā·ik ˌäp·ə'rā·shən }

algebraic set A set made up of all zeros of some specified set of polynomials in n variables with coefficients in a specified field F, in a specified extension field of F. { ¦al·jə¦brā·ik 'set }

algebraic sum **1.** The result of the addition of two or more quantities, with the addition of a negative quantity equivalent to subtraction of the corresponding positive quantity. **2.** For two fuzzy sets A and B, with membership functions m_A and m_B, that fuzzy set whose membership function m_{A+B} satisfies the equation $m_{A+B}(x) = m_A(x) + m_B(x) - [m_A(x) \cdot m_B(x)]$ for every element x. { ¦al·jə¦brā·ik 'səm }

algebraic surface A subset S of a complex n-space which consists of the set of complex solutions of a system of polynomial equations in n variables such that S is a complex two-manifold in the neighborhood of most of its points. { ¦al·jə¦brā·ik 'sər·fəs }

algebraic symbol A letter that represents a number or a symbol indicating an algebraic operation. { ¦al·jə¦brā·ik 'sim·bəl }

algebraic term In an expression, a term that contains only numbers and algebraic symbols. { ¦al·jə¦brā·ik 'tərm }

algebraic topology The study of topological properties of figures using the methods of abstract algebra; includes homotopy theory, homology theory, and cohomology theory. { ¦al·jə¦brā·ik tə'päl·ə·jē }

algebra of subsets An algebra of subsets of a set S is a family of subsets of S that contains the null set, the complement (relative to S) of each of its members, and the union of any two of its members. { ¦al·jə·brə əv 'səb,sets }

algebra with identity An algebra which has an element, not equal to 0 and denoted by 1, such that, for any element x in the algebra, $x1 = 1x = x$. { ¦al·jə·brə with i'den·ə·tē }

algorithm A set of well-defined rules for the solution of a problem in a finite number of steps. { 'al·gə,rith·əm }

aliasing Introduction of error into the computed amplitudes of the lower frequencies in a Fourier analysis of a function carried out using discrete time samplings whose interval does not allow the proper analysis of the higher frequencies present in the analyzed function. { 'āl·yəs·iŋ }

alignment chart *See* nomograph. { ə'līn·mənt ,chärt }

aliquant A divisor that does not divide a quantity into equal parts. { 'al·ə,kwänt }

aliquot A divisor that divides a quantity into equal parts with no remainder. { 'al·ə,kwät }

allometry A relation between two variables x and y that can be written in the form $y = ax^n$, where a and n are constants. { ə'läm·ə·trē }

almost every A proposition concerning the points of a measure space is said to be true at almost every point, or to be true almost everywhere, if it is true for every point in the space, with the exception at most of a set of points which form a measurable set of measure zero. { ¦ȯl,mōst 'ev·rē }

almost-perfect number An integer that is 1 greater than the sum of all its factors other than itself. { ¦ȯl,mōst ¦pər·fik 'nəm·bər }

almost-periodic function A continuous function $f(x)$ such that for any positive number ϵ there is a number M so that for any real number x, any interval of length M contains a nonzero number t such that $f(x + t) - f(x) < \epsilon$. { 'ȯl,mōst ,pir·ē'äd·ik 'fəŋk·shən }

alpha rule *See* renaming rule. { 'al·fə ,rül }

alt *See* altitude.

alternate angles A pair of nonadjacent angles that a transversal forms with each of two lines; they lie on opposite sides of the transversal, and are both interior, or both exterior, to the two lines. { 'ȯl·tər·nət 'aŋ·gəlz }

alternating form A bilinear form f which changes sign under interchange of its independent variables; that is, $f(x,y) = -f(y,x)$ for all values of the independent variables x and y. { 'ȯl·tər,nād·iŋ 'fȯrm }

alternating function A function in which the interchange of two independent variables causes the dependent variable to change sign. { 'ȯl·tər·nād·iŋ 'fəŋk·shən }

alternating group A group made up of all the even permutations of n objects. { 'ȯl·tər·nād·iŋ 'grüp }

alternating series Any series of real numbers in which consecutive terms have opposite signs. { 'ȯl·tər·nād·iŋ 'sir·ēz }

altitude The perpendicular distance from the base to the top (a vertex or parallel line) of a geometric figure such as a triangle or parallelogram. Abbreviated alt. { 'al·tə,tüd }

alysoid *See* catenary. { 'al·ə,sȯid }

ambiguous case **1.** For the solution of a plane triangle, the case in which two sides and the angle opposite one of them is given, and there are two distinct solutions. **2.** For the solution of a spherical triangle, the case in which two sides and the angle opposite one of them is given, or two angles and the side opposite one of them is given, and there are two distinct solutions. { am¦big·yə·wəs 'kās }

amicable numbers Two numbers such that the exact divisors of each number (except the number itself) add up to the other number. { 'am·ə·kə·bəl 'nəm·bərz }

amplitude The angle between a vector representing a specified complex number on an Argand diagram and the positive real axis. Also known as argument. { 'am·plə,tüd }

anallagmatic curve A curve that is its own inverse curve with respect to some circle. { ə¦nal·ig¦mad·ik 'kərv }

analysis The branch of mathematics most explicitly concerned with the limit process or the concept of convergence; includes the theories of differentiation, integration and measure, infinite series, and analytic functions. Also known as mathematical analysis. { ə'nal·ə·səs }

analytic continuation The process of extending an analytic function to a domain larger than the one on which it was originally defined. { ,an·əl'id·ik kən·tin·yü'ā·shən }

analytic curve A curve whose parametric equations are real analytic functions of the same real variable. { ,an·əl'id·ik 'kərv }

analytic function A function which can be represented by a convergent Taylor series. Also known as holomorphic function. { ,an·əl'id·ik 'fuŋk·shən }

analytic geometry The study of geometric figures and curves using a coordinate system and the methods of algebra. Also known as cartesian geometry. { ,an·əl 'id·ik jē'äm·ə·trē }

analytic hierarchy A systematic procedure for representing the elements of any problem which breaks down the problem into its smaller constituents and then calls for only simple pairwise comparison judgments to develop priorities at each level. { ,an·əl'id·ik 'hī·ər,är·kē }

analytic number theory The study of problems concerning the discrete domain of integers by means of the mathematics of continuity. { ˌan·əl′id·ik ′nəm·bər ˌthē·ə·rē }

analytic structure A covering of a locally Euclidean topological space by open sets, each of which is homeomorphic to an open set in Euclidean space, such that the coordinate transformation (in both directions) between the overlap of any two of these sets is given by analytic functions. { an·əl′id·ik ′strək·chər }

analytic trigonometry The study of the properties and relations of the trigonometric functions. { ˌan·əl′id·ik ˌtrig·ə′näm·ə·trē }

AND function An operation in logical algebra on statements P, Q, R, such that the operation is true if all the statements P, Q, R, ... are true, and the operation is false if at least one statement is false. { ′and ˌfuŋk·shən }

angle The geometric figure, arithmetic quantity, or algebraic signed quantity determined by two rays emanating from a common point or by two planes emanating from a common line. { ′aŋ·gəl }

angle bisection The division of an angle by a line or plane into two equal angles. { ′aŋ·gəl bī′sek·shən }

angle of contingence For two points on a plane curve, the angle between the tangents to the curve at those points. { ′aŋ·gəl əv kən′tin·jəns }

angle of geodesic contingence For two points on a curve on a surface, the angle of intersection of the geodesics tangent to the curve at those points. { ′aŋ·gəl əv ˌjē·ə¦des·ik kən′tin·jəns }

angular distance **1.** For two points, the angle between the lines from a point of observation to the points. **2.** The angular difference between two directions, numerically equal to the angle between two lines extending in the given directions. **3.** The arc of the great circle joining two points, expressed in angular units. { ′an·gyə·lər ′dis·təns }

angular radius For a circle drawn on a sphere, the smaller of the angular distances from one of the two poles of the circle to any point on the circle. { ′aŋ·gyə·lər ′rād·ē·əs }

annihilator For a set S, the class of all functions of specified type whose value is zero at each point of S. { ə′nī·əˌlād·ər }

annular solid A solid generated by rotating a closed plane curve about a line which lies in the plane of the curve and does not intersect the curve. { ′an·yə·lər ′säl·əd }

annulus The ringlike figure that lies between two concentric circles. { ′an·yə·ləs }

annulus conjecture For dimension n, the assertion that if f and g are locally flat embeddings of the $(n - 1)$ sphere, S^{n-1}, in real n space, R^n, with $f(S^{n-1})$ in the bounded component of $R^n - g(S^{n-1})$, then the closed region in R^n bounded by $f(S^{n-1})$ and $g(S^{n-1})$ is homeomorphic to the direct product of S^{n-1} and the closed interval [0,1]; it is established for $n \neq 4$. { ′an·yə·ləs kən′jek·chər }

antecedent **1.** The numerator of a ratio. **2.** The first of the two statements in an implication. { ′an·təˌsēd·ənt }

anticlastic Having the property of a surface or portion of a surface whose two principal curvatures at each point have opposite signs, so that one normal section is concave and the other convex. { ¦an·tē¦klas·tik }

anticommutator The anticommutator of two operators, A and B, is the operator $AB + BA$. { ˌan·tē'käm·yəˌtād·ər }

anticommute Two operators anticommute if their anticommutator is equal to zero. { ˌan·tē·kə'myüt }

antiderivative *See* indefinite integral. { ¦an·tē·di¦riv·əd·iv }

antilog *See* antilogarithm of a number. { 'an·tiˌläg }

antilogarithm For a number x, a second number whose logarithm equals x. Abbreviated antilog. Also known as inverse logarithm. { ¦an·ti'läg·əˌrith·əm }

antipodal points The points at opposite ends of a diameter of a sphere. { an¦tip·əd·əl 'pȯins }

antisymmetric dyadic A dyadic equal to the negative of its conjugate. { ¦an·tē·si¦me·trik dī'ad·ik }

antisymmetric matrix A matrix which is equal to the negative of its transpose. Also known as skew-symmetric matrix. { ¦an·tē·si¦me·trik 'mā·triks }

antisymmetric relation A relation, which may be denoted $\geq$, among the elements of a set such that if $a \geq b$ and $b \geq a$ then $a = b$. { ˌant·i·si¦me·trik ri'lā·shən }

antisymmetric tensor A tensor in which interchanging two indices of an element changes the sign of the element. { ¦an·tē·si¦me·trik 'ten·sər }

apex **1.** The vertex of a triangle opposite the side which is regarded as the base. **2.** The vertex of a cone or pyramid. { 'āˌpeks }

Apollonius' problem The problem of constructing a circle that is tangent to three given circles. { ˌap·ə¦lōn·ē·əs ¦präb·ləm }

apothem The perpendicular distance from the center of a regular polygon to one of its sides. Also known as short radius. { 'ap·əˌthem }

applicable surfaces Surfaces such that there is a length-preserving map of one onto the other. { ¦ap·lə·kə·bəl 'sər·fəs·əz }

approximate **1.** To obtain a result that is not exact but is near enough to the correct result for some specified purpose. **2.** To obtain a series of results approaching the correct result. { ə'präk·sə·mət (adjective) *or* ə'präk·səˌmāt (verb) }

approximate reasoning The process by which a possibly imprecise conclusion is deduced from a collection of imprecise premises. { ə¦präks·ə·mət 'rēz·ən·iŋ }

approximation **1.** A result that is not exact but is near enough to the correct result for some specified purpose. **2.** A procedure for obtaining such a result. { ə¦präk·sə¦mā·shən }

a priori Pertaining to deductive reasoning from assumed axioms or supposedly self-evident principles, supposedly without reference to experience. { ¦ā prē¦ȯr·ē }

a priori probability *See* mathematical probability. { ¦ā prē¦ȯr·ē ˌpräb·ə'bil·əd·ē }

arabic numerals The numerals 0, 1, 2, 3, 4, 5, 6, 7, 8, and 9. { 'ar·ə·bik 'nüm·rəlz }

arc **1.** A continuous piece of the circumference of a circle. **2.** *See* edge. { ärk }

arc cosecant **1.** For a number x, any angle whose cosecant equals x. **2.** For a number x, the angle between $-\pi/2$ radians and $\pi/2$ radians whose cosecant equals x; it is the value at x of the inverse of the restriction of the cosecant function to the interval between $-\pi/2$ and $\pi/2$. { 'ärk kō'sē‚kant }

arc cosine **1.** For a number x, any angle whose cosine equals x. **2.** For a number x, the angle between 0 radians and π radians whose sine equals x; it is the value at x of the inverse of the restriction of the cosine function to the interval between 0 and π. { 'ärk 'kō‚sīn }

arc cotangent **1.** For a number x, any angle whose cotangent equals x. **2.** For a number x, the angle between 0 radians and π radians whose cotangent equals x; it is the value at x of the inverse of the restriction of the cotangent function to the interval between 0 and π. { 'ärk kō'tan·jənt }

Archimedean ordered field A field with a linear order that satisfies the axiom of Archimedes. { ‚ärk·ə¦mē·dē·ən ¦ȯrd·ərd 'fēld }

Archimedean solid One of 13 possible solids whose faces are all regular polygons, though not necessarily all of the same type, and whose polyhedral angles are all equal. { ¦är·kə¦mēd·ē·ən 'säl·əd }

Archimedean spiral A plane curve whose equation in polar coordinates (r, θ) is $r^m = a^m\theta$, where a and m are a positive or negative integer. { ¦är·kə¦mēd·ē·ən 'spī·rəl }

Archimedes' axiom *See* axiom of Archimedes. { ¦är·kə¦mēd‚ēz 'ak·sē·əm }

Archimedes' problem The problem of dividing a hemisphere into two parts of equal volume with a plane parallel to the base of the hemisphere; it cannot be solved by Euclidean methods. { ¦är·kə¦mēd‚ēz 'präb·ləm }

Archimedes' spiral *See* spiral of Archimedes. { ¦är·kə¦mēd'ēz 'spī·rəl }

arc hyperbolic function An inverse function of one of the hyperbolic functions. { ¦ärk ‚hī·pər'bäl·ik 'fəŋk·shən }

arcmin *See* minute.

arc secant **1.** For a number x, any angle whose secant equals x. **2.** For a number x, the angle between 0 radians and π radians whose secant equals x; it is the value at x of the inverse of the restriction of the secant function to the interval between 0 and π. { ¦ärk 'sē‚kant }

arc sine **1.** For a number x, any angle whose sine equals x. **2.** For a number x, the angle between $-\pi/2$ radians and $\pi/2$ radians whose sine equals x; it is the value at x of the inverse of the restriction of the sine function to the interval between $-\pi/2$ and $\pi/2$. { ¦ärk ¦sīn }

arc tangent **1.** For a number x, any angle whose tangent equals x. **2.** For a number x, the angle between $-\pi/2$ radians and $\pi/2$ radians whose tangent equals x; it is the value at x of the inverse of the restriction of the tangent function to the interval between $-\pi/2$ and $\pi/2$. { ¦ärk 'tan·jənt }

arcwise connected set A set in which each pair of points can be joined by a simple arc whose points are all in the set. { 'ärk‚wīz kə‚nek·təd 'set }

area A measure of the size of a two-dimensional surface, or of a region on such a surface. { 'er·ē·ə }

Argand diagram A two-dimensional cartesian coordinate system for representing the complex numbers, the number $x + iy$ being represented by the point whose coordinates are x and y. { 'är‚gän 'di·ə‚gram }

Arguesian plane *See* Desarguesian plane. { är¦gesh·ən 'plān }

argument *See* amplitude; independent variable. { 'är·gyə·mənt }

arithlog paper Graph paper marked with a semilogarithmic coordinate system. { ə'rith‚läg ‚pā·pər }

arithmetic Addition, subtraction, multiplication, and division, usually of integers, rational numbers, real numbers, or complex numbers. { ə'rith·mə‚tik }

arithmetical addition The addition of positive numbers or of the absolute values of signed numbers. { ¦a·rith¦med·ə·kəl ə'dish·ən }

arithmetic-geometric mean For two positive numbers a_1 and b_1, the common limit of the sequences $\{a_n\}$ and $\{b_n\}$ defined recursively by the equations $a_{n+1} = {}^1/_2(a_n + b_n)$ and $b_{n+1} = (a_n b_n)^{1/2}$. { ¦a·rith¦med·ik ‚jē·ə¦me·trik 'mēn }

arithmetic mean The average of a collection of numbers obtained by dividing the sum of the numbers by the quantity of numbers. Also known as average (av). { ¦a·rith¦med·ik 'mēn }

arithmetic progression A sequence of numbers for which there is a constant d such that the difference between any two successive terms is equal to d. { ¦a·rith¦med·ik prə'gresh·ən }

arithmetic series A series whose terms form an arithmetic progression. { ¦a·rith¦med·ik ‚sir‚ēz }

arithmetic sum **1.** The result of the addition of two or more positive quantities. **2.** The result of the addition of the absolute values of two or more quantities. { ¦a·rith¦med·ik 'səm }

arithmetization **1.** The study of various branches of higher mathematics by methods that make use of only the basic concepts and operations of arithmetic. **2.** Representation of the elements of a finite or denumerable set by nonnegative integers. Also known as Gödel numbering. { ə‚rith·məd·ə'zā·shən }

arm A side of an angle. { ärm }

articulation point *See* cut point. { är‚tik·yə'lā·shən ‚pȯint }

ascending series **1.** A series each of whose terms is greater than the preceding term. **2.** *See* power series. { ə'send·iŋ 'sir·ēz }

Ascoli's theorem The theorem that a set of uniformly bounded, equicontinuous, real-valued functions on a closed set of a real euclidean n-dimensional space contains a sequence of functions which converges uniformly on compact subsets. { as'kō‚lēz ‚thir·əm }

associate curve *See* Bertrand curve. { ə'sō·sē·ət ˌkərv }

associated prime ideal A prime ideal I in a commutative ring R is said to be associated with a module M over R if there exists an element x in M such that I is the annihilator of x. { ə'sō·sēˌād·əd 'prīm ˌī·dēl }

associated radii of convergence For a power series in n variables, $z_1, \ldots, z_n$, any set of numbers, $r_1, \ldots, r_n$, such that the series converges when $|z_i| < r_i$, $i = 1, \ldots, n$, and diverges when $|z_i| > r_i$, $i = 1, \ldots, n$. { əˈsō·sēˌād·əd ˈrād·dēˌī əv kən'vər·jəns }

associated tensor A tensor obtained by taking the inner product of a given tensor with the metric tensor, or by performing a series of such operations. { ə'sō·sēˌād·əd 'ten·sər }

associate matrix *See* Hermitian conjugate. { ə'sō·sē·ət 'mā·triks }

associate operator *See* adjoint operator. { ə'sō·sē·ət 'äp·əˌrād·ər }

associates Two elements x and y in a commutative ring with identity such that $x = ay$, where a is a unit. Also known as equivalent elements. { ə'sō·sē·ətz }

associative algebra An algebra in which the vector multiplication obeys the associative law. { ə'sō·sēˌād·iv 'al·jə·brə }

associative law For a binary operation that is designated $\circ$, the relationship expressed by $a \circ (b \circ c) = (a \circ b) \circ c$. { ə'sō·sēˌād·iv 'lȯ }

astroid A hypocycloid for which the diameter of the fixed circle is four times the diameter of the rolling circle. { 'aˌstrȯid }

asymptote **1.** A line approached by a curve in the limit as the curve approaches infinity. **2.** The limit of the tangents to a curve as the point of contact approaches infinity. { 'as·əmˌtōt }

asymptotic curve A curve on a surface whose osculating plane at each point is the same as the tangent plane to the surface. { āˌsim'täd·ik 'kərv }

asymptotic directions For a hyperbolic point on a surface, the two directions in which the normal curvature vanishes; equivalently, the directions of the asymptotic curves passing through the point. { ˌā·sim'täd·ik də'rek·shənz }

asymptotic expansion A series of the form $a_0 + (a_1/x) + (a_2/x^2) + \cdots + (a_n/x_n) + \cdots$ is an asymptotic expansion of the function $f(x)$ if there exists a number N such that for all nN the quantity $x_n[f(x) - S_n(x)]$ approaches zero as x approaches infinity, where $S_n(x)$ is the sum of the first n terms in the series. Also known as asymptotic series. { āˌsim'täd·ik ik'span·shən }

asymptotic formula A statement of equality between two functions which is not a true equality but which means the ratio of the two functions approaches 1 as the variable approaches some value, usually infinity. { āˌsim'täd·ik 'fȯr·myə·lə }

asymptotic series *See* asymptotic expansion. { āˌsim'täd·ik 'sir·ēz }

asymptotic stability The property of a vector differential equation which satisfies the conditions that (1) whenever the magnitude of the initial condition is sufficiently small, small perturbations in the initial condition produce small perturbations in the solution; and (2) there is a domain of attraction such that whenever the initial condition belongs to this domain the solution approaches zero at large times. { ā‚sim'täd·ik stə'bil·əd·ē }

atlas An atlas for a manifold is a collection of coordinate patches that covers the manifold. { 'at·ləs }

atom An element, A, of a measure algebra, other than the zero element, which has the property that any element which is equal to or less than A is either equal to A or equal to the zero element. { 'ad·əm }

augend A quantity to which another quantity is added. { 'ȯ‚jənd }

augmented matrix The matrix of the coefficients, together with the constant terms, in a system of linear equations. { 'ȯg·men·təd 'mā·triks }

autocorrelation function For a specified function $f(t)$, the average value of the product $f(t)f(t-\tau)$, where τ is a time-delay parameter; more precisely, the limit as T approaches infinity of $1/(2T)$ times the integral from $-T$ to T of $F(t)f(t-\tau)\,dt$. { ¦ȯd·ō‚kär·ə'lā·shən ‚fuŋk·shən }

automata theory A theory concerned with models used to simulate objects and processes such as computers, digital circuits, nervous systems, cellular growth and reproduction. { ȯ'täm·əd·ə 'thē·ə·rē }

automorphism An isomorphism of an algebraic structure with itself. { ¦ȯd·ō'mȯr ‚fiz·əm }

autoregressive series A function of the form $f(t) = a_1f(t-1) + a_2f(t-2) + \cdots + a_mf(t-m) + k$, where k is any constant. { ¦ȯd·ō·ri¦gres·iv 'sir·ēz }

auxiliary equation The equation that is obtained from a given linear differential equation by replacing with zero the term that involves only the independent variable. Also known as reduced equation. { ȯg¦zil·yə·re i'kwā·zhən }

av *See* arithmetic mean.

average *See* arithmetic mean. { 'av·rij }

average curvature For a given arc of a plane curve, the ratio of the change in inclination of the tangent to the curve, over the arc, to the arc length. { ¦av·rij 'kərv·ə·chər }

average deviation In statistics, the average or arithmetic mean of the deviation, taken without regard to sign, from some fixed value, usually the arithmetic mean of the data. Abbreviated AD. Also known as mean deviation. { 'av·rij ‚dē·vē'ā·shən }

axial symmetry Property of a geometric configuration which is unchanged when rotated about a given line. { 'ak·sē·əl 'sim·ə·trē }

axiom Any of the assumptions upon which a mathematical theory (such as geometry, ring theory, and the real numbers) is based. Also known as postulate. { 'ak·sē·əm }

axiom of Archimedes The postulate that if x is any real number, there exists an integer n such that n is greater than x. Also known as Archimedes' axiom. { ¦ak·sē·əm əv ‚ärk·ə'mē‚dēz }

axiom of choice The axiom that for any family A of sets there is a function that assigns to each set S of the family A a member of S. { ¦ak·sē·əm əv 'chȯis }

axis 1. In a coordinate system, the line determining one of the coordinates, obtained by setting all other coordinates to zero. **2.** A line of symmetry for a geometric figure. **3.** For a cone whose base has a center, a line passing through this center and the vertex of the cone. { 'ak·səs }

axis of abscissas The horizontal or x axis of a two-dimensional cartesian coordinate system, parallel to which abscissas are measured. { 'ak·səs əv ab'sis·əz }

axis of ordinates The vertical or y axis of a two-dimensional cartesian coordinate system, parallel to which ordinates are measured. { 'ak·səs əv 'ȯrd·nəts }

B

backward difference One of a series of quantities obtained from a function whose values are known at a series of equally spaced points by repeatedly applying the backward difference operator to these values; used in interpolation and numerical calculation and integration of functions. { ¦bak·wərd 'dif·rəns }

backward difference operator A difference operator, denoted ∇, defined by the equation $\nabla f(x) = f(x) - f(x - h)$, where h is a constant denoting the difference between successive points of interpolation or calculation. { ¦bak·wərd ¦dif·rəns 'äp·ə‚rād·ər }

Baire function The smallest class of functions on a topological space which contains the continuous functions and is closed under pointwise limits. { 'ber ‚fəŋk·shən }

Baire measure A measure defined on the class of all Baire sets such that the measure of any closed, compact set is finite. { 'ber ‚mezh·ər }

Baire's category theorem The theorem that a complete metric space is of second category; equivalently, the intersection of any sequence of open dense sets in a complete metric space is dense. { ¦berz 'kad·ə‚gör·ē ‚thir·əm }

Baire set A member of the smallest sigma algebra containing all closed, compact subsets of a topological space. { 'ber ‚set }

balanced digit system A number system in which the allowable digits in each position range in value from $-n$ to n, where n is some positive integer, and $n + 1$ is greater than one-half the base. { 'bal·ənst 'dij·ət ‚sis·təm }

balanced incomplete block design For positive integers b, ν, r, k, and λ, an arrangement of ν elements into b subsets or blocks so that each block contains exactly k distinct elements, each element occurs in r blocks, and every combination of two elements occurs together in exactly λ blocks. Also known as (b,ν,r,k,λ)-design. { ¦bal·ənst ‚iŋ·kəm‚plēt 'bläk di‚zīn }

balanced set A set S in a real or complex vector space X such that if x is in S and $a \leq 1$, then ax is in S. { 'bal·ənst ‚set }

balance equation An equation expressing a balance of quantities in the sense that the local or individual rates of change are zero. { 'bal·əns i'kwā·zhən }

Banach algebra An algebra which is a Banach space satisfying the property that for every pair of vectors, the norm of the product of those vectors does not exceed the product of their norms. { 'bä‚näk 'al·jə·brə }

Banach's fixed-point theorem A theorem stating that if a mapping f of a metric space

E into itself is a contraction, then there exists a unique element x of E such that $fx = x$. Also known as Caccioppoli-Banach principle. { ¦bä,näks ,fikst ,pȯint 'thir·əm }

Banach space A real or complex vector space in which each vector has a non-negative length, or norm, and in which every Cauchy sequence converges to a point of the space. Also known as complete normed linear space. { 'bä,näk ,spās }

Banach-Steinhaus theorem If a sequence of bounded linear transformations of a Banach space is pointwise bounded, then it is uniformly bounded. { ¦bä,näk ¦stīn,hau̇s ,thir·əm }

barycenter The center of mass of a system of finitely many equal point masses distributed in euclidean space in such a way that their position vectors are linearly independent. { 'bar·ə,sen·tər }

barycentric coordinates The coefficients in the representation of a point in a simplex as a linear combination of the vertices of the simplex. { ,bar·ə'sen·trik kō'ȯrd·ən,əts }

base **1.** A side or face upon which the altitude of a geometric configuration is thought of as being constructed. **2.** For a logarithm, the number of which the logarithm is the exponent. **3.** For a number system, the number whose powers determine place value. **4.** For a topological space, a collection of sets, unions of which form all the open sets of the space. { bās }

base angle Either of the two angles of a triangle that have the base for a side. { 'bās ,aŋ·gəl }

base for the neighborhood system *See* local base. { ¦bās fər thə 'nā·bər,hu̇d ,sis·təm }

base notation *See* radix notation. { 'bās nō'tā·shən }

base space of a bundle The topological space B in the bundle (E,p,B). { ¦bās ,spās əv ə 'bən·dəl }

base vector One of a set of linearly independent vectors in a vector space such that each vector in the space is a linear combination of vectors from the set; that is, a member of a basis. { 'bās ,vek·tər }

basic solution In bifurcation theory, a simple, explicitly known solution of a nonlinear equation, in whose neighborhood other solutions are studied. { 'bā·sik sə'lü·shən }

basis A set of linearly independent vectors in a vector space such that each vector in the space is a linear combination of vectors from the set. { 'bā·səs }

Bayes' theorem A theorem stating that the probability of a hypothesis, given the original data and some new data, is proportional to the probability of the hypothesis, given the original data only, and the probability of the new data, given the original data and the hypothesis. Also known as inverse probability principle. { ¦bāz 'thir·əm }

bei function One of the functions that is defined by $\mathrm{ber}_n(z) \pm i\,\mathrm{bei}_n(z) = J_n(ze^{\pm 3\pi i/4})$, where J_n is the nth Bessel function. { 'bī ,fəŋk·shən }

ber function One of the functions that is defined by $\mathrm{ber}_n(z) \pm i\,\mathrm{bei}_n(z) = J_n(ze^{\pm 3\pi i/4})$, where J_n is the nth Bessel function. { 'ber ,fəŋk·shən }

Bernoulli differential equation *See* Bernoulli equation. { ber′nü·lē *or* ¦ber·nü¦yē ′dif·ə′ren·chəl i′kwā·zhən }

Bernoulli equation A nonlinear first-order differential equation of the form $(dy/dx) + yf(x) = y^n g(x)$, where n is a number different from unity and f and g are given functions. Also known as Bernoulli differential equation. { ber′nü·lē i′kwā·zhən }

Bernoulli number The numerical value of the coefficient of $x_n^2/(2n)!$ is the expansion of $xe_x/(e^x-1)$. { ber′nü·lē ′nəm·bər }

Bernoulli polynomial The nth such polynomial is

$$\sum_{k=0}^{n} \binom{n}{k} B_k Z^{n-k}$$

where $\binom{n}{k}$ is a binomial coefficient, and B_k is a Bernoulli number. { ber′nü·lē ′päl·ə′nō·mē·əl }

Bernoulli's lemniscate A curve shaped like a figure eight whose equation in rectangular coordinates is expressed as $(x^2 + y^2)^2 = a^2(x^2 - y^2)$. { ber′nü·lēz lem′nis·kət }

Bertrand curve One of a pair of curves having the same principal normals. Also known as associate curve; conjugate curve. { ′ber′tränd ′kərv }

Bertrand's postulate The proposition that there exists at least one prime number between any integer greater than three and twice the integer minus two. { ′ber′tränz ′päs·chə·lət }

Bessel equation The differential equation $z^2f''(z) + zf'(z) + (z^2 - n^2)f(z) = 0$. { ′bes·əl i′kwā·zhən }

Bessel function A solution of the Bessel equation. Also known as cylindrical function. Symbolized $J_n(z)$. { ′bes·əl ′fəŋk·shən }

Bessel inequality The statement that the sum of the squares of the inner product of a vector with the members of an orthonormal set is no larger than the square of the norm of the vector. { ′bes·əl ′in·ē′kwäl·əd·ē }

Bessel transform *See* Hankel transform. { ′bes·əl ′tranz′fȯrm }

beta function A function of two positive variables, defined by

$$\mathrm{B}(m,n) = \int_0^1 x^{m-1}(1-x)^{n-1}dx$$

{ ′bād·ə ′fənk·shən }

Betti group *See* homology group. { ′bāt·tē ′grüp }

Betti number *See* connectivity number. { ′bāt·tē ′nəm·bər }

Bézout domain An integral domain in which all finitely generated ideals are principal. { ′bā′zō dō′mān }

Bianchi identity A differential identity satisfied by the Riemann curvature tensor: the antisymmetric first covariant derivative of the Riemann tensor vanishes identically. { ′byäŋ·kē ī′den·əd·ē }

bicompact set *See* compact set. { bī′käm‚pakt ¦set }

biconditional operation A logic operator on two statements P and Q whose result is true if P and Q are both true or both false, and whose result is false otherwise. Also known as if and only if operation; match. { ¦bī·kən‚dish·ən·əl ‚äp·ə′rā·shən }

biconnected graph A connected graph in which two points must be removed to disconnect the graph. { ¦bī·kə′nek·təd ′graf }

bicontinuous function *See* homeomorphism. { ¦bī·kən′tin·yə·wəs ′fəŋk·shən }

bicorn A plane curve whose equation in cartesian coordinates x and y is $(x^2 + 2ay - a^2)^2 = y^2(a^2 - x^2)$, where a is a constant. { ′bī‚kȯrn }

Bieberbach conjecture The conjecture that if a function $f(z)$ is analytic and univalent in the unit disk, and if it has the power series expansion $z + a_2z^2 + a_3z^3 + \cdots$, then, for all n ($n = 2, 3, \ldots$), the absolute value of a_n is equal to or less than n. { ′bē·bə‚bäk kən‚jek·chər }

bifurcation The appearance of qualitatively different solutions to a nonlinear equation as a parameter in the equation is varied. { bī·fər′kā·shən }

bifurcation theory The study of the local behavior of solutions of a nonlinear equation in the neighborhood of a known solution of the equation; in particular, the study of solutions which appear as a parameter in the equation is varied and which at first approximate the known solution, thus seeming to branch off from it. Also known as branching theory. { ‚bī·fər′kā·shən ‚thē·ə·rē }

bigit *See* bit. { ′bij·ət }

bigraded module A collection of modules $E_{s,t}$, indexed by pairs of integers s and t, with each module over a fixed principal ideal domain. { ¦bī‚grād·əd ‚mäj·əl }

biharmonic function A solution to the partial differential equation $\Delta^2 u(x,y,z) = 0$, where Δ is the Laplacian operator; occurs frequently in problems in electrostatics. { ¦bī·här′män·ik ′fəŋk·shən }

bijection A mapping f from a set A onto a set B which is both an injection and a surjection; that is, for every element b of B there is a unique element a of A for which $f(a) = b$. Also known as bijective mapping. { ′bī‚jek·shən }

bijective mapping *See* bijection. { ‚bī′jek·tiv ′map·iŋ }

bilateral Laplace transform A generalization of the Laplace transform in which the integration is done over the negative real numbers as well as the positive ones. { bī′lad·ə·rəl lə′pläs ′tranz‚fȯrm }

bilinear concomitant An expression $B(u,v)$, where u, v are functions of x, satisfying $vL(u) - u\bar{L}(v) = (d/dx)\cdot B(u,v)$, where L, $\bar{L}$ are given adjoint differential equations. { bī′lin·ē·ər kən′käm·ə·tənt }

bilinear expression An expression which is linear in each of two variables separately. { bī′lin·ē·ər ik′spresh·ən }

bilinear form **1.** A polynomial of the second degree which is homogeneous of the first degree in each of two sets of variables; thus, it is a sum of terms of the form $a_{ij}x_iy_j$, where $x_1, \ldots, x_m$ and $y_1, \ldots, y_n$ are two sets of variables and the a_{ij} are constants. **2.** More generally, a mapping $f(x, y)$ from $E \times F$ into R, where R is a commutative ring and $E \times F$ is the cartesian product of two modules E and F over R, such that for each x in E the function which takes y into $f(x, y)$ is linear, and for each y in F the function which takes x into $f(x, y)$ is linear. { ¦bī‚lin·ē·ər 'fȯrm }

bilinear transformations *See* Möbius transformations. { bī'lin·ē·ər tranz·fər'mā·shənz }

billion 1. The number 10^9. **2.** In British usage, the number 10^{12}. { 'bil·yən }

binary digit *See* bit. { 'bīn·ə·rē 'dij·ət }

binary notation *See* binary number system. { 'bīn·ə·rē nō'tā·shən }

binary number A number expressed in the binary number system of positional notation. { 'bīn·ə·rē 'nəm·bər }

binary number system A representation for numbers using only the digits 0 and 1 in which successive digits are interpreted as coefficients of successive powers of the base 2. Also known as binary notation; binary system. { 'bīn·ə·rē 'nəm·bər ‚sis·təm }

binary numeral One of the two digits 0 and 1 used in writing a number in binary notation. { 'bī‚ner·ē 'nüm·rəl }

binary operation A rule for combining two elements of a set to obtain a third element of that set, for example, addition and multiplication. { 'bīn·ə·rē äp·ə'rā·shən }

binary system *See* binary number system. { 'bīn·ə·rē 'sis·təm }

binary-to-decimal conversion The process of converting a number written in binary notation to the equivalent number written in ordinary decimal notation. { 'bīn·ə·rē tə 'des·məl kən'vər·zhən }

binary tree A rooted tree in which each vertex has a maximum of two successors. { 'bīn·ə·rē 'trē }

binomial A polynomial with only two terms. { bī'nō·mē·əl }

binomial array *See* Pascal's triangle. { bī'nō·mē·əl ə'rā }

binomial coefficient A coefficient in the expansion of $(x + y)_n$, where n is a positive integer; the $(k + 1)$st coefficient is equal to the number of ways of choosing k objects out of n without regard for order. Symbolized $\binom{n}{k}$; ${}_nC_k$; $C(n,k)$; C_k^n. { bī'nō·mē·əl kō·ə'fish·ənt }

binomial differential A differential of the form $x^p(a + bx^q)^r dx$, where p, q, r are integers. { bī'nō·mē·əl ‚dif·ə'ren·chəl }

binomial equation An equation having the form $x^n - a = 0$. { bī'nō·mē·əl i'kwā·zhən }

binomial expansion *See* binomial series. { bī'nō·mē·əl ik'span·shən }

binomial law The probability of an event occurring r times in n Bernoulli trials is equal to $\binom{n}{r}p^r(1 - p)^{n-r}$, where p is the probability of the event. { bī'nō·mē·əl ‚lȯ }

binomial series The expansion of $(x + y)^n$ when n is neither a positive integer nor zero. Also known as binomial expansion. { bī'nō·mē·əl 'sir·ēz }

binomial surd A sum of two roots of rational numbers, at least one of which is an irrational number. { bī'nō·mē·əl 'sərd }

binomial theorem The rule for expanding $(x + y)^n$. { bī'nō·mē·əl 'thir·əm }

binormal A vector on a curve at a point so that, together with the positive tangent and principal normal, it forms a system of right-handed rectangular cartesian axes. { bī'nȯr·məl }

bipartite cubic The points satisfying the equation $y^2 = x(x - a)(x - b)$. { bī'pär‚tīt 'kyü·bik }

bipartite graph A linear graph (network) in which the nodes can be partitioned into two groups G_1 and G_2 such that for every arc (i,j) node i is in G_1 and node j in G_2. { bī'pär‚tīt 'graf }

bipolar coordinate system **1.** A two-dimensional coordinate system defined by the family of circles that pass through two common points, and the family of circles that cut the circles of the first family at right angles. **2.** A three-dimensional coordinate system in which two of the coordinates depend on the x and y coordinates in the same manner as in a two-dimensional bipolar coordinate system and are independent of the z coordinate, while the third coordinate is proportional to the z coordinate. { ¦bī‚pō·lər kō'ȯrd·ən·ət ‚sis·təm }

biquadratic Any fourth-degree algebraic expression. Also known as quartic. { ¦bī·kwə'drad·ik }

biquadratic equation *See* quartic equation. { ¦bī·kwə'drad·ik i'kwā·zhən }

biquinary abacus An abacus in which the frame is divided into two parts by a bar which separates each wire into two- and five-counter segments. { bī'kwin·ə·rē 'ab·ə·kəs }

biquinary notation A mixed-base notation system in which the first of each pair of digits counts 0 or 1 unit of five, and the second counts 0, 1, 2, 3, or 4 units. Also known as biquinary number system. { bī'kwin·ə·rē nō'tā·shən }

biquinary number system *See* biquinary notation. { bī'kwin·ə·rē 'nəm·bər ‚sis·təm }

birectangular Property of a geometrical object that has two right angles. { ¦bī·rek'taŋ·gyə·lər }

bisection algorithm A procedure for determining the root of a function to any desired accuracy by repeatedly dividing a test interval in half and then determining in which half the value of the function changes sign. { 'bī‚sek·shən 'al·gə‚rith·əm }

bisector The ray dividing an angle into two equal angles. { ‚bī'sek·tər }

bit In a pure binary numeration system, either of the digits 0 or 1. Also known as bigit; binary digit. { bit }

bitangent *See* double tangent. { bī'tan·jənt }

biunique correspondence A correspondence that is one to one in both directions. { ¦bī·yü‚nēk ‚kär·ə'spän·dəns }

blurring An operation that decreases the value of the membership function of a fuzzy set if it is greater than 0.5, and increases it if it is less than 0.5. { 'blər·iŋ }

body of revolution A symmetrical body having the form described by rotating a plane curve about an axis in its plane. { 'bäd·ē əv rev·ə'lü·shən }

Bolyai geometry *See* Lobachevski geometry. { 'bȯl·yī jē'äm·ə·trē }

Bolzano's theorem The theorem that a single-valued, real-valued, continuous function of a real variable is equal to zero at some point in an interval if its values at the end points of the interval have opposite sign. { ˌbōl'tsän·ōz ˌthir·əm }

Bolzano-Weierstrass theorem The theorem that every bounded, infinite set in finite dimensional euclidean space has a cluster point. { ˌbōl'tsän·ō 'vī·ərˌshträs ˌthir·əm }

Boolean algebra An algebraic system with two binary operations and one unary operation important in representing a two-valued logic. { 'bü·lē·ən 'al·jə·brə }

Boolean calculus Boolean algebra modified to include the element of time. { 'bü·lē·ən 'kal·kyə·ləs }

Boolean determinant A function defined on Boolean matrices which depends on the elements of the matrix in a manner analogous to the manner in which an ordinary determinant depends on the elements of an ordinary matrix, with the operation of multiplication replaced by intersection and the operation of addition replaced by union. { ¦bül·ē·ən di'tər·mə·nənt }

Boolean function A function $f(x,y,...,z)$ assembled by the application of the operations AND, OR, NOT on the variables x, y,..., z and elements whose common domain is a Boolean algebra. { 'bü·lē·ən 'fəŋk·shən }

Boolean matrix A rectangular array of elements each of which is a member of a Boolean algebra. { ¦bül·ē·ən 'māˌtriks }

Boolean operation table A table which indicates, for a particular operation on a Boolean algebra, the values that result for all possible combination of values of the operands; used particularly with Boolean algebras of two elements which may be interpreted as "true" and "false." { ¦bül·ē·ən ˌäp·ə'rā·shən ˌtā·bəl }

Boolean operator A logic operator that is one of the operators AND, OR, or NOT, or can be expressed as a combination of these three operators. { ¦bül·ē·ən 'äp·əˌrād·ər }

Boolean ring A commutative ring with the property that for every element a of the ring, $a \times a$ and $a + a = 0$; it can be shown to be equivalent to a Boolean algebra. { ¦bül·ē·ən 'riŋ }

bordering For a determinant, the procedure of adding a column and a row, which usually have unity as a common element and all other elements equal to zero. { 'bȯrd·ər·iŋ }

Borel measurable function 1. A real-valued function such that the inverse image of the set of real numbers greater than any given real number is a Borel set. **2.** More generally, a function to a topological space such that the inverse image of any open set is a Borel set. { bȯ·rel ¦mezh·rə·bəl 'fənk·shən }

Borel measure A measure defined on the class of all Borel sets of a topological space such that the measure of any compact set is finite. { bə'rel ˌmezh·ər }

Borel set A member of the smallest σ-algebra containing the compact subsets of a topological space. { bȯ·rel ¦set }

Borel sigma algebra The smallest sigma algebra containing the compact subsets of a topological space. { bȯ·rel ¦sig·mə 'al·jə·brə }

borrow An arithmetically negative carry; it occurs in direct subtraction by raising the low-order digit of the minuend by one unit of the next-higher-order digit; for example, when subtracting 67 from 92, a tens digit is borrowed from the 9, to raise the 2 to a factor of 12; the 7 of 67 is then subtracted from the 12 to yield 5 as the units digit of the difference; the 6 is then subtracted from 8, or 9 − 1, yielding 2 as the tens digit of the difference. { 'bä·rō }

boundary *See* frontier. { 'bau̇n·drē }

boundary condition A requirement to be met by a solution to a set of differential equations on a specified set of values of the independent variables. { 'bau̇n·drē kən'dish·ən }

boundary point In a topological space, a point of a set with the property that every neighborhood of the point contains points of both the set and its complement. { 'bau̇n·drē ˌpȯint }

boundary value problem A problem, such as the Dirichlet or Neumann problem, which involves finding the solution of a differential equation or system of differential equations which meets certain specified requirements, usually connected with physical conditions, for certain values of the independent variable. { 'bau̇n·drē ˌval·yü ˌpräb·ləm }

bounded difference For two fuzzy sets A and B, with membership functions m_A and m_B, the fuzzy set whose membership function $m_{A \ominus B}$ has the value $m_A(x) - m_B(x)$ for every element x for which $m_A(x) \geq m_B(x)$, and has the value 0 for every element x for which $m_A(x) \leq m_B(x)$. { 'bau̇nd·əd 'dif·rəns }

bounded function **1.** A function whose image is a bounded set. **2.** A function of a metric space to itself which moves each point no more than some constant distance. { ¦bau̇n·dəd 'fəŋk·shən }

bounded growth The property of a function f defined on the positive real numbers which requires that there exist numbers M and a such that the absolute value of $f(t)$ is less than Ma^t for all positive values of t. { ¦bau̇n·dəd 'grōth }

bounded linear transformation A linear transformation T for which there is some positive number A such that the norm of $T(x)$ is equal to or less than A times the norm of x for each x. { ¦bau̇n·dəd ¦lin·ē·ər tranz·fər'mā·shən }

bounded product For two fuzzy sets A and B, with membership functions m_A and m_B, the fuzzy set whose membership function $m_{A \odot B}$ has the value $m_A(x) + m_B(x) - 1$ for every element x for which $m_A(x) + m_B(x) \geq 1$, and has the value 0 for every element x for which $m_A(x) + m_B(x) \leq 1$. { 'bau̇nd·əd 'präd·əkt }

bounded sequence A sequence whose members form a bounded set. { 'bau̇nd·əd 'sē·kwəns }

bounded set **1.** A collection of numbers whose absolute values are all smaller than some constant. **2.** A set of points, the distance between any two of which is smaller than some constant. { ¦bau̇n·dəd 'set }

bounded sum For two fuzzy sets A and B, with membership functions m_A and m_B, the fuzzy set whose membership function $m_{A \oplus B}$ has the value $m_A(x) + m_B(x)$ for every element x for which $m_A(x) + m_B(x) \leq 1$, and has the value 1 for every element x for which $m_A(x) + m_B(x) \geq 1$. { ¦bau̇n·dəd 'səm }

bounded variation A real-valued function is of bounded variation on an interval if its total variation there is bounded. { ¦baůn·dəd ver·ē′ā·shən }

bound variable In logic, a variable that occurs within the scope of a quantifier, and cannot be replaced by a constant. { ¦baůnd ′ver·ē·ə·bəl }

boxcar function A function whose value is zero except for a finite interval of its argument, for which it has a constant nonzero value. { ′bäks‚kär ‚fəŋk·shən }

branch 1. A complex function which is analytic in some domain and which takes on one of the values of a multiple-valued function in that domain. 2. A section of a curve that is separated from other sections of the curve by discontinuities, singular points, or other special points such as maxima and minima. { branch }

branch cut A line or curve of singular points used in defining a branch of a multiple-valued complex function. { ′branch ‚kət }

branching diagram In bifurcation theory, a graph in which a parameter characterizing solutions of a nonlinear equation is plotted against a parameter that appears in the equation itself. { ′branch·iŋ ‚dī·ə‚gram }

branching theory *See* bifurcation theory. { ′branch·iŋ ‚thē·ə·rē }

branch point 1. A point at which two or more sheets of a Riemann surface join together. 2. In bifurcation theory, a value of a parameter in a nonlinear equation at which solutions branch off from the basic solution. { ′branch ‚pȯint }

Brianchon's theorem The theorem that if a hexagon circumscribes a conic section, the three lines joining three pairs of opposite vertices are concurrent (or are parallel). { ¦brē·ən¦känz ‚thir·əm }

bridge A line whose removal disconnects a component of a graph. Also known as isthmus. { brij }

bridging The operation of carrying in addition or multiplication. { ′brij·iŋ }

Briggs' logarithm *See* common logarithm. { ¦brigz ′log·ə‚rith·əm }

broken line A line which is composed of a series of line segments lying end to end, and which does not form a continuous line. { ¦brō·kən ‚līn }

Bromwich contour A path of integration in the complex plane running from $c - i\infty$ to $c + i\infty$, where c is a real, positive number chosen so that the path lies to the right of all singularities of the analytic function under consideration. { ′bräm‚wich ‚kän‚tůr }

Brouwer's theorem A fixed-point theorem stating that for any continuous mapping f of the solid n-sphere into itself there is a point x such that $f(x) = x$. { ′braů·ərz ‚thir·əm }

Budan's theorem The theorem that the number of roots of an nth-degree polynomial lying in an open interval equals the difference in the number of sign changes induced by n differentiations at the two ends of the interval. { ′bü‚dänz ‚thir·əm }

bullet nose A plane curve whose equation in cartesian coordinates x and y is $(a^2/x^2) - (b^2/y^2) = 1$, where a and b are constants. { ′bůl·ət ‚nōz }

bundle A triple (E, p, B), where E and B are topological spaces and p is a continuous map of E onto B; intuitively E is the collection of inverse images under p of points from B glued together by the topology of X. { 'bən·dəl }

bundle of planes The set of all planes which pass through a given point. { ¦bən·dəl əv 'plānz }

Buniakowski's inequality *See* Cauchy-Schwarz inequality. { ˌbu̇n·yə'ko̍f·skēz ˌin·i'kwäl·əd·ē }

Burali-Forti paradox The order-type of the set of all ordinals is the largest ordinal, but that ordinal plus one is larger. { bu̇'räl·ē 'fo̍r·tē 'par·əˌdäks }

(b,v,r,k,λ)-design *See* balanced incomplete block design. { ¦bē ¦vē ¦är ¦kā 'lam·də diˌzīn }

C

Caccioppoli-Banach principle *See* Banach's fixed-point theorem. { ˌkä·chē′äp·ə·lē ′bäˌnäk ˌprin·sə·pəl }

Calabi conjecture If the volume of a certain type of surface, defined in a higher dimensional space in terms of complex numbers, is known, then a particular kind of metric can be defined on it; the conjecture was subsequently proved to be correct. { kə′lä·bē kənˌjek·chər }

calculus The branch of mathematics dealing with differentiation and integration and related topics. { ′kal·kyə·ləs }

calculus of enlargement *See* calculus of finite differences. { ′kal·kyə·ləs əv in′lärj·mənt }

calculus of finite differences A method of interpolation that makes use of formal relations between difference operators which are, in turn, defined in terms of the values of a function on a set of equally spaced points. Also known as calculus of enlargement. { ′kal·kyə·ləs əv ′fīˌnīt ′dif·rən·səs }

calculus of residues The application of the Cauchy residue theorem and related theorems to compute the residues of a meromorphic function at simple poles, evaluate contour integrals, expand meromorphic functions in series, and carry out related calculations. { ′kal·kyə·ləs əv ′rez·əˌdüz }

calculus of tensors The branch of mathematics dealing with the differentiation of tensors. { ′kal·kyə·ləs əv ′ten·sərs }

calculus of variations The study of problems concerning maximizing or minimizing a given definite integral relative to the dependent variables of the integrand function. { ′kal·kyə·ləs əv ˌver·ē′ā·shənz }

calculus of vectors That branch of calculus concerned with differentiation and integration of vector-valued functions. { ′kal·kyə·ləs əv ′vek·tərz }

canal surface The envelope of a family of spheres of equal radii whose centers are on a given space curve. { kə′nal ˌsər·fəs }

cancellation law A rule which allows formal division by common factors in equal products, even in systems which have no division, as integral domains; $ab = ac$ implies that $b = c$. { kan·sə′lā·shən ˌlȯ }

canonical coordinates Any set of generalized coordinates of a system together with their conjugate momenta. { kə′nän·ə·kəl kō′ȯrd·ən·əts }

canonical matrix A member of an equivalence class of matrices that has a particularly simple form, where the equivalence classes are determined by one of the relations defining equivalent, similar, or congruent matrices. { kə'nän·ə·kəl 'mā,triks }

canonical transformation Any function which has a standard form, depending on the context. { kə'nän·ə·kəl ,tranz·fər'mā·shən }

Cantor diagonal process A technique of proving statements about infinite sequences, each of whose terms is an infinite sequence by operation on the *n*th term of the *n*th sequence for each *n*; used to prove the uncountability of the real numbers. { 'kän·tȯr dī'ag·ən·əl ,präs·əs }

Cantor function A real-valued nondecreasing continuous function defined on the closed interval [0,1] which maps the Cantor ternary set onto the interval [0,1]. { 'kän·tȯr ,fəŋk·shən }

Cantor's axiom The postulate that there exists a one-to-one correspondence between the points of a line extending indefinitely in both directions and the set of real numbers. { 'kan·tərz 'ak·sē·əm }

Cantor ternary set A perfect, uncountable, totally disconnected subset of the real numbers having Lebesgue measure zero; it consists of all numbers between 0 and 1 (inclusive) with ternary representations containing no ones. { 'kän·tȯr 'tər·nə·rē ,set }

Cantor theorem A theorem that there is no one-to-one correspondence between a set and the collection of its subsets. { 'kän·tȯr 'thir·əm }

cap The symbol ∩, which indicates the intersection of two sets. { kap }

Carathéodory outer measure A positive, countably subadditive set function defined on the class of all subsets of a given set; used for defining measures. { ,kär·ə¦tā·ə'dȯr·ē ¦au̇d·ər 'mezh·ər }

cardinal number The number of members of a set; usually taken as a particular well-ordered set representative of the class of all sets which are in one-to-one correspondence with one another. { 'kärd·nəl 'nəm·bər }

cardioid A heart-shaped curve generated by a point of a circle that rolls without slipping on a fixed circle of the same diameter. { 'kärd·ē,ȯid }

carry An arithmetic operation that occurs in the course of addition when the sum of the digits in a given position equals or exceeds the base of the number system; a multiple *m* of the base is subtracted from this sum so that the remainder is less than the base, and the number *m* is then added to the next-higher-order digit. { 'kar·ē }

cartesian axis One of a set of mutually perpendicular lines which all pass through a single point, used to define a cartesian coordinate system; the value of one of the coordinates on the axis is equal to the directed distance from the intersection of axes, while the values of the other coordinates vanish. { kär'tē·zhən 'ak·səs }

cartesian coordinates **1.** The set of numbers which locate a point in space with respect to a collection of mutually perpendicular axes. **2.** *See* rectangular coordinates. { kär'tē·zhən kō'ȯrd·nəts }

cartesian coordinate system A coordinate system in *n* dimensions where *n* is any integer made by using *n* number axes which intersect each other at right angles at an origin, enabling any point within that rectangular space to be identified by the distances from the n lines. Also known as rectangular cartesian coordinate system. { kär'tē·zhən kō'ȯrd·nət ˌsis·təm }

cartesian geometry *See* analytic geometry. { kär'tē·zhan jē'äm·ə·trē }

cartesian oval A plane curve consisting of all points P such that $aFP + bF''P = c$, where F and F'' are fixed points and a, b, and c are constants which are not necessarily positive. { kär'tē·zhən 'ō·vəl }

cartesian plane A plane whose points are specified by cartesian coordinates. { kär'tēzh·ən 'plān }

cartesian product In reference to the product of P and Q, the set $P \times Q$ of all pairs (p, q), where p belongs to P and q belongs to Q. { kär'tē·zhan 'präd·əkt }

cartesian surface A surface obtained by rotating the curve $n_0(x^2 + y^2)^{1/2} \pm n_1[(x - a)^2 + y^2]^{1/2} = c$ about the x axis. { kär'tē·zhan 'sər·fəs }

cartesian tensor The aggregate of the functions of position in a tensor field in an n-dimensional cartesian coordinate system. { kär'tē·zhan 'ten·sər }

Cassinian oval *See* oval of Cassini. { kə'sin·ē·ən 'ō·vəl }

casting-out nines A method of checking the correctness of elementary arithmetical operations, based on the fact that an integer yields the same remainder as the sum of its decimal digits, when divided by 9. { ¦kast·iŋ ˌau̇t 'nīnz }

catastrophe theory A theory of mathematical structure in which smooth continuous inputs lead to discontinuous responses. { kə'tas·trə·fē ˌthē·ə·rē }

category A class of objects together with a set of morphisms for each pair of objects and a law of composition for morphisms; sets and functions form an important category, as do groups and homomorphisms. { 'kad·əˌgȯr·ē }

catenary The curve obtained by suspending a uniform chain by its two ends; the graph of the hyperbolic cosine function. Also known as alysoid; chainette. { 'kat·əˌner·ē }

caterer problem A linear programming problem in which it is required to find the optimal policy for a caterer who must choose between buying new napkins and sending them to either a fast or a slow laundry service. { 'kād·ə·rər ˌpräb·ləm }

Cauchy boundary conditions The conditions imposed on a surface in euclidean space which are to be satisfied by a solution to a partial differential equation. { kō·shē 'bau̇n·drē kənˌdish·ənz }

Cauchy condensation test A monotone decreasing series of positive terms Σa_n converges or diverges, as does $\Sigma p^n a_{p^n}$, for any positive integer p. { kō·shē ˌkän·den'sā·shən ˌtest }

Cauchy formula An expression for the value of an analytic function f at a point z in terms of a line integral

$$f(z) = \frac{1}{2\pi i} \int_C \frac{f(\zeta)}{\zeta - z}\, d\zeta$$

where C is a simple closed curve containing z. Also known as Cauchy integral formula. { kō·shē ˌfȯr·myə·lə }

Cauchy-Hadamard theorem The theorem that the radius of convergence of a Taylor series in the complex variable z is the reciprocal of the limit superior, as n approaches infinity, of the nth root of the absolute value of the coefficient of z^n. { kō·shē 'had·ə·mär ˌthir·əm }

Cauchy inequality The square of the sum of the products of two variables for a range of values is less than or equal to the product of the sums of the squares of these two variables for the same range of values. { kō·shē ˌin·i'kwäl·əd·ē }

Cauchy integral formula *See* Cauchy formula. { kō·shē ¦in·tə·grəl ¦fȯr·myə·lə }

Cauchy integral test *See* Cauchy's test for convergence. { kō·shē 'in·tə·grəl ˌtest }

Cauchy integral theorem The theorem that if γ is a closed path in a region R satisfying certain topological properties, then the integral around γ of any function analytic in R is zero. { kō·shē 'in·tə·grəl ˌthir·əm }

Cauchy mean The Cauchy mean-value theorem for the ratio of two continuous functions. { kō·shē ˌmēn }

Cauchy mean-value theorem If f and g are functions satisfying certain conditions on an interval $[a,b]$, then there is a point x in the interval at which the ratio of derivatives $f'(x)/g'(x)$ equals the ratio of the net change in f, $f(b) - f(a)$, to that of g. { kō·shē ¦mēn ¦val·yü ˌthir·əm }

Cauchy principal value Also known as principal value. **1.** The Cauchy principal value of

$$\int_{-\infty}^{\infty} f(x)dx \quad \text{is} \quad \lim_{s\to\infty} \int_{-s}^{s} f(x)dx$$

provided the limit exists. **2.** If a function f is bounded on an interval (a,b) except in the neighborhood of a point c, the Cauchy principal value of

$$\int_{a}^{b} f(x)dx \quad \text{is} \quad \lim_{\delta\to 0}\left[\int_{a}^{c-\delta} f(x)dx + \int_{c+\delta}^{b} f(x)dx\right]$$

provided the limit exists. { kō·shē ¦prin·sə·pəl ¦val·yü }

Cauchy problem The problem of determining the solution of a system of partial differential equation of order m from the prescribed values of the solution and of its derivatives of order less than m on a given surface. { kō·shē ˌpräb·ləm }

Cauchy product A method of multiplying two absolutely convergent series to obtain a series which converges absolutely to the product of the limits of the original series:

$$\left(\sum_{n=0}^{\infty} a_n\right)\left(\sum_{n=0}^{\infty} b_n\right) = \sum_{n=0}^{\infty} c_n \quad \text{where} \quad c_n = \sum_{k=0}^{n} a_k b_{n-k}$$

{ kō·shē ˌpräd·əkt }

Cauchy radical test A test for convergence of series of positive terms: if the nth root of the nth term is less than some number less than unity, the series converges; if it remains equal to or greater than unity, the series diverges. { kō·shē 'rad·i·kəl ˌtest }

Cauchy ratio test A series of nonnegative terms converges if the limit, as n approaches infinity, of the ratio of the $(n + 1)$st to nth term is smaller than 1, and diverges if it is greater than 1; the test fails if this limit is 1. Also known as ratio test. { kō·shē 'rā·shō ˌtest }

Cauchy residue theorem The theorem expressing a line integral around a closed curve of a function which is analytic in a simply connected domain containing the curve, except at a finite number of poles interior to the curve, as a sum of residues of the function at these poles. { kō·shē 'rez·ə‚dü ‚thir·əm }

Cauchy-Riemann equations A pair of partial differential equations that is satisfied by the real and imaginary parts of a complex function $f(z)$ if and only if the function is analytic: $\partial u/\partial x = \partial v/\partial y$ and $\partial u/\partial y = -\partial v/\partial x$, where $f(z) = u + iv$ and $z = x + iy$. { kō·shē 'rē‚män i'kwā·zhənz }

Cauchy-Schwarz inequality The square of the inner product of two vectors does not exceed the product of the squares of their norms. Also known as Buniakowski's inequality; Schwarz' inequality. { kō·shē 'shwȯrts in·i'kwäl·əd·ē }

Cauchy sequence A sequence with the property that the difference between any two terms is arbitrarily small provided they are both sufficiently far out in the sequence; more precisely stated: a sequence $\{a_n\}$ such that for every $\epsilon > 0$ there is an integer N with the property that if n and m are both greater than N, then $a_n - a_m < \epsilon$. Also known as fundamental sequence. { kō·shē 'sē·kwəns }

Cauchy's mean-value theorem *See* second mean-value theorem. { kō·shēz ‚mēn ‚val·yü 'thir·əm }

Cauchy's test for convergence **1.** A series is absolutely convergent if the limit as n approaches infinity of its nth term raised to the $1/n$ power is less than unity. **2.** A series a_n is convergent if there exists a monotonically decreasing function f such that $f(n) = a_n$ for n greater than some fixed number N, and if the integral of $f(x)dx$ from N to ∞ converges. Also known as Cauchy integral test; Maclaurin-Cauchy test. { kō·shēz ‚test fər kən'vər·jəns }

Cauchy transcendental equation An equation whose roots are characteristic values of a certain type of Sturm-Liouville problem: $\tan \sigma\pi = (k + K)/(\sigma^2 - kK)$, where k and K are given, and σ is to be determined. { kō·shē ¦trans‚en¦dent·əl i'kwā·zhən }

Cavalieri's theorem The theorem that two solids have the same volume if their altitudes are equal and all plane sections parallel to their bases and at equal distances from their bases are equal. { ‚kav·ə'lyer·ēz ‚thir·əm }

Cayley algebra The nonassociative division algebra consisting of pairs of quaternions; it may be identified with an eight-dimensional vector space over the real numbers. { 'kā·lē ‚al·jə·brə }

Cayley-Hamilton theorem The theorem that a linear transformation or matrix is a root of its own characteristic polynomial. Also known as Hamilton-Cayley theorem. { ¦kāl·ē ¦ham·əl·tən ‚thir·əm }

Cayley-Klein parameters A set of four complex numbers used to describe the orientation of a rigid body in space, or equivalently, the rotation which produces that orientation, starting from some reference orientation. { ¦kāl·ē ¦klīn pə‚ram·əd·ərz }

Cayley numbers The division algebra consisting of pairs of quaternions; may be identified with an eight-dimensional vector space over the real numbers. Also known as octonions. { 'kāl·ē ‚nəm·bərz }

Cayley's sextic A plane curve with the equation $r = 4a \cos^3(\theta/3)$, where r and θ are radial and angular polar coordinates and a is a constant. { 'kā·lēz 'sek·stik }

Cayley's theorem A theorem that any group G is isomorphic to a subgroup of the group of permutations on G. { 'kā‚lēz ‚thir·əm }

ceiling The smallest integer that is equal to or greater than a given real number a; symbolized $\lceil a \rceil$. { 'sē·liŋ }

cell **1.** The homeomorphic image of the unit ball. **2.** One of the $(n - 1)$-dimensional polytopes that enclose a given n-dimensional polytope. { sel }

cell complex A topological space which is the last term of a finite sequence of spaces, each obtained from the previous by sewing on a cell along its boundary. { 'sel ‚käm‚pleks }

cellular automaton A mathematical construction consisting of a system of entities, called cells, whose temporal evolution is governed by a collection of rules, so that its behavior over time may appear highly complex or chaotic. { 'sel·yə·lər ȯ'täm·ə·tən }

center **1.** The point that is equidistant from all the points on a circle or sphere. **2.** The point (if it exists) about which a curve (such as a circle, ellipse, or hyperbola) is symmetrical. **3.** The point (if it exists) about which a surface (such as a sphere, ellipsoid, or hyperboloid) is symmetrical. **4.** For a regular polygon, the center of its circumscribed circle. **5.** The subgroup consisting of all elements that commute with all other elements in a given group. **6.** The subring consisting of all elements a such that $ax = xa$ for all x in a given ring. { 'sen·tər }

center of area For a plane figure, the center of mass of a thin uniform plate having the same boundaries as the plane figure. Also known as center of figure; centroid. { 'sen·tər əv 'er·ē·ə }

center of curvature At a given point on a curve, the center of the osculating circle of the curve at that point. { 'sen·tər əv 'kər·və·chər }

center of figure *See* center of area; center of volume. { 'sen·tər əv 'fig·yər }

center of geodesic curvature For a curve on a surface at a given point, the center of curvature of the orthogonal projection of the curve onto a plane tangent to the surface at the point. { ¦sen·tər əv ‚jē·ə'des·ik ‚kərv·ə·chər }

center of inversion The point O with respect to which an inversion is defined, so that every point P is mapped by the inversion into a point Q that is collinear with O and P. { 'sen·tər əv in'vər·zhən }

center of normal curvature For a given point on a surface and for a given direction, the normal section of the surface through the given point and in the given direction. { 'sen·tər əv 'nȯrm·əl 'kər·və·chər }

center of perspective The point specified by Desargues' theorem, at which lines passing through corresponding vertices of two triangles are concurrent. { 'sen·tər əv pər'spek·tiv }

center of principal curvature For a given point on a surface, the center of normal curvature at the point in one of the two principal directions. { 'sen·tər əv 'prin·sə·pəl ‚kər·və·chər }

center of similitude A point of intersection of lines that join the ends of parallel radii of coplanar circles. { 'sen·tər əv si'mil·ə‚tüd }

center of spherical curvature The center of the osculating sphere at a specified point on a space curve. { ¦sen·tər əv ¦sfer·ə·kəl 'kər·və·chər }

center of volume For a three-dimensional figure, the center of mass of a homogeneous solid having the same boundaries as the figures. Also known as center of figure; centroid. { 'sen·tər əv 'väl·yəm }

centrad A unit of plane angle equal to 0.01 radian or to about 0.573 degree. { 'sent‚rad }

central angle In a circle, an angle whose sides are radii of the circle. { 'sen·trəl 'aŋ·gəl }

central conic A conic that has a center, namely, a circle, ellipse, or hyperbola. { 'sen·trəl 'kän·ik }

central difference One of a series of quantities obtained from a function whose values are known at a series of equally spaced points by repeatedly applying the central difference operator to these values; used in interpolation or numerical calculation and integration of functions. { 'sen·trəl 'dif·rəns }

central difference operator A difference operator, denoted ∂, defined by the equation $\partial f(x) = f(x + h/2) - f(x - h/2)$, where h is a constant denoting the difference between successive points of interpolation or calculation. { ¦sen·trəl ¦dif·rəns 'äp·ə‚rād·ər }

centralizer The subgroup consisting of all elements which commute with a given element of a group. { 'sen·tra‚līz·ər }

central mean operator A difference operator, denoteed μ, defined by the equation $\mu f(x) = [f(x + h/2) + f(x - h/2)]/2$, where h is a constant denoting the difference between successive points of interpolation or calculation. Also known as averaging operator. { ¦sen·trəl ¦mēn 'äp·ə‚rād·ər }

central plane For a fixed ruling of a ruled surface, the plane tangent to the surface at the central point of the ruling. { 'sen·trəl 'plān }

central point For a fixed ruling L on a ruled surface, the limiting position, as a variable ruling L' approaches L, of the foot on L of the common perpendicular to L and L'. { 'sen·trəl 'pȯint }

central projection A mapping of a configuration into a plane that associates with any point of the configuration the intersection with the plane of the line passing through the point and a fixed point. { 'sen·trəl prə'jek·shən }

central quadric A quadric surface that has a center, namely, a sphere, ellipsoid, or hyperboloid. { 'sen·trəl 'kwä·drik }

centroid *See* center of area; center of volume. { 'sen‚trȯid }

centroids of areas and lines Points positioned identically with the centers of gravity of corresponding thin homogeneous plates or thin homogeneous wires; involved in the analysis of certain problems of mechanics such as the phenomenon of bending. { 'sen‚trȯidz əv ¦er·ē·əz ən 'līnz }

Cesáro equation An equation which relates the arc length along a plane curve and the radius of curvature. { chā'zä·rō i‚kwā·zhən }

Cesáro summation A method of attaching sums to certain divergent sequences and series by taking averages of the first n terms and passing to the limit. { chā'zä·rō sə'mā·shən }

Ceva's theorem The theorem that if three concurrent straight lines pass through the vertices *A*, *B*, and *C* of a triangle and intersect the opposite sides, produced if necessary at *D*, *E*, and *F*, then the product $AF \cdot BD \cdot CE$ of the lengths of three alternate segments equals the product $FB \cdot DC \cdot EA$ of the other three. { 'chā·vəz ˌthir·əm }

cevian A straight line that passes through a vertex of a triangle or tetrahedron and intersects the opposite side or face. { 'chāv·ē·ən }

chain *See* linearly ordered set. { chān }

chain complex A sequence $\{C_n\}$, $-\infty < n < \infty$, of Abelian groups together with a sequence of boundary homomorphisms d_n: $C_n \rightarrow C_{n-1}$ such that $d_{n-1} \circ d_n = 0$ for each n. { 'chān ˌkämˌpleks }

chainette *See* catenary. { chā'net }

chain homomorphism A sequence of homomorphisms f_n: $C_n \rightarrow D_n$ between the groups of two chain complexes such that $f_{n-1}d_n = \overline{d}_n f_n$ where d_n and $\overline{d}_n$ are the boundary homomorphisms of $\{C_n\}$ and $\{D_n\}$ respectively. { 'chān ˌhō·mō'mȯrˌfiz·əm }

chain of simplices A member of the free Abelian group generated by the simplices of a given dimension of a simplicial complex. { 'chān əv 'sim·pləˌsēz }

chain rule A rule for differentiating a composition of functions: $(d/dx) f(g(x)) = f'(g(x)) \cdot g'(x)$. { 'chān ˌrül }

chance variable *See* random variable. { ¦chans 'ver·ē·ə·bəl }

character group The set of all continuous homomorphisms of a topological group onto the group of all complex numbers with unit norm. { 'kar·ik·tər ˌgrüp }

characteristic **1.** That part of the logarithm of a number which is the integral (the whole number) to the left of the decimal point in the logarithm. **2.** For a family of surfaces that depend continuously on a parameter, the limiting curve of intersection of two members of the family as the two values of the parameter determining them approach a common value. { ˌkar·ik·tə'ris·tik }

characteristic cone A conelike region important in the study of initial value problems in partial differential equations. { ˌkar·ik·tə'ris·tik 'kōn }

characteristic curve **1.** One of a pair of conjugate curves in a surface with the property that the directions of the tangents through any point of the curve are the characteristic directions of the surface. **2.** A curve plotted on graph paper to show the relation between two changing values. **3.** A characteristic curve of a one-parameter family of surfaces is the limit of the curve of intersection of two neighboring surfaces of the family as those surfaces approach coincidence. { ˌkar·ik·tə'ris·tik 'kərv }

characteristic directions For a point *P* on a surface *S*, the pair of conjugate directions which are symmetric with respect to the directions of the lines of curvature on *S* through *P*. { ˌkar·ik·tə'ris·tik də'rek·shənz }

characteristic equation **1.** Any equation which has a solution, subject to specified boundary conditions, only when a parameter occurring in it has certain values. **2.** Specifically, the equation $A\mathbf{u} = \lambda\mathbf{u}$, which can have a solution only when the parameter λ has certain values, where A can be a square matrix which multiplies the vector $\mathbf{u}$, or a linear differential or integral operator which operates on the function $\mathbf{u}$, or in general, any linear operator operating on the vector $\mathbf{u}$ in a finite or infinite dimensional vector space. Also known as eigenvalue equation. **3.** An equation which sets the characteristic polynomial of a given linear transformation on a finite dimensional vector space, or of its matrix representation, equal to zero. { ˌkar·ik·tə'ris·tik i'kwā·zhən }

characteristic form A means of classifying partial differential equations. { ˌkar·ik·tə'ris·tik 'fȯrm }

characteristic function **1.** The function χ_A defined for any subset A of a set by setting $\chi_A = (x) = 1$ if x is in A and $\chi_A = (x)\ 0$ if x is not in A. Also known as indicator function. **2.** *See* eigenfunction. { ˌkar·ik·tə'ris·tik 'fəŋk·shən }

characteristic manifold **1.** A surface used to study the problem of existence of solutions to partial differential equations. **2.** The linear set of eigenvectors corresponding to a given eigenvalue of a linear transformation. { ˌkar·ik·tə'ris·tik 'man·əˌfōld }

characteristic number *See* eigenvalue. { ˌkar·ik·tə'ris·tik 'nəm·bər }

characteristic point The characteristic point of a one-parameter family of surfaces corresponding to the value u_0 of the parameter is the limit of the point of intersection of the surfaces corresponding to the values u_0, u_1, and u_2 of the parameter as u_1 and u_2 approach u_0 independently. { ˌkar·ik·tə'ris·tik 'pȯint }

characteristic polynomial The polynomial whose roots are the eigenvalues of a given linear transformation on a finite dimensional vector space. { ˌkar·ik·tə'ris·tik ˌpäl·ə'nō·mē·əl }

characteristic ray For a differential equation, an integral curve which generates all the others. { ˌkar·ik·tə'ris·tik 'rā }

characteristic root *See* eigenvalue. { ˌkar·ik·tə'ris·tik 'rüt }

characteristic value *See* eigenvalue. { ˌkar·ik·tə'ris·tik 'val·yü }

characteristic vector *See* eigenvector. { ˌkar·ik·tə'ris·tik 'vek·tər }

Charlier polynomials Families of polynomials which are orthogonal with respect to Poisson distributions. { shärˈlyā ˌpäl·ə'nō·mē·əlz }

Charpit's method A method for finding a complete integral of the general first-order partial differential equation in two independent variables; it involves solving a set of five ordinary differential equations. { 'chärˌpits ˌmeth·əd }

chart An n-chart is a pair (U,h), where U is an open set of a topological space and h is a homeomorphism of U onto an open subset of n-dimensional euclidean space. { chärt }

Chebyshev approximation *See* min-max technique.

Chebyshev polynomials A family of orthogonal polynomials which solve Chebyshev's differential equation. { 'cheb·ə·shəf ˌpäl·i'nō·mē·əlz }

Chebyshev's differential equation A special case of Gauss' hypergeometric second-order differential equation: $(1 - x^2)f''(x) - xf'(x) + n^2f(x) = 0$. { 'cheb·ə·shəfs dif·ə'ren·chəl i'kwā·zhən }

Choquet theorem Let K be a compact convex set in a locally convex Hausdorff real vector space and assume that either (1) the set of extreme points of K is closed or (2) K is metrizable; then for every point x in K there is at least one Radon probability measure m on X, concentrated on the set of extreme points of K, such that x is the centroid of m. { shō'kā ˌthir·əm }

chord A line segment which intersects a curve or surface only at the endpoints of the segment. { kȯrd }

Christoffel symbols Symbols that represent particular functions of the coefficients and their first-order derivatives of a quadratic form. Also known as three-index symbols. { 'kris·tȯf·əl ˌsim·bəlz }

chromatic number For a specified surface, the smallest number n such that for any decomposition of the surface into regions the regions can be colored with n colors in such a way that no two adjacent regions have the same color. { krō'mad·ik 'nəm·bər }

Church-Rosser theorem If for a lambda expression there is a terminating reduction sequence yielding a reduced form B, then the leftmost reduction sequence will yield a reduced form that is equivalent to B up to renaming. { ¦chərch ¦rȯs·ər ¦thir·əm }

Church's thesis The claim that a function is computable in the intuitive sense if and only if it is computable by a Turing machine. Also known as Turing's thesis. { ¦chərch·əz ¦thē·səs }

circle **1.** The set of all points in the plane at a given distance from a fixed point. **2.** A unit of angular measure, equal to one complete revolution, that is, to 2π radians or 360°. Also known as turn. { 'sər·kəl }

circle graph *See* pie chart. { 'sər·kəl ˌgraf }

circle of convergence The region in which a power series possesses a limit. { 'sər·kəl əv kən'vər·jəns }

circle of curvature The circle tangent to a curve on the concave side and having the same curvature at the point of tangency as does the curve. { 'sər·kəl əv 'kər·və·chər }

circle of inversion A circle with respect to which two specified curves are inverse curves. { 'sər·kəl əv in'vər·zhən }

circulant determinant A determinant in which the elements of each row are the same as those of the previous row moved one place to the right, with the last element put first. { 'sər·kyə·lənt də'tər·mə·nənt }

circulant matrix A matrix in which the elements of each row are those of the previous row moved one place to the right. { 'sər·kyə·lənt 'māˌtriks }

circular cone A cone whose base is a circle. { 'sər·kyə·lər 'kōn }

circular conical surface The lateral surface of a right circular cone. { ¦sər·kyə·lər ¦kän·ə·kəl 'sər·fəs }

circular cylinder A solid bounded by two parallel planes and a cylindrical surface whose intersections with planes perpendicular to the straight lines forming the surface are circles. { 'sər·kyə·lər 'sil·ən·dər }

circular functions *See* trigonometric functions. { 'sər·kyə·lər 'fəŋk·shənz }

circular helix A curve that lies on a right circular cylinder and intersects all the elements of the cylinder at the same angle. { 'sər·kyə·lər 'hē,liks }

circular nomograph A chart with concentric circular scales for three variables, laid out so that any straight line passes through values of the variables satisfying a given equation. { 'sər·kyə·lər 'nō·mə,graf }

circular point A point on a surface at which the normal curvature is the same in all directions. { 'sər·kyə·lər 'pȯint }

circular point at infinity In projective geometry, one of two points at which every circle intersects the ideal line. { ¦sər·kyə·lər ¦pȯint at in'fin·əd·ē }

circular segment Portion of circle cut off from the main body of the circle by a straight line (chord) through the circle. { 'sər·kyə·lər 'seg·mənt }

circular slide rule A slide rule in a circular form whose advantages over a straight slide rule are its precision, because it is equivalent to a straight slide rule many times longer than the circular slide rule's diameter, and ease of multiplication, because the scale is continuous. { 'sər·kyə·lər 'slīd ,rül }

circulation For the circulation of a vector field around a closed path, the line integral of the field vector around the path. { ,sər·kyə·'lā·shən }

circumcenter For a triangle or a regular polygon, the center of the circle that is circumscribed about the triangle or polygon. { ¦sər·kəm¦sen·tər }

circumcircle A circle that passes through all the vertices of a given polygon, if such a circle exists. { 'sər·kəm,sər·kəl }

circumference **1.** The length of a circle. **2.** For a sphere, the length of any great circle on the sphere. { sər'kəm·fə·rəns }

circumradius The radius of a circle that is circumscribed about a polygon. { ¦sər·kəm'rād·ē·əs }

circumscribed **1.** A closed curve (or surface) is circumscribed about a polygon (or polyhedron) if every vertex of the polygon (or polyhedron) is incident upon the curve (or surface) and the polygon (or polyhedron) is contained in the curve (or surface). **2.** A polygon (or polyhedron) is circumscribed about a closed curve (or surface) if every side of the polygon (or face of the polyhedron) is tangent to the curve (or surface) and the curve (or surface) is contained within the polygon (or polyhedron). { 'sər·kəm,skrībd }

cissoid A plane curve consisting of all points which lie on a variable line passing through a fixed point, and whose distance from the fixed point is equal to the distance between the intersections of the line with two given curves. { 'sis,ȯid }

cissoid of Diocles The cissoid of a circle and a tangent line with respect to a fixed point on the circumference of the circle diametrically opposite the point of tangency. { 'si,sȯid əv 'dī·ə·klēz }

class **1.** A set that consists of all the sets having a specified property. **2.** The class of a plane curve is the largest number of tangents that can be drawn to the curve from any point in the plane that is not on the curve. { klas }

class C^n The class of all functions that are continuous on a given domain and have continuous derivatives of all orders up to and including the nth. { ˌklas ′sē ′en }

class formula A formula which states that the order of a finite group G is equal to the sum, over a set of representatives x_i of the distinct conjugacy classes of G, of the index of the normalizer of x_i in G. { ′klas ˌfȯr·myə·lə }

classical canonical matrix A form to which any matrix can be reduced by a collineatory transformation, with zeros except for a sequence of Jordan matrices siutated along the principal diagonal. { ′klas·ə·kəl kə′nän·ə·kəl ′mā·triks }

closed covering A closed covering of a set S in a topological space is a collection of closed sets whose union contains S. { ¦klōzd ′kəv·ər·iŋ }

closed curve A curve that has no end points. { ¦klōzd ′kərv }

closed graph theorem If T is a linear transformation on Banach space X to Banach space Y whose domain $D(T)$ is closed and whose graph, that is, the set of pairs (x,Tx) for x in $D(T)$, is closed in $X \times Y$, then T is bounded (and hence continuous). { ¦klōzd ¦graf ′thir·əm }

closed intervals A closed interval of real numbers, denoted by $[a,b]$, consists of all numbers equal to or greater than a and equal to or less than b. { ¦klōzd ′in·tər·vəlz }

closed linear manifold A topologically closed vector subspace of a topological vector space. { ¦klōzd ¦lin·ē·ər ′man·ə·fōld }

closed linear transformation A linear transformation T such that the set of points of the form $[x, T(x)]$ is closed in the cartesian product $\bar{D} \times \bar{R}$ of the closure of the domain D and the closure of the range R of T. { ¦klōzd ¦lin·ē·ər ˌtranz·fər′mā·shən }

closed orthonormal set *See* complete orthonormal set. { ¦klōzd ¦ȯr·thō¦nȯr·məl ′set }

closed map A function between two topological spaces which sends each closed set of one into a closed set of the other. { ¦klōzd ′map }

closed *n*-cell A set that is homeomorphic with the set of points in n-dimensional euclidean space ($n = 1, 2, \ldots$) whose distance from the origin is equal to or less than unity. { ¦klōzd ′en ˌsel }

closed operator A linear transformation f whose domain A is contained in a normed vector space X satisfying the condition that if $\lim x_n = x$ for a sequence x_n in A, and $\lim f(x_n) = y$, then x is in A and $f(x) = y$. { ¦klōzd ′äp·əˌrād·ər }

closed set A set of points which contains all its cluster points. Also known as topologically closed set. { ¦klōzd ′set }

closed surface A surface that has no bounding curve. { ¦klōzd ′sər·fəs }

closure **1.** The union of a set and its cluster points; the smallest closed set containing the set. **2.** Property of a mathematical set such that a specified mathematical operation that is applied to elements of the set produces only elements of the same set { ′klō·zhər }

clothoid *See* Cornu's spiral. { 'klȯth·ȯid }

cluster point A cluster point of a set in a topological space is a point p whose neighborhoods all contain at least one point of the set other than p. Also known as accumulation point; limit point. { 'kləs·tər ˌpȯint }

coaxial circles Family of circles such that any pair have the same radical axis. { kō'ak·sē·əl 'sər·kəlz }

coaxial cylinders Two cylinders whose cylindrical surfaces consist of the lines that pass through concentric circles in a given plane and are perpendicular to this plane. { kō'ak·sē·əl 'sil·ən·dərz }

coaxial planes Planes that pass through the same straight line. Also known as collinear planes. { kō'ak·sē·əl 'planz }

coboundary An image under the coboundary operator. { kō'bau̇n·drē }

coboundary operator If $\{C^n\}$ is a sequence of Abelian groups, coboundary operators are homomorphisms $\{\delta^n\}$ such that δ^n: $C^n \rightarrow C^{n+1}$ and $\delta^{n+1} - \delta^n = 0$. { kō'bau̇n·drē 'äp·əˌrād·ər }

cochain complex A sequence of Abelian groups C^n, $-\infty < n < \infty$, together with coboundary homomorphisms δ^n: $C^n \rightarrow C^{n+1}$ such that $\delta^{n+1} - \delta^n = 0$. { 'kōˌchān 'kämˌpleks }

cochleoid A plane curve whose equation in polar coordinates is $r\theta = a \sin \theta$. { 'käk·lēˌȯid }

cocycle A chain of simplices whose coboundary is 0. { 'kōˌsī·kəl }

coefficient A factor in a product. { ¦kō·ə'fish·ənt }

coefficient of strain Multiplier used in transformations to elongate or compress configurations in a direction parallel to an axis. { ¦kō·ə'fish·ənt əv 'strān }

cofactor *See* minor. { 'kōˌfak·tər }

cofinal A subset C of a directed set D is cofinal if for each element of D there is a larger element in C. { kō'fīn·əl }

cohomology group One of a series of Abelian groups $H^n(K)$ that are used in the study of a simplicial complex K and are closely related to homology groups, being associated with cocycles and coboundaries in the same manner as homology groups are associated with cycles and boundaries. { 'kō·hə'mäl·ə·jē ˌgrüp }

cohomology theory A theory which uses algebraic groups to study the geometric properties of topological spaces; closely related to homology theory. { kō·hō'mäl·ə·jē 'thē·ə·rē }

collinear Lying on a single straight line. { kə'lin·ē·ər }

collinear planes *See* coaxial planes. { ˌkō'lin·ē·ər 'plānz }

collinear vectors Two vectors, one of which is a non-zero scalar multiple of the other. { kə'lin·ē·ər 'vek·tərz }

collineation A mapping which transforms points into points, lines into lines, and planes into planes. Also known as collineatory transformation. { kəˌlin·ē'ā·shən }

collineatory transformation *See* collineation. { kə'lin·yə,tȯr·ē ,tranz·fər'mā·shən }

colog *See* cologarithm. { 'kō,läg }

cologarithm The cologarithm of a number is the logarithm of the reciprocal of that number. Abbreviated colog. { ¦kō'läg·ə,rith·əm }

color class In a given coloring of a graph, the set of vertices which are assigned the same color. { 'kəl·ər ,klas }

coloring An assignment of colors to the vertices of a graph so that adjacent vertices are assigned different colors. { 'kəl·ər·iŋ }

column *See* place. { 'käl·əm }

column operations A set of rules for manipulating the columns of a matrix so that the image of the corresponding linear transformation remains unchanged. { 'käl·əm ,äp·ə'rā·shənz }

column rank The number of linearly independent columns of a matrix; the dimension of the image of the corresponding linear transformation. { 'käl·əm ,raŋk }

column space The vector space spanned by the columns of a matrix. { 'käl·əm ,spās }

column vector A matrix consisting of only one column. { 'käl·əm ,vek·tər }

combescure transformation A one-to-one continuous mapping of one space curve onto another space curve so that tangents to corresponding points are parallel. { 'kōm·sə,kyu̇r tranz·fər'mā·shən }

combination A selection of one or more of the elements of a given set without regard to order. { ,käm·bə'nā·shən }

combinatorial analysis **1.** The determination of the number of possible outcomes in ideal games of chance by using formulas for computing numbers of combinations and permutations. **2.** The study of large finite problems. { kəm,bī·nə'tȯr·ē·əl ə'nal·ə·səs }

combinatorial theory The branch of mathematics which studies the arrangements of elements into sets. { kəm,bī·nə'tȯr·ē·əl 'thē·ə·rē }

combinatorial topology The study of polyhedrons, simplicial complexes, and generalizations of these. Also known as piecewise linear topology. { kəm,bī·nə'tȯr·ē·əl tə'päl·ə·jē }

combinatorics Combinatorial topology which studies geometric forms by breaking them into simple geometric figures. { ,kəm·bə·nə'tȯr·iks }

common denominator Any common multiple of the denominators of a collection of fractions. { ¦käm·ən də'näm·ə,nād·ər }

common difference The fixed difference between any term in an arithmetic progression and the preceding term. { 'käm·ən 'dif·rəns }

common divisor For a set of integers, an integer c such that each of the integers in the set is divisible by c. { 'käm·ən di'vīz·ər }

common fraction A fraction whose numerator and denominator are both integers. { 'käm·ən 'frak·shən }

common logarithm The exponent in the representation of a number as a power of 10. Also known as Briggs' logarithm. { ¦käm·ən 'läg·ə‚rith·əm }

common multiple A quantity (polynomial number) divisible by all quantities in a given set. { ¦käm·ən 'məl·tə·pəl }

commutative algebra An algebra in which the multiplication operation obeys the commutative law. { ¦käm·yə‚tād·iv 'al·jə·brə }

commutative diagram A diagram in which any two mappings between the same pair of sets, formed by composition of mappings represented by arrows in the diagram, are equal. { ¦käm·yə‚tād·iv 'dī·ə‚gram }

commutative group *See* Abelian group. { ¦käm·yə‚tād·iv ‚grüp }

commutative law A rule which requires that the result of a binary operation be independent of order; that is, $ab = ba$. { ¦käm·yə‚tād·iv ‚lȯ }

commutative operation A binary operation that obeys a commutative law, such as addition and multiplication on the real or complex numbers. Also known as Abelian operation. { ¦käm·yə‚tād·iv ‚äp·ə'rā·shən }

commutative ring A ring in which the multiplication obeys the commutative law. Also known as Abelian ring. { ¦käm·yə‚tād·iv ‚riŋ }

commutator The commutator of a and b is the element c of a group such that $bac = ab$. { 'käm·yə‚tād·ər }

commutator subgroup The subgroup of a given group G consisting of all products of the form $g_1g_2 \ldots g_n$, where each g_i is the commutator of some pair of elements in G. { 'käm·yə‚tād·ər 'səb‚grüp }

compactification For a topological space X, a compact topological space that contains X. { käm'pak·tə·fe‚kā·shən }

compact-open topology A topology on the space of all continuous functions from one topological space into another; a subbase for this topology is given by the sets $W(K,U) = \{f:f(K) \subset U\}$, where K is compact and U is open. { ¦käm‚pakt ¦ō·pən tə'päl·ə·jē }

compact operator A linear transformation from one normed vector space to another, with the property that the image of every bounded set has a compact closure. { ¦käm‚pakt 'äp·ə‚rād·ər }

compact set A set in a topological space with the property that every open cover has a finite subset which is also a cover. Also known as bicompact set. { ¦käm‚pakt 'set }

compact space A topological space which is a compact set. { ¦käm‚pakt 'spās }

compactum A topological space that is metrizable and compact. { käm'pak·təm }

comparable functions Two real-valued functions with a common domain of definition such that the values of one of the functions are equal to or greater than the values of the other for all the points in this domain. { 'käm·prə·bəl 'fəŋk·shənz }

comparison test A simple test for the convergence of an infinite series, according to which a series converges if the absolute values of each of its terms are equal to or less than the corresponding term of a series that is known to converge, and diverges if each of its terms is equal to or greater than the absolute value of the corresponding term of a series that is known to diverge. { kəm'par·ə·sən ˌtest }

complement 1. The complement of a number A is another number B such that the sum $A + B$ will produce a specified result. 2. For a subset of a set, the collection of all members of the set which are not in the given subset. 3. For a fuzzy set A with membership function m_A, the complement of A is the fuzzy set $\bar{A}$ whose membership function mA has the value $1 - m_A(x)$ for every element x. 4. *See* radix complement. { 'käm·plə·mənt }

complementary angle One of a pair of angles whose sum is 90°. { ˌkäm·plə'men·trē 'aŋ·gəl }

complementary function Any solution of the equation obtained from a given linear differential equation by replacing the inhomogeneous term with zero. { ˌkäm·plə'men·trē 'fəŋk·shən }

complementary minor *See* minor. { ˌkäm·plə'men·trē 'mī·nər }

complementary operation An operation on a Boolean algebra of two elements (labeled "true" and "false") whose result is the negation of a given operation; for example, NAND is complementary to the AND function. { ˌkäm·plə'men·trē ˌäp·ə'rā·shən }

complementation The act of replacing a set by its complement. { ˌkäm·plə·mən'tā·shən }

complemented lattice A lattice with distinguished elements a and b, and with the property that corresponding to each point x of the lattice, there is a y such that the greatest lower bound of x and y is a, and the least upper bound of x and y is b. { 'käm·pləˌment·əd 'lad·əs }

complete graph A graph with exactly one edge connecting each pair of distinct vertices and no loops. { kəm¦plēt 'graf }

complete induction *See* mathematical induction. { kəm'plēt in'dək·shən }

complete integral 1. A solution of an nth order ordinary differential equation which depends on n arbitrary constants as well as the independent variable. Also known as complete primitive. 2. A solution of a first-order partial differential equation with n independent variables which depends upon n arbitrary parameters as well as the independent variables. { kəm'plēt 'in·tə·grəl }

complete lattice A partially ordered set in which every subset has both a supremum and an infimum. { kəm'plēt 'lad·əs }

completely normal space A topological space with the property that any pair of sets with disjoint closures can be separated by open sets. { kəm'plēt·lē ¦nór·məl 'spās }

completely reducible representation A representation of a group as a family of linear operators of a vector space V such that V is the direct sum of subspaces $V_1, \ldots,$ V_n which are invariant under these operators, but $V_1, \ldots, V_n$ do not have any proper closed subspaces which are also invariant under these operators. Also known as semisimple representation. { kəm¦plēt·lē ri¦düs·ə·bəl ˌrep·ri·zen'tā·shən }

completely regular space A topological space X where for every point x and neighborhood U of x there is a continuous function from X to [0,1] with $f(x) = 1$ and $f(y) = 0$, if y is not in U. Also known as Tychonoff space. { kəm'plēt·lē ¦reg·yə·lər 'spās }

complete metric space A metric space in which every Cauchy sequence converges to a point of the space. Also known as complete space. { kəm'plēt ¦me·trik 'spās }

complete normed linear space *See* Banach space. { kəm'plēt ¦nȯrmd ¦lin·ē·ər 'spas }

complete order *See* linear order. { kəm'plēt 'ȯrd·ər }

complete orthonormal set A set of mutually orthogonal unit vectors in a (possibly infinite dimensional) vector space which is contained in no larger such set, that is no nonzero vector is perpendicular to all the vectors in the set. Also known as closed orthonormal set. { kəm'plēt ¦ȯr·thō¦nȯr·məl 'set }

complete primitive *See* complete integral. { kəm'plēt 'prim·əd·iv }

complete quadrangle A plane figure consisting of a quadrangle and its two diagonals. Also known as complete quadrilateral. { kəm'plēt 'kwä,draŋ·gəl }

complete quadrilateral *See* complete quadrangle. { kəm'plēt ,kwä·drə'lad·ə·rəl }

complete residue system modulo *n* A set of integers that includes one and only one member of each number class modulo n. { kəm¦plēt ¦rez·ə·dü ¦sis·təm ¦mäj·ə,lō 'en }

complete space *See* complete metric space. { kəm'plēt 'spās }

completing the square A method of solving quadratic equations, consisting of moving all terms to the left side of the equation, dividing through by the coefficient of the square term, and adding to both sides a number sufficient to make the left side a perfect square. { kəm'plēd·iŋ thə 'skwer }

completion For a metric space X, a complete metric space obtained from X by formally adding limits to Cauchy sequences. { kəm'plē·shən }

complex A space which is represented as a union of simplices which intersect only on their faces. { 'käm,pleks }

complex conjugate **1.** One of a pair of complex numbers with identical real parts and with imaginary parts differing only in sign. Also known as conjugate. **2.** The matrix whose elements are the complex conjugates of the corresponding elements of a given matrix. { 'käm,pleks 'kän·jə·gət }

complex fraction A fraction whose numerator or denominator is a fraction. { 'käm,pleks 'frak·shən }

complex integer *See* Gaussian integer. { ¦käm,pleks 'int·ə·jər }

complex number Any number of the form $a + bi$, where a and b are real numbers, and $i^2 = -1$. { 'käm,pleks 'nəm·bər }

complex plane A plane whose points are assigned the real and imaginary parts of complex numbers for coordinates. { ¦käm,pleks 'plān }

complex sphere *See* Riemann sphere. { 'käm,pleks 'sfir }

complex unit Any complex number, $x + iy$, whose absolute value, $\sqrt{(x^2 + y^2)}$, equals 1. { 'käm,pleks 'yü·nət }

complex variable A variable which assumes complex numbers for values. { 'käm,pleks 'ver·ē·ə·bəl }

component **1.** In a graph system, a connected subgraph which is not a subgraph of any other connected subgraph. **2.** For a set S, a connected subset of S that is not a subset of any other connected subset of S. { kəm'pō·nənt }

component vectors Vectors parallel to specified (usually perpendicular) axes whose sum equals a given vector. { kəm'pō·nənt ,vek·tərz }

composite function A function of one or more independent variables that are themselves functions of one or more other independent variables. { kəm'päz·ət 'fəŋk·shən }

composite group A group that contains normal subgroups other than the identity element and the whole group. { kəm'päz·ət 'grüp }

composite number Any positive integer which is not prime. Also known as composite quantity. { kəm'päz·ət 'nəm·bər }

composite quantity *See* composite number. { kəm'päz·ət 'kwän·əd·ē }

composition **1.** The composition of two mappings, f and g, denoted $g \circ f$, where the domain of g includes the range of f, is the mapping which assigns to each element x in the domain of f the element $g(y)$, where $y = f(x)$. **2.** *See* addition. { ,käm·pə'zish·ən }

composition series A normal series G_1, G_2, ..., of a group, where each G_i is a proper normal subgroup of G_{i-1} and no further normal subgroups both contain G_i and are contained in G_{i-1}. { ,käm·pə'zish·ən 'sir,ēz }

compositum Let E and F be fields, both contained in some field L; the compositum of E and F, denoted EF, is the smallest subfield of L containing E and F. { kəm'päz·əd·əm }

compound curve A curve made up of two arcs of differing radii whose centers are on the same side, connected by a common tangent; used to lay out railroad curves because curvature goes from nothing to a maximum gradually, and vice versa. { 'käm,paủnd 'kərv }

compound number A quantity which is expressed as the sum of two or more quantities in terms of different units, for example, 3 feet 10 inches, or 2 pounds 5 ounces. { 'käm ,paủnd 'nəm·bər }

computable function A function whose value can be calculated by some Turing machine in a finite number of steps. Also known as effectively computable function. { kəm'pyüd·ə·bəl 'fəŋk·shən }

computation **1.** The act or process of calculating. **2.** The result so obtained. { ,käm·pyə'tā·shən }

concave function A function $f(x)$ is said to be concave over the interval a,b if for any three points x_1, x_2, x_3 such that $ax_1x_2x_3b$, $f(x_2) \geq L(x_2)$, where $L(x)$ is the equation of the straight line passing through the points $[x_1,f(x_1)]$ and $[x_3,f(x_3)]$. { 'känˌkāv 'fəŋk·shən }

concave polygon A polygon at least one of whose angles is greater than 180°. { 'känˌkāv 'päl·əˌgän }

concave polyhedron A polyhedron for which there is at least one plane that contains a face of the polyhedron and that is such that parts of the polyhedron are on both sides of the plane. { 'kän¦kāv ˌpäl·ə'hē·drən }

concentrated A measure (or signed measure) m is concentrated on a measurable set A if any measurable set B with nonzero measure has a nonnull intersection with A. { 'kän·sənˌtrād·əd }

concentration An operation that provides a relatively sharp boundary to a fuzzy set; for a fuzzy set A with membership function m_A, a concentration of A is a fuzzy set whose membership function has the value $[m_A(x)]^\alpha$ for every element x, where α is a fixed number that is greater than 1. { ˌkän·sən'trā·shən }

concentric circles A family of coplanar circles with the same center. { kən'sen·trik 'sər·kəlz }

conchoid A plane curve consisting of the locus of both ends of a line segment of constant length on a line which rotates about a fixed point, while the midpoint of the segment remains on a fixed curve which does not contain the fixed point. { 'käŋˌkȯid }

conchoid of Nicomedes The conchoid of a straight line with respect to a fixed point that does not lie on the line. { 'käŋˌkȯid əv ˌnik·ə'mē·dēz }

concurrent line One of two or more lines that have a point in common. { kən'kər·ənt ˌlīn }

concurrent plane One of three or more planes that have a point in common. { kən'kər·ənt 'plān }

condensation point For a set in a topological space, a point whose neighborhoods all contain uncountably many points of the set. { ˌkän·dən'sā·shən ˌpȯint }

condition The product of the norm of a matrix and of its inverse. { kən'dish·ən }

conditional implication *See* implication. { kən¦dish·ən·əl ˌim·plə'kā·shən }

conditional convergence The property of a series that is convergent but not absolutely convergent. { kən'dish·ən·əl kən'vər·jəns }

conditional expectation If X is a random variable on a probability space (Ω,F,P), the conditional expectation of X with respect to a given sub σ-field F' of F is an F'-measurable random variable whose expected value over any set in F' is equal to the expected value of X over this set. { kən'dish·ən·əl ˌekˌspek'tā·shən }

conditionally compact set A set whose closure is compact. Also known as relatively compact set. { kən'dish·ən·əl·ē ¦kämˌpakt ˌset }

cone A solid bounded by a region enclosed in a closed curve on a plane and a surface formed by the segments joining each point of the closed curve to a point which is not in the plane. { kōn }

cone of revolution The surface obtained by rotating a line around another line which it intersects, using the intersection point as a pivot. { 'kōn əv rev·ə'lü·shən }

configuration An arrangement of geometric objects. { kən,fig·yə'rā·shən }

confluent hypergeometric function A solution to differential equation $z(d^2w/dz^2) + (\rho - z)(dw/dz) - \alpha w = 0$. { kən'flü·ənt ¦hī·pər,jē·ə¦me,trik 'fəŋk·shən }

confocal conics **1.** A system of ellipses and hyperbolas that have the same pair of foci. **2.** A system of parabolas that have the same focus and the same axis of symmetry. { kän'fō·kəl 'kän·iks }

confocal coordinates Coordinates of a point in the plane with norm greater than 1 in terms of the system of ellipses and hyperbolas whose foci are at (1,0) and (−1,0). { kän'fō·kəl ,kō'ȯrd·ən·əts }

confocal quadrics Quadrics that have the same principal planes and whose sections by any one of these planes are confocal conics. { kän'fō·kəl 'kwäd·riks }

conformable matrices Two matrices which can be multiplied together; this is possible if and only if the number of columns in the first matrix equals the number of rows in the second. { kən'fȯr·mə·bəl 'mā·trə,sēz }

conformal mapping An angle-preserving analytic function of a complex variable. { kən'fȯr·məl 'map·iŋ }

congruence **1.** The property of geometric figures that can be made to coincide by a rigid transformation. **2.** The property of two integers having the same remainder on division by another integer. { kən'grü·əns }

congruence transformation Also known as transformation. **1.** A mapping which associates with each real quadratic form on a set of coordinates the quadratic form that results when the coordinates are subjected to a linear transformation. **2.** A mapping which associates with each square matrix A the matrix $B = SAT$, where S and T are nonsingular matrices, and T is the transpose of S; if A represents the coefficients of a quadratic form, then this definition is equivalent to definition 1. { kən'grü·əns ,tranz·fər,mā·shən }

congruent matrices Two matrices A and B related by the transformation $B = SAT$, where S and T are nonsingular matrices and T is the transpose of S. { kən'grü·ənt 'mā·trə,sēz }

congruent numbers Two numbers having the same remainder when divided by a given quantity called the modulus. { kən'grü·ənt 'nəm·bərz }

conic A curve which may be represented as the intersection of a cone with a plane; the four types of conics are circle, ellipse, parabola, and hyperbola. Also known as conic section. { 'kän·ik }

conical helix A curve that lies on a cone and cuts all the elements of the cone at the same angle. { 'kän·ə·kəl 'hē·liks }

conical projection A projection which associates with each point P in a plane Q the point p in a second plane q which is collinear with O and P, where O is a fixed point lying outside Q. { 'kän·ə·kəl prə'jek·shən }

conical surface A surface formed by the lines which pass through each of the points of a closed plane curve and a fixed point which is not in the plane of the curve. { 'kän·ə·kəl 'sər·fəs }

conicoid A quadric surface (ellipsoid, paraboloid, or hyperboloid) other than a limiting (degenerate) case of such a surface. { 'kän·ə‚kȯid }

conjugate 1. An element y of a group related to a given element x by $y = z^{-1}xz$ or $zy = xz$, where z is another element of the group. Also known as transform. **2.** *See* complex conjugate. { 'kän·jə·gət }

conjugate angles Two angles whose sum is 360° or 2π radians. Also known as explementary angles. { 'kän·jə·gət 'aŋ·gəlz }

conjugate arcs Two arcs of a circle whose sum is the complete circle. { 'kän·jə·gət 'ärks }

conjugate axis For a hyperbola whose equation in cartesian coordinates has the standard form $(x^2/a^2) - (y^2/b^2) = 1$, the portion of the y axis from $(0,-b)$ to $(0,b)$. { 'kän·jə·gət 'ak·səs }

conjugate binomial surds *See* conjugate radicals. { 'kän·jə·gət bī'nōm·ē·əl 'sərdz }

conjugate convex functions Two functions $f(x)$ and $g(y)$ are conjugate convex functions if the derivative of $f(x)$ is 0 for $x = 0$ and constantly increasing for $x > 0$, and the derivative of $g(y)$ is the inverse of the derivative of $f(x)$. { 'kän·jə·gət 'kän‚veks 'fəŋk·shənz }

conjugate curve 1. A member of one of two families of curves on a surface such that exactly one member of each family passes through each point P on the surface, and the directions of the tangents to these two curves at P are conjugate directions. **2.** *See* Bertrand curve. { 'kän·jə·gət 'kərv }

conjugate diameters 1. For a conic section, any pair of straight lines either of which bisects all the chords that are parallel to the other. **2.** For an ellipsoid or hyperboloid, any three lines passing through the point of symmetry of the surface such that the plane containing the conjugate diameters (first definition) of one of the lines also contains the other two lines. { 'kän·jə·gət dī'am·əd·ərz }

conjugate diametral planes A pair of diametral planes, each of which is parallel to the chords that define the other. { 'kän·jə·gət ‚dī·ə'me·trəl 'plānz }

conjugate directions For a point on a surface, a pair of directions, one of which is the direction of a curve on the surface through the point, while the other is the direction of the characteristic of the planes tangent to the surface at points on the curve. { 'kän·jə·gət di'rek·shənz }

conjugate elements 1. Two elements a and b in a group G for which there is an element x in G such that $ax = xb$. **2.** Two elements of a determinant that are interchanged if the rows and columns of the determinant are interchanged. { 'kän·jə·gət 'el·ə·mənts }

conjugate hyperbolas Two hyperbolas having the same asymptotes with semiaxes interchanged. { 'kän·jə·gət hī'pər·bə·ləz }

conjugate lines 1. For a conic section, two lines each of which passes through the intersection of the tangents to the conic at its points of intersection with the other line. **2.** For a quadric surface, two lines each of which intersects the polar line of the other. { 'kän·jə·gət 'līnz }

conjugate planes For a quadric surface, two planes each of which contains the pole of the other. { 'kän·jə·gət 'plānz }

conjugate points For a conic section, two points either of which lies on the line that passes through the points of contact of the two tangents drawn to the conic from the other. { 'kän·jə·gət 'pȯins }

conjugate radicals Binomial surds that are of the type $a\sqrt{b} + c\sqrt{d}$ and $a\sqrt{b} - c\sqrt{d}$, where a, b, c, d are rational but $\sqrt{b}$ and $\sqrt{d}$ are not both rational. Also known as conjugate binomial surds. { 'kän·jə·gət 'rad·ə·kəlz }

conjugate roots Conjugate complex numbers which are roots of a given equation. { 'kän·jə·gət 'rüts }

conjugate ruled surface The ruled surface whose rulings are the lines that are tangent to a given ruled surface at the points of its line of striction and are perpendicular to the rulings of the given ruled surface at these points. { 'kän·jə·gət ¦rüld 'sər·fəs }

conjugate space The set of all continuous linear functionals defined on a normed linear space. { 'kän·jə·gət 'spās }

conjugate subgroups Two subgroups A and B of a group G for which there exists an element x in G such that B consists of the elements of the form xax^{-1}, where a is in A. { 'kän·jə·gət 'səb‚grüps }

conjugate system of curves Two one-parameter families of curves on a surface such that a unique curve of each family passes through each point of the surface, and the directions of the tangents to these two curves at any point on the surface are the conjugate directions at that point. { 'kän·jə·gət ¦sis·təm əv 'kərvz }

conjugate triangles Two triangles in which the poles of the sides of each with respect to a given curve are the vertices of the other. { 'kän·jə·gət 'trī‚aŋ·gəlz }

conjunction The connection of two statements by the word "and." { kən'jəŋk·shən }

conjunctive matrices Two matrices A and B related by the transformation $B = SAT$, where S and T are nonsingular matrices and S is the Hermitian conjugate of I. { kən'jəŋk·tiv 'mā·trə‚sēz }

conjunctive transformation The transformation $B = SAT$, where S is the Hermitian conjugate of T, and matrices A and B are equivalent. { kən'jəŋk·tiv ‚tranz·fər 'mā·shən }

connected graph A graph in which each pair of points is connected by a path. { kə'nek·təd 'graf }

connected set A set in a topological space which is not the union of two nonempty sets A and B for which both the intersection of the closure of A with B and the intersection of the closure of B with A are empty; intuitively, a set with only one piece. { kə'nek·təd 'set }

connected space A topological space which cannot be written as the union of two nonempty disjoint open subsets. { kə'nek·təd 'spās }

connected surface A surface between any two points of which there is a continuous path that does not cross the surface's boundary. { kə'nek·təd 'sər·fəs }

connectivity number 1. The number of points plus 1 which can be removed from a curve without separating the curve into more than one piece. **2.** The number of closed cuts or cuts joining points of previous cuts (or joining points on the boundary) plus 1 which can be made on a surface without separating the surface. Also known as Betti number. **3.** In general, the n-dimensional connectivity number of a topological space X is the number of infinite cyclic groups whose direct sum with the torsion group $G_n(X)$ forms the homology group $H_n(X)$. { kə͵nek′tiv·əd·ē ͵nəm·bər }

consequent **1.** The second term or denominator of a ratio. **2.** The second of the two statements in an implication. { ′kän·sə·kwənt }

consistency condition The requirement that a mathematical theory be free from contradiction. { kən′sis·tən·sē kən′dish·ən }

consistent equations Two or more equations that are all satisfied by at least one set of values of the variables. { kən′sis·tənt i′kwā·zhənz }

constant of integration An arbitrary constant that must be added to an indefinite integral of a function to obtain all the indefinite integrals of that function. Also known as integration constant. { ′kän·stənt əv ͵in·tə′grā·shən }

constant term A term that does not contain a variable. Also known as absolute term. { ′kän·stənt ′tərm }

constrained optimization problem A nonlinear programming problem in which there are constraint functions. { kən′strānd äp·tə·mə′zā·shən ͵präb·ləm }

constraint function A function defining one of the prescribed conditions in a nonlinear programming problem. { kən′strānt ͵fəŋk·shən }

construction The process of drawing with suitable instruments a geometrical figure satisfying certain specified conditions. { kən′strək·shən }

content *See* Jordan content. { ′kän͵tent }

contiguous functions Any pair of hypergeometric functions in which one of the parameters differs by unity and the other two are equal. { kən′tig·yə·wəs ′fəŋk·shənz }

continuant The determinant of a continuant matrix. { kən′tin·yə·wənt }

continuant matrix A square matrix all of whose nonzero elements lie on the principal diagonal or the diagonals immediately above and below the principal diagonal. Also known as triple-diagonal matrix. { kən′tin·yə·wənt ′mā·triks }

continued equality An expression in which three or more quantities are set equal by means of two or more equality signs. { kən′tin·yüd i′kwäl·əd·ē }

continued fraction The sum of a number and a fraction whose denominator is the sum of a number and a fraction, and so forth; it may have either a finite or an infinite number of terms. { kən′tin·yüd ′frak·shən }

continued-fraction expansion **1.** An expansion of a driving-point function about infinity (or zero) in a continued fraction, in which the terms are alternately constants and multiples of the complex frequency (or multiples of the reciprocal of the complex frequency). **2.** A representation of a real number by a continued fraction, in a manner similar to the representation of real numbers by a decimal expansion. { kən′tin·yüd ′frak·shən ik′span·shən }

continued product A product of three or more factors, or of an infinite number of factors. { kən'tin·yüd 'präd·əkt }

continuous at a point A function f is continuous at a point x if for every sequence $\{x_n\}$ whose limit is x, the sequence $f(x_n)$ converges to $f(x)$; in a general topological space, for every neighborhood W of $f(x)$, there is a neighborhood N of x such that $f^{-1}(W)$ is contained in N. { kən¦tin·yə·wəs ad ə 'point }

continuous deformation A transformation of an object that magnifies, shrinks, rotates, or translates portions of the object in any manner without tearing. { kən¦tin·yə·wəs ˌdē·fȯr'mā·shən }

continuous extension A continuous function which is equal to another continuous function defined on a smaller domain. { kən¦tin·yə·wəs ik'sten·shən }

continuous function A function which is continuous at each point of its domain. Also known as continuous transformation. { kən¦tin·yə·wəs 'fəŋk·shən }

continuous geometry A generalization of projective geometry. { kən¦tin·yə·wəs jē'äm·ə·trē }

continuous image The image of a set under a continuous function. { kən¦tin·yə·wəs 'im·ij }

continuous operator A linear transformation of Banach spaces which is continuous with respect to their topologies. { kən¦tin·yə·wəs 'äp·əˌrād·ər }

continuous spectrum The portion of the spectrum of a linear operator which is a continuum. { kən¦tin·yə·wəs 'spek·trəm }

continuous transformation *See* continuous function. { kən¦tin·yə·wəs tranz·fər'mā·shən }

continuum A compact, connected set. { kən'tin·yə·wəm }

contour integral A line integral of a complex function, usually over a simple closed curve. { 'känˌtu̇r ˌin·tə·grəl }

contracted curvature tensor A symmetric tensor of second order, obtained by summation on two indices of the Riemann curvature tensor which are not antisymmetric. Also known as contracted Riemann-Christoffel tensor; Ricci tensor. { kən'trak·təd 'kər·və·chər ˌten·sər }

contracted Riemann-Christoffel tensor *See* contracted curvature tensor. { kən'trak·təd ¦rē·män kris'tȯf·əl ˌten·sər }

contraction **1.** A continuous function of a metric space to itself which moves each pair of points closer together. **2.** The operation of setting one of the contravariant indices of a tensor equal to one of the covariant indices and summing over this index, yielding a tensor of order two less than that of the original tensor. { kən'trak·shən }

contraction semigroup A strongly continuous semigroup all of whose elements have norms which are equal to or less than a constant which is, in turn, less than 1. { kən'trak·shən 'sem·iˌgrüp }

contrapositive The contrapositive of the statement "if p, then q" is the equivalent statement "if not q, then not p." { ¦kän·trə'päz·əd·iv }

contravariant functor A functor which reverses the sense of morphisms. { ¦kän·trə'ver·ē·ənt 'fəŋk·tər }

contravariant index A tensor index such that, under a transformation of coordinates, the procedure for obtaining a component of the transformed tensor for which this index has the value p involves taking a sum over q of the product of a component of the original tensor for which the index has the value q times the partial derivative of the pth transformed coordinate with respect to the qth original coordinate; it is written as a superscript. { ¦kän·trə'ver·ē·ənt 'in,deks }

contravariant tensor A tensor with only contravariant indices. { ¦kän·trə'ver·ē·ənt 'ten·sər }

contravariant vector A contravariant tensor of degree 1, such as the tensor whose components are differentials of the coordinates. { ¦kän·trə'ver·ē·ənt 'vek·tər }

convergence The property of having a limit for infinite series, sequences, products, and so on. { kən'vər·jəns }

convergence in measure A sequence of functions $f_n(x)$ converges in measure to $f(x)$ if given any $\epsilon > 0$, the measure of the set of points at which $f_n(x) - f(x) > \epsilon$ is less than ϵ, provided n is sufficiently large. { kən'vər·jəns in 'mezh·ər }

convergent One of the continued fractions that is obtained from a given continued fraction by terminating after a finite number of terms. { kən'vər·jənt }

convergent integral An improper integral which has a finite value. { kən'vər·jənt 'in·tə·grəl }

convergent sequence A sequence which has a limit. { kən'vər·jənt 'sē·kwəns }

convergent series A series whose sequence of partial sums has a limit. { kən'vər·jənt 'sir,ēz }

converse The converse of the statement "if p, then q" is the statement "if q, then p." { 'kän,vərs }

conversion factor The numerical factor by which one must multiply (or divide) a quantity that is expressed in terms of a certain unit to express the quantity in terms of another unit. Also known as unit conversion factor. { kən'vər·zhən ,fak·tər }

conversion ratio *See* conversion factor. { kən'vər·zhən ,rā·shō }

convex angle A polyhedral angle that lies entirely on one side of each of its faces. { 'kän,veks 'aŋ·gəl }

convex body A convex set that has at least one interior point. { 'kän,veks 'bäd·ē }

convex combination A linear combination of vectors in which the sum of the coefficients is 1. { 'kän,veks ,käm·bə'nā·shən }

convex curve A plane curve for which any straight line that crosses the curve crosses it at just two points. { 'kän,veks 'kərv }

convex function A function $f(x)$ is considered to be convex over the interval a,b if for any three points x_1, x_2, x_3 such that $ax_1x_2x_3b$, $f(x_2) \leq L(x_2)$, where $L(x)$ is the equation of the straight line passing through the points $[x_1, f(x_1)]$ and $[x_3, f(x_3)]$. { 'kän,veks 'fəŋk·shən }

convex function in the sense of Jensen A function $f(x)$ over an interval a,b such that, for any two points x_1 and x_2 satisfying $a < x_1 < x_2 < b$, $f[(x_1 + x_2)/2] \leqq (1/2)\cdot[f(x_1) + f(x_2)]$. { ¦kän‚veks ¦fəŋk·shən in thə ‚sens əv 'jen·sən }

convex hull The smallest convex set containing a given collection of points in a real linear space. Also known as convex linear hull. { 'kän‚veks 'həl }

convex linear combination A linear combination in which the scalars are nonnegative real numbers whose sum is 1. { ¦kän‚veks ‚lin·ē·ər ‚käm·bə'nā·shən }

convex linear hull *See* convex hull. { 'kän‚veks 'lin·ē·ər ‚həl }

convex polygon A polygon all of whose interior angles are less than or equal to 180°. { 'kän‚veks 'päl·i‚gän }

convex polyhedron A polyhedron in the plane which is a convex set, for example, any regular polyhedron. { 'kän‚veks ¦päl·i¦hē·drən }

convex programming Nonlinear programming in which both the function to be maximized or minimized and the constraints are appropriately chosen convex or concave functions of the independent variables. { 'kän‚veks 'prō‚gram·iŋ }

convex sequence A sequence of numbers, $a_1, a_2, \ldots$, such that $a_{i+1} \leqq (1/2)(a_i + a_{i+2})$ for all $i \geqq 1$ (or for all i satisfying $1 \leqq i < n - 2$ if the sequence is a finite sequence with n terms). { 'kän‚veks 'sē·kwəns }

convex set A set which contains the entire line segment joining any pair of its points. { 'kän‚veks 'set }

convolution The convolution of the functions f and g is the function F, defined by

$$F(x) = \int_0^x f(t)g(x - t)\, dt$$

{ ‚kän·və'lü·shən }

convolution family *See* faltung. { ‚kän·və'lü·shən ‚fam·lē }

convolution theorem A theorem stating that, under specified conditions, the integral transform of the convolution of two functions is equal to the product of their integral transforms. { ‚kän·və'lü·shən ‚thir·əm }

coordinate axes One of a set of lines or curves used to define a coordinate system; the value of one of the coordinates uniquely determines the location of a point on the axis, while the values of the other coordinates vanish on the axis. { kō'ȯrd·ən·ət 'ak‚sēz }

coordinate basis A basis for tensors on a manifold induced by a set of local coordinates. { kō'ȯrd·ən·ət 'bā·səs }

coordinates A set of numbers which locate a point in space. { kō'ȯrd·ən·əts }

coordinate systems A rule for designating each point in space by a set of numbers. { kō'ȯrd·ən·ət ‚sis·təmz }

coordinate transformation A mathematical or graphic process of obtaining a modified set of coordinates by performing some nonsingular operation on the coordinate axes, such as rotating or translating them. { kō'ȯrd·ən·ət tranz·fər'mā·shən }

Cornu's spiral A plane curve whose curvature is proportional to its arc length, and whose cartesian coordinates are given in parametric form by the Fresnel integrals. Also known as clothoid; Euler's spiral. { 'kȯr·nüz ¦spī·rəl }

correlation curve *See* correlogram. { ˌkär·ə'lā·shən ˌkərv }

correlogram A curve showing the assumed correlation between two mathematical variables. Also known as correlation curve. { kə'rel·əˌgram }

corresponding angles For two lines, l_1 and l_2, cut by a transversal t, a pair of angles such that (1) one of the angles has sides l_1 and t while the other has sides l_2 and t; (2) both angles are on the same side of t; and (3) the angles are on the same sides of l_1 and l_2, respectively. { 'kär·əˌspänd·iŋ 'aŋ·gəlz }

cos *See* cosine function.

cosecant The reciprocal of the sine. Denoted csc. { kō'sēˌkant }

coset For a subgroup of a group, a set consisting of all elements of the form xh or of all elements of the form hx, where h is an element of the subgroup and x is a fixed element of the group. { 'kōˌset }

cosh *See* hyperbolic cosine.

cosine function In a right triangle with an angle θ, the cosine function gives the ratio of adjacent side to hypotenuse; more generally, it is the function which assigns to any real number θ the abscissa of the point on the unit circle obtained by moving from (1,0) counterclockwise θ units along the circle, or clockwise θ units if θ is less than 0. Denoted cos. { 'kōˌsīn ˌfəŋk·shən }

cosine series A Fourier series that contains only terms that are even in the independent variable, that is, the constant term and terms involving the cosine function. { 'kōˌsīn ˌsir·ēz }

cot *See* cotangent.

cotangent The reciprocal of the tangent. Denoted cot; ctn. { kō'tan·jənt }

coterminal angles Two angles that have the same initial line and the same terminal line and therefore differ by a multiple of 2π radians or 360°. { ¦kō¦tərm·ən·əl 'aŋ·gəlz }

coth *See* hyperbolic cotangent.

countability axioms Two conditions which are satisfied by a euclidean space and one or the other of which is often assumed in the study of a general topological space; the first states that any point in the topological space has a countable local base, while the second states that the topological space has a countable base. { ˌkau̇n·tə'bil·əd·ē ˌax·sē·əmz }

countable Either finite or denumerable. Also known as enumerable. { 'kau̇nt·ə·bəl }

countably additive Given a measure m, and a sequence of pairwise disjoint measurable sets, the property that the measure of the union is equal to the sum of the measures of the sets. { 'kau̇nt·ə·blē 'ad·əd·iv }

countably additive set function A real-value function defined on a class of sets such that the value of the function on the union of any pairwise disjoint sequence of sets is equal to the sum of the sequence of the values of the function on the sets. { ¦kaůn·tə·blē ¦ad·əd·iv 'set ˌfəŋk·shən }

countably compact set A set with the property that every cover with countably many open sets contains a finite number of sets which is also a cover. { 'kaůnt·ə·blē ˌkämˌpakt 'set }

countably infinite set *See* denumerable set. { 'kaůnt·ə·blē ˌin·fə·nət 'set }

countably subadditive A set function m is countably subadditive if, given any sequence of sets, the measure of the union is less than or equal to the sum of the measures of the sets. { kaůnt·ə·blē səb'ad·əd·iv }

countably subadditive set function A real-valued function defined on a class of sets such that the value of the function on the union of any sequence of sets is equal to or less than the sum of the sequence of the values of the function on the sets. { ¦kaůn·tə·blē səb¦ad·əd·iv 'set ˌfəŋk·shən }

covariant components Vector or tensor components which, in a transformation from one set of basis vectors to another, transform in the same manner as the basis vectors. { kō'ver·ē·ənt kəm'pō·nəns }

covariant derivative For a tensor field at a point P of an affine space, a new tensor field equal to the difference between the derivative of the original field defined in the ordinary manner and the derivative of a field whose value at points close to P are parallel to the value of the original field at P as specified by the affine connection. { kō'ver·ē·ənt də'riv·əd·iv }

covariant functor A functor which does not change the sense of morphisms. { kō'ver·ē·ənt 'fəŋk·tər }

covariant index A tensor index such that, under a transformation of coordinates, the procedure for obtaining a component of the transformed tensor for which this index has value p involves taking a sum over q of the product of a component of the original tensor for which the index has the value q times the partial derivative of the qth original coordinate with respect to the pth transformed coordinate; it is written as a subscript. { kō'ver·ē·ənt 'inˌdeks }

covariant tensor A tensor with only covariant indices. { kō'ver·ē·ənt 'ten·sər }

covariant vector A covariant tensor of degree 1, such as the gradient of a function. { kō'ver·ē·ənt 'vek·tər }

cover *See* covering. { 'kəv·ər }

covering For a set A, a collection of sets whose union contains A. Also known as cover. { 'kəv·riŋ }

covers *See* coversed sine.

coversed sine The coversed sine of A is $1 - \text{sine } A$. Denoted covers. Also known as coversine; versed cosine. { ¦kōˌvərst 'sīn }

coversine *See* coversed sine. { ˌkōˌvər'sīn }

cracovian An object which is the same as a matrix except that the product of cracovians A and B is equal to the matrix product $A'B$, where A' is the transpose of A. { krə'kō·vē·ən }

Cramer's rule The method of solving a system of linear equations by means of determinants. { 'krā·mərz ˌrül }

critical function A function satisfying the Euler equations in the calculus of variations. { 'krid·ə·kəl 'fəŋk·shən }

critical point A point at which the first derivative of a function is either 0 or does not exist. { 'krid·ə·kəl 'pȯint }

critical table A table, usually for a function that varies slowly, which gives only values of the argument near which changes in the value of the function, as rounded to the number of decimal places displayed in the table, occur. { 'krid·ə·kəl ˌtā·bəl }

critical value The value of the dependent variable at a critical point of a function. { 'krid·ə·kəl 'val·yü }

cross-cap The self-intersecting surface that results when a Möbius band is deformed so that its boundary is a circle. { 'krȯs ˌkap }

cross curve A plane curve whose equation in cartesian coordinates x and y is $(a^2/x^2) + (b^2/y^2) = 1$, where a and b are constants. Also known as cruciform curve. { 'krȯs ˌkərv }

cross multiplication Multiplication of the numerator of each of two fractions by the denominator of the other, as when eliminating fractions from an equation. { ¦krȯs ˌməl·tə·plə'kā·shən }

crossover length A length characteristic of a fractal network such that at scales which are small compared with this length the fractal nature of the structure is manifest in its dynamics, whereas at scales which are large compared with this length the dynamics resemble those of a crystalline structure. { 'krȯsˌō·vər ˌleŋkth }

cross product 1. An anticommutative multiplication on the vectors of euclidean three-dimensional space. Also known as vector product. **2.** The product of the two mean terms of a proportion, or the product of the two extreme terms; in the proportion $a/b = c/d$, it is ad or bc. { 'krȯs ˌprä·dəkt }

cross ratio For four collinear points, A, B, C, and D, the ratio $(AB)(CD)/(AD)(CB)$, or one of the ratios obtained from this quantity by a permutation of A, B, C, and D. { 'krȯs ˌrā·shō }

cross section 1. The intersection of an n-dimensional geometric figure in some euclidean space with a lower dimensional hyperplane. **2.** A right inverse for the projection of a fiber bundle. { 'krȯs ˌsek·shən }

Crout reduction Modification of the Gauss procedure for numerical solution of simultaneous linear equations; adapted for use on desk calculators and digital computers. { 'krau̇t ri'dək·shən }

cruciform curve *See* cross curve. { 'krü·səˌfȯrm ˌkərv }

crunode A point on a curve through which pass two branches of the curve with different tangents. Also known as node. { ¦krü¦nōd }

csc *See* cosecant.

csch *See* hyperbolic cosecant.

ctn *See* cotangent.

cubature The numerical integration of a function of two variables. { 'kyüb·ə·chər }

cube **1.** Regular polyhedron whose faces are all square. **2.** For a number a, the new number obtained by taking the threefold product of a with itself: $a \times a \times a$. { kyüb }

cube root Another number whose cube is the original number. { 'kyüb 'rüt }

cubical parabola A plane curve whose equation in cartesian coordinates x and y is $y = x^3$. { 'kyüb·ə·kəl pə'rab·ə·lə }

cubic curve A plane curve which has an equation of the form $f(x, y) = 0$, where $f(x, y)$ is a polynomial of degree three in x and y. { 'kyü·bik 'kərv }

cubic determinant A mathematical form analogous to an ordinary determinant, with the elements forming a cube instead of a square. { 'kyü·bik di'tər·mə·nənt }

cubic equation A polynomial equation with no exponent larger than 3. { 'kyü·bik i'kwā·zhən }

cubic polynomial A polynomial in which all exponents are no greater than 3. { 'kyü·bik ˌpäl·ə'nō·mē·əl }

cubic spline One of a collection of cubic polynomials used in interpolating a function whose value is specified at each of a collection of distinct ordered values, X_i ($i = 1, ..., n$), and whose slope is specified at X_1 and X_n; one cubic polynomial is found for each interval, such that the interpolating system has the prescribed values at each of the X_i, the prescribed slope at X_1 and X_n, and a continuous slope at each of the X_i. { 'kyü·bik 'splīn }

cubic surd A cube root of a rational number that is itself an irrational number. { 'kyü·bik 'sərd }

cuboctahedron A polyhedron whose faces consist of six equal squares and eight equal equilateral triangles, and which can be formed by cutting the corners off a cube; it is one of the 13 Archimedean solids. Also spelled cubooctahedron. { ˌkyü¦bäk·tə'hē·drən }

cubooctahedron *See* cuboctahedron. { ¦kyü·bōˌäk·tə'hē·drən }

Cullen number A number having the form $C_n = (n \cdot 2^n) + 1$ for $n = 0, 1, 2, \ldots$. { 'kəl·ən ˌnəm·bər }

cup The symbol $\cup$, which indicates the union of two sets. { kəp }

cup product A multiplication defined on cohomology classes; it gives cohomology a ring structure. { 'kəp ˌpräd·əkt }

curl The curl of a vector function is a vector which is formally the cross product of the del operator and the vector. Also known as rotation (rot). { kərl }

curtate cycloid A trochoid in which the distance from the center of the rolling circle to the point describing the curve is less than the radius of the rolling circle. { 'kərˌtāt 'sīˌklȯid }

curvature The reciprocal of the radius of the circle which most nearly approximates a curve at a given point; the rate of change of the unit tangent vector to a curve with respect to arc length of the curve. { 'kər·və·chər }

curvature tensor *See* Riemann-Christoffel tensor. { 'kər·və·chər ˌten·sər }

curve The continuous image of the unit interval. { kərv }

curved surface A surface having no part that is a plane surface. { 'kərvd 'sər·fəs }

curve tracing The method of graphing a function by plotting points and analyzing symmetries, derivatives, and so on. { 'kərv ˌtrās·iŋ }

curvilinear coordinates Any linear coordinates which are not cartesian coordinates; frequently used curvilinear coordinates are polar coordinates and cylindrical coordinates. { 'kər·və'lin·ē·ər kō'ȯrd·ən·əts }

curvilinear solid A solid whose surfaces are not planes. { ˌkər·və'lin·ē·ər 'säl·əd }

curvilinear transformation A transformation from one coordinate system to another in which the coordinates in the new system are arbitrary twice-differentiable functions of the coordinates in the old system. { 'kər·və'lin·ē·ər tranz·fər'mā·shən }

cusp A singular point of a curve at which the limits of the tangents of the portions of the curve on either side of the point coincide. Also known as spinode. { kəsp }

cuspidal cubic A cubic curve that has one cusp, one point of inflection, and no node. { 'kəs·pəd·əl 'kyü·bik }

cuspidal locus A curve consisting of the cusps of some family of curves. { 'kəs·pəd·əl 'lō·kəs }

cusp of the first kind A cusp such that the two portions of the curve adjacent to the cusp lie on opposite sides of the limiting tangent to the curve at the cusp. Also known as simple cusp. { 'kəsp əv thə 'fərst ˌkīnd }

cusp of the second kind A cusp such that the two portions of the curve adjacent to the cusp lie on the same side of the limiting tangent to the curve at the cusp. { 'kəsp əv thə 'sek·ənd kīnd }

cut A subset of a given set whose removal from the original set leaves a set that is not connected. { kət }

cut capacity For a network whose points have been partitioned into two specified classes, C_1 and C_2, the sum of the capacities of all the segments directed from a point in C_1 to a point in C_2. Also known as cut value. { 'kət kə'pas·əd·ē }

cut point A point in a component of a graph whose removal disconnects that component. Also known as articulation point. { 'kət ˌpȯint }

cut value *See* cut capacity. { 'kət ˌval·yü }

cycle **1.** A member of the kernel of a boundary homomorphism. **2.** A closed path in a graph that does not pass through any vertex more than once and passes through at least three vertices. **3.** *See* cyclic permutation. { 'sī·kəl }

cyclic curve **1.** A curve (such as a cycloid, cardioid, or epicycloid) generated by a point of a circle that rolls (without slipping) on a given curve. **2.** The intersection of a quadric surface with a sphere. Also known as spherical cyclic curve. **3.** The stereographic projection of a spherical cyclic curve. Also known as plane cyclic curve. { 'sīk·lik 'kərv }

cyclic extension A Galois extension whose Galois group is cyclic. { 'sīk·lik ik'sten·chən }

cyclic group A group that has an element a such that any element in the group can be expressed in the form a^n, where n is an integer. { 'sīk·lik ˌgrüp }

cyclic identity The principle that the sum of any component of the Riemann-Christoffel tensor and two other components obtained from it by cyclic permutation of any three indices, while the fourth is held fixed, is zero. { 'sīk·lik īˌden·təd·ē }

cyclic permutation A permutation of an ordered set of symbols which sends the first to the second, the second to the third, ..., the last to the first. Also known as cycle. { 'sīk·lik pər·myə'tā·shən }

cyclindroid 1. A cylindrical surface generated by the lines perpendicular to a plane that pass through an ellipse in the plane. **2.** A surface that is generated by a straight line that moves so as to intersect two curves and remain parallel to a given plane. { si'klinˌdrȯid }

cycloid The curve traced by a point on the circumference of a circle as the circle rolls along a straight line. { 'sīˌklȯid }

cyclosymmetric function A function whose value is unchanged under a cyclic permutation of its variables. { ˌsi·klō·si¦me·trik 'fəŋk·shən }

cyclotomic equation An equation which has the form $x^{n-1} + x^{n-2} + \cdots + x + 1 = 0$, where n is a prime number. { ¦sī·klō¦täm·ik i'kwā·zhən }

cyclotomic field The extension field of a given field K which is the smallest extension field of K that includes the nth roots of unity for some integer n. { ˌsī·klə¦täm·ik 'fēld }

cyclotomy Theory of dividing the circle into equal parts or constructing regular polygons or, analytically, of finding the nth roots of unity. { sī'kläd·ə·mē }

cylinder 1. A solid bounded by a cylindrical surface and two parallel planes, or the surface of such a solid. **2.** *See* cylindrical surface. { 'sil·ən·dər }

cylinder function Any solution of the Bessel equation, including Bessel functions, Neumann functions, and Hankel functions. { 'sil·ən·dər ˌfəŋk·shən }

cylindrical coordinates A system of curvilinear coordinates in which the position of a point in space is determined by its perpendicular distance from a given line, its distance from a selected reference plane perpendicular to this line, and its angular distance from a selected reference line when projected onto this plane. { sə'lin·drə·kəl ˌkō'ȯrd·ən·əts }

cylindrical function *See* Bessel function. { sə'lin·dri·kəl ˌfəŋk·shən }

cylindrical helix A curve lying on a cylinder which intersects the elements of the cylinder at a constant angle. { sə'lin·drə·kəl 'hēˌliks }

cylindrical surface A surface consisting of each of the straight lines which are parallel to a given straight line and pass through a given curve. Also known as cylinder. { sə'lin·drə·kəl 'sər·fəs }

D

d'Alembertian A differential operator in four-dimensional space,

$$\frac{\partial^2}{\partial x^2} + \frac{\partial^2}{\partial y^2} + \frac{\partial^2}{\partial z^2} - \frac{1}{c^2}\frac{\partial^2}{\partial t^2}$$

which is used in the study of relativistic mechanics. { ¦dal·əm¦bər·shən }

d'Alembert's test for convergence A series Σa_n converges if there is an N such that the absolute value of the ratio a_n/a_{n-1} is always less than some fixed number smaller than 1, provided n is at least N, and diverges if the ratio is always greater than 1. { ¦dal·əm¦bərz ˌtest fər kən'vər·jəns }

Darboux's monodromy theorem The proposition that, if the function $f(z)$ of the complex variable z is analytic in a domain D bounded by a simple closed curve C, and $f(z)$ is continuous in the union of D and C and is injective for z on C, then $f(z)$ is injective for z in D. { 'där·büz ¦män·əˌdrä·mē ˌthir·əm }

decagon A 10-sided polygon. { 'dek·əˌgän }

decahedron A polyhedron that has 10 faces. { ˌdek·ə'hē·drən }

decidable predicate A predicate for which there exists an algorithm which, for any given value of its independent variables, provides a definite answer as to whether or not it is true. { di'sīd·ə·bəl 'pred·ə·kət }

decimal A number expressed in the scale of tens. { 'des·məl }

decimal fraction Any number written in the form: an integer followed by a decimal point followed by a (possibly infinite) string of digits. { ¦des·məl ¦frak·shən }

decimal number A number signifying a decimal fraction by a decimal point to the left of the numerator with the number of figures to the right of the point equal to the power of 10 of the denominator. { ¦des·məl ¦nəm·bər }

decimal number system A representational system for the real numbers in which place values are read in powers of 10. { ¦des·məl 'nəm·bər ˌsis·təm }

decimal place Reference to one of the digits following the decimal point in a decimal fraction; the kth decimal place registers units of 10^{-k}. { ¦des·məl ¦plās }

decimal point A dot written either on or slightly above the line; used to mark the point at which place values change from positive to negative powers of 10 in the decimal number system. { 'des·məl ˌpȯint }

decimal system A number system based on the number 10; in theory, each unit is 10 times the next smaller one. { 'des·məl ˌsis·təm }

decomposable process A process which can be reduced to several basic events. { dē·kəm′pō·zə·bəl ′präs·əs }

decomposition The expression of a fraction as a sum of partial fractions. { dē‚käm·pə′zish·ən }

decreasing function 1. A function of x whose value gets smaller as x gets larger, that is, if $x < y$, then $f(x) > f(y)$. **2.** *See* monotone nonincreasing function. { di′krēs·iŋ ‚fəŋk·shən }

decrement The quantity by which a variable is decreased. { ′dek·rə·mənt }

Dedekind cut A set of rational numbers satisfying certain properties, with which a unique real number may be associated; used to define the real numbers as an extension of the rationals. { ′dā·də·kint ‚kət }

Dedekind test If the series

$$\sum_i (b_i - b_{i+1})$$

converges absolutely, the b_i converge to zero, and the series

$$\sum_i a_i$$

has bounded partial sums, then the series

$$\sum_i a_i b_i$$

converges. { ′dā·də·kint ‚test }

deduction The process of deriving a statement from certain assumed statements by applying the rules of logic. { di′dək·shən }

defective equation An equation that has fewer roots than another equation from which it has been derived. { di′fekt·iv i′kwā·zhən }

defective number *See* deficient number. { di′fek·tiv ′nəm·bər }

deficiency index For a curve or equation involving two complex variables this is the genus of the Riemann surface associated to the equation. { də′fish·ən·sē ‚in‚deks }

deficient number A positive integer the sum of whose divisors, including 1 but excluding itself, is less than itself. Also known as defective number. { də′fish·ənt ′nəm·bər }

definite Riemann integral A number associated with a function defined on an interval $[a,b]$ which is

$$\lim_{N\to\infty} \sum_{k=0}^{N-1} f\left(a + \frac{k}{N}\right) \cdot \frac{b-a}{N}$$

if f is bounded and continuous; denoted by

$$\int_a^b f(x)dx;$$

if f is a positive function, the definite integral measures the area between the graph of f and the x axis. { ¦def·ə·nət ′rē‚män ‚in·tə·grəl }

deformation A homotopy of the identity map to some other map. { ˌdef·ər'mā·shən }

degeneracy The condition in which two characteristic functions of an operator have the same characteristic value. { di'jen·ə·rə·sē }

degenerate conic A straight line, a pair of straight lines, or a point, which is a limiting form of a conic. { di'jen·ə·rət 'kän·ik }

degenerate simplex A modification of a simplex in which the points $p_0, \ldots, p_n$ on which the simplex is based are linearly dependent. { di'jen·ə·rət 'simˌpleks }

degree **1.** A unit for measurement of plane angles, equal to 1/360 of a complete revolution, or 1/90 of a right angle. Symbolized °. **2.** For a term in one variable, the exponent of that variable. **3.** For a term in several variables, the sum of the exponents of its variables. **4.** For a polynomial, the degree of the highest-degree term. **5.** For a differential equation, the greatest power to which the highest-order derivative occurs. **6.** For an algebraic curve defined by the polynomial equation $f(x,y) = 0$, the degree of the polynomial $f(x,y)$. **7.** For a vertex in a graph, the number of arcs which have that vertex as an end point. { di'grē }

degree of degeneracy The number of characteristic functions of an operator having the same characteristic value. Also known as order of degeneracy. { di'grē əv di'jen·ə·rə·sē }

de Gua's rule The rule that if, in a polynomial equation $f(x) = 0$, a group of r consecutive terms is missing, then the equation has at least r imaginary roots if r is even, or the equation has at least $r + 1$ or $r - 1$ imaginary roots if r is odd (depending on whether the terms immediately preceding and following the group have like or unlike signs). { də'gwäz ˌrül }

Delambre analogies *See* Gauss formulas. { də'lam·brə əˌnal·ə·jēz }

del operator The rule which replaces the function f of three variables, x, y, z, by the vector valued function whose components in the x, y, z directions are the respective partial derivatives of f. Written ∇f. Also known as nabla. { 'del ˌäp·əˌrād·ər }

delta function A distribution δ such that

$$\int_{-\infty}^{\infty} f(t)\delta(x - t)dt$$

is $f(x)$. Also known as Dirac delta function; unit impulse. { 'del·tə ˌfəŋk·shən }

deltoid The plane curve traced by a point on a circle while the circle rolls along the inside of another circle whose radius is three times as great. Also known as Steiner's hypocycloid tricuspid. { 'delˌtȯid }

De Moivre's theorem The nth power of the quantity $\cos\theta + i \sin\theta$ is $\cos n\theta + i \sin n\theta$ for any integer n. { də'mwäv·rəz ˌthir·əm }

De Morgan's rules The complement of the union of two sets equals the intersection of their respective complements; the complement of the intersection of two sets equals the union of their complements. { də'mȯr·gənz ˌrülz }

De Morgan's test A series with term u_n, for which u_{n+1}/u_n converges to 1, will converge absolutely if there is $c > 0$ such that the limit superior of $n(|u_{n+1}/u_n|-1)$ equals $-1-c$. { də'mȯr·gənz ˌtest }

denominator In a fraction, the term that divides the other term (called the numerator), and is written below the line. { də'näm·ə,nād·ər }

dense-in-itself set A set every point of which is an accumulation point; a set without any isolated points. { 'dens in it'self ,set }

dense subset A subset of a topological space whose closure is the entire space. { ¦dens 'səb,set }

density For an increasing sequence of integers, the greatest lower bound of the quantity $F(n)/n$, where $F(n)$ is the number of integers in the sequence (other than zero) equal to or less than n. { 'den· səd·ē }

density function A density function for a measure m is a function which gives rise to m when it is integrated with respect to some other specified measure. { 'den·səd·ē ,fəŋk·shən }

denumerable set A set which may be put in one-to-one correspondence with the positive integers. Also known as countably infinite set. { də'nüm·rə·bəl 'set }

dependent variable If y is a function of x, that is, if the function assigns a single value of y to each value of x, then y is the dependent variable. { di'pen·dənt 'ver·ē·ə·bəl }

depressed equation An equation that results from reducing the number of roots in a given equation with one unknown by dividing the original equation by the difference of the unknown and a root. { di'prest i'kwā·zhən }

derangement numbers The numbers D_n, $n = 1, 2, 3, \ldots$, giving the number of permutations of a set of n elements that carry no element of the set into itself. { di'rānj·mənt ,nəm·bərz }

derivation **1.** The process of deducing a formula. **2.** A function D on an algebra which satisfies the equation $D(uv) = uD(v) + vD(u)$. { ,der·ə'vā·shən }

derivative The slope of a graph $yf(x)$ at a given point c; more precisely, it is the limit as h approaches zero of $f(c + h) - f(c)$ divided by h. Also known as differential coefficient; rate of change. { də'riv·əd·iv }

derived curve A curve whose ordinate, for each value of the abscissa, is equal to the slope of some given curve. Also known as first derived curve. { də'rīvd 'kərv }

derived set The set of cluster points of a given set. { də'rīvd 'set }

derogatory matrix A matrix whose order is greater than the order of its reduced characteristic equation. { də'räg·ə,tȯr·ē 'mā·triks }

Desarguesian plane Any projective plane in which points and lines satisfy Desargues' theorem. Also known as Arguesian plane. { dā·zär¦gā·zē·ən 'plān }

Desargues' theorem If the three lines passing through corresponding vertices of two triangles are concurrent, then the intersections of the three pairs of corresponding sides lie on a straight line, and conversely. { dā'zärgz ,thir·əm }

Descartes' rule of signs A polynomial with real coefficients has at most k real positive roots, where k is the number of sign changes in the polynomial. { dā 'kärts 'rül əv 'sīnz }

descriptive geometry The application of graphical methods to the solution of three-dimensional space problems. { di'skrip·tiv jē'äm·ə·trē }

determinant A certain real-valued function of the column vectors of a square matrix which is zero if and only if the matrix is singular; used to solve systems of linear equations and to study linear transformations. { də'tər·mə·nənt }

determinant tensor A tensor whose components are each equal to the corresponding component of the Levi-Civita tensor density times the square root of the determinant of the metric tensor, and whose contravariant components are each equal to the corresponding component of the Levi-Civita density divided by the square root of the metric tensor. Also known as permutation tensor. { də'tər·mə·nənt 'ten·sər }

developable surface A surface that can be obtained from a plane sheet by deformation, without stretching or shrinking. { di¦vel·əp·ə·bəl 'sər·fəs }

devil on two sticks *See* devil's curve. { 'dev·əl ȯn ˌtü 'stiks }

devil's curve A plane curve whose equation in cartesian coordinates x and y is $y^4 - a^2y^2 = x^4 - b^2x^2$, where a and b are constants. Also known as devil on two sticks. { 'dev·əlz 'kərv }

dextrorse curve *See* right-handed curve. { 'dekˌstrȯrs ˌkərv }

dextrorsum *See* right-handed curve. { dek'strȯr·səm }

diagonal **1.** The set of points all of whose coordinates are equal to one another in an n-dimensional coordinate system. **2.** A line joining opposite vertices of a polygon with an even number of sides. { dī'ag·ən·əl }

diagonalize To convert a square matrix to a diagonal matrix, usually by multiplying it on the left by a second matrix A of the same order, and on the right by the inverse of A. { dī'ag·ən·əˌlīz }

diagonally dominant matrix A matrix in which the absolute value of each diagonal element is either greater than the sum of the absolute values of the off-diagonal elements of the same row or greater than the sum of the off-diagonal elements in the same column. { dī'ag·ən·əl·ē 'däm·ə·nənt 'māˌtriks }

diagonal matrix A matrix whose nonzero entries all lie on the principal diagonal. { dī'ag·ən·əl 'mā·triks }

diagram A picture in which sets are represented by symbols and mappings between these sets are represented by arrows. { 'dī·əˌgram }

diakoptics A piecewise approach to the solution of large-scale interconnected systems, in which the large system is first broken up into several small pieces or subdivisions, the subdivisions are solved separately, and finally the effect of interconnection is determined and added to each subdivision to yield the complete solution of the system. { ˌdī·ə'käp·tiks }

diameter **1.** A line segment which passes through the center of a circle, and whose end points lie on the circle. **2.** The length of such a line. **3.** For a conic, any straight line that passes through the midpoints of all the chords of the conic that are parallel to a given chord. **4.** For a set, the smallest number that is greater than or equal to the distance between every pair of points of the set. { dī'am·əd·ər }

diametral curve A curve that passes through the midpoints of a family of parallel chords of a given curve. { dī'am·ə·trəl 'kərv }

diametral plane 1. A plane that passes through the center of a sphere. 2. A plane that passes through the mid-points of a family of parallel chords of a quadric surface that are parallel to a given chord. { dī'am·ə·trəl 'plān }

diametral surface A surface that passes through the midpoints of a family of parallel chords of a given surface that are parallel to a given chord. { dī'am·ə·trəl 'sər·fəs }

Dido's problem The problem of finding the curve, with a given perimeter, that encloses the greatest possible area; the curve is a circle. { 'dē‚dōz ‚präb·ləm }

difference 1. The result of subtracting one number from another. 2. The difference between two sets A and B is the set consisting of all elements of A which do not belong to B; denoted $A - B$. { 'dif·rəns }

difference equation An equation expressing a functional relationship of one or more independent variables, one or more functions dependent on these variables, and successive differences of these functions. { 'dif·rəns i'kwā·zhən }

difference methods Versions of the predictor-corrector methods of calculating numerical solutions of differential equations in which the prediction and correction formulas express the value of the solution function in terms of finite differences of a derivative of the function. { 'dif·rəns ‚meth·ədz }

difference operator One of several operators, such as the displacement operator, forward difference operator, or central mean operator, which can be used to conveniently express formulas for interpolation or numerical calculation or integration of functions and can be manipulated as algebraic quantities. { 'dif·rəns ‚äp·ə‚rād·ər }

difference quotient The increment of the value of a function divided by the increment of the independent variable; for the function $y = f(x)$, it is $\Delta y/\Delta x = [f(x + \Delta x) - f(x)]\Delta x$, where Δx and Δy are the increments of x and y. { 'dif·rəns ‚kwō·shənt }

differentiable atlas A family of embeddings $h_i{:}E^n \rightarrow M$ of euclidean space into a topological space M with the property that $h_i^{-1}h_j{:}E^n \rightarrow E^n$ is a differentiable map for each pair of indices, i, j. { ‚dif·ə'ren·chə·bəl 'at·ləs }

differentiable function A function which has a derivative at each point of its domain. { ‚dif·ə'ren·chə·bəl 'fəŋk·shən }

differentiable manifold A topological space with a maximal differentiable atlas; roughly speaking, a smooth surface. { ‚dif·ə'ren·chə·bəl 'man·ə‚fōld }

differential 1. The differential of a real-valued function $f(x)$, where x is a vector, evaluated at a given vector c, is the linear, real-valued function whose graph is the tangent hyperplane to the graph of $f(x)$ at $x = c$; if x is a real number, the usual notation is $df = f'(c)dx$. 2. *See* total differential. { ‚dif·ə'ren·chəl }

differential calculus The study of the manner in which the value of a function changes as one changes the value of the independent variable; includes maximum-minimum problems and expansion of functions into Taylor series. { ‚dif·ə'ren·chəl 'kal·kyə·ləs }

differential coefficient *See* derivative. { ˌdif·əˈren·chəl ˌkō·iˈfish·ənt }

differential equation An equation expressing a relationship between functions and their derivatives. { ˌdif·əˈren·chəl iˈkwā·zhən }

differential form A homogeneous polynomial in differentials. { ˌdif·əˈren·chəl ˈfȯrm }

differential game A game in which the describing equations are differential equations. { ˌdif·əˈren·chəl ˈgām }

differential geometry The study of curves and surfaces using the methods of differential calculus. { ˌdif·əˈren·chəl jēˈäm·ə·trē }

differential operator An operator on a space of functions which maps a function f into a linear combination of higher-order derivatives of f. { ˌdif·əˈren·chəl ˈäp·əˌrād·ər }

differential topology The branch of mathematics dealing with differentiable manifolds. { ˌdif·əˈren·chəl təˈpäl·ə·jē }

differentiation The act of taking a derivative. { ˌdif·əˌren·chēˈā·shən }

digit A character used to represent one of the nonnegative integers smaller than the base of a system of positional notation. Also known as numeric character. { ˈdij·ət }

digit place *See* digit position. { ˈdij·ət ˌplās }

digit position The position of a particular digit in a number that is expressed in positional notation, usually numbered from the lowest significant digit of the number. Also known as digit place. { ˈdij·ət pəˌzish·ən }

digraph *See* directed graph. { ˈdīˌgraf }

dihedral *See* dihedron. { dīˈhē·drəl }

dihedral angle The angle between two planes; it is said to be zero if the planes are parallel; if the planes intersect, it is the plane angle between two lines, one in each of the planes, which pass through a point on the line of intersection of the two planes and are perpendicular to it. { dīˈhē·drəl ˌaŋ·gəl }

dihedron A geometric figure formed by two half planes that are bounded by the same straight line. Also known as dihedral. { dīˈhē·drən }

dilation **1.** A transformation which changes the size, and only the size, of a geometric figure. **2.** An operation that provides a relatively flexible boundary to a fuzzy set; for a fuzzy set A with membership function m_A, a dilation of A is a fuzzy set whose membership function has the value $m_A(x)^\beta$ for every element x, where β is a fixed number that is greater than 0 and less than 1. { dəˈlā·shən }

dimension **1.** The number of coordinates required to label the points of a geometrical object. **2.** For a vector space, the number of vectors in any basis of the vector space. **3.** For a simplex, one less than the number of vertices of the simplex. **4.** For a simplicial complex, the largest of the dimensions of the simplices that make up the complex. { dəˈmen·chən }

dimensionless number A ratio of various physical properties (such as density or heat capacity) and conditions (such as flow rate or weight) of such nature that the resulting number has no defining units of weight, rate, and so on. Also known as nondimensional parameter. { də'men·chən·ləs 'nəm·bər }

dimension theory The study of abstract notions of dimension, which are topological invariants of a space. { də'men·chən ˌthē·ə·rē }

dioctahedral Having 16 faces. { ˌdīˌäk·tə'hē·drəl }

diophantine analysis A means of determining integer solutions for certain algebraic equations. { ¦dī·ə¦fant·ən ə'nal·ə·səs }

diophantine equations Equations with more than one independent variable and with integer coefficients for which integer solutions are desired. { ¦dī·ə¦fant·ən i'kwā·zhənz }

Dirac delta function *See* delta function. { di'rak 'del·tə ˌfəŋk·shən }

Dirac spinor *See* spinor. { di'rak 'spin·ər }

directed angle An angle for which one side is designated as initial, the other as terminal. { də'rek·təd 'aŋ·gəl }

directed graph A graph in which a direction is shown for every arc. Also known as digraph. { də'rek·təd 'graf }

directed line A line on which a positive direction has been specified. { də'rek·təd 'līn }

directed number A number together with a sign. { də'rek·təd 'nəm·bər }

directed set A partially ordered set with the property that for every pair of elements a,b in the set, there is a third element which is larger than both a and b. { də'rek·təd 'set }

directional derivative The rate of change of a function in a given direction; more precisely, if f maps an n-dimensional euclidean space into the real numbers, and $x = (x_1, ..., x_n)$ is a vector in this space, and $u = (u_1, ..., u_n)$ is a unit vector in the space (that is, $u_1^2 + \cdots + u_n^2 = 1$), then the directional derivative of f at x in the direction of u is the limit as h approaches zero of $[f(x + hu) - f(x)]/h$. { də'rek·shən·əl də'riv·əd·iv }

direction angles The three angles which a line in space makes with the positive x, y, and z axes. { də'rek·shən 'aŋ·gəlz }

direction cosine The cosine of one of the direction angles of a line in space. { də'rek·shən 'kōˌsīn }

direction numbers Any three numbers proportional to the direction cosines of a line in space. Also known as direction ratios. { di'rek·shən ˌnəm·bərz }

direction ratios *See* direction numbers. { di'rek·shən ˌrā·shōz }

director circle A circle consisting of the points of intersection of pairs of perpendicular tangents to an ellipse or hyperbola. { di'rek·tər 'sər·kəl }

direct product Given a finite family of sets $A_1, ..., A_n$, the direct product is the set of all n-tuples $(a_1, ..., a_n)$, where a_i belongs to A_i for $i = 1, ..., n$. { də'rekt 'präd·əkt }

direct proportion A statement that the ratio of two variable quantities is equal to a constant. { də'rekt prə'pȯr·shən }

directrix 1. A fixed line used in one method of defining a conic; the distance from this line divided by the distance from a fixed point (called the focus) is the same for all points on the conic. **2.** A curve through which a line generating a given ruled surface always passes. { də'rek·triks }

direct sum If each of the sets in a finite direct product of sets has a group structure, this structure may be imposed on the direct product by defining the composition "componentwise"; the resulting group is called the direct sum. { də¦rekt 'səm }

direct variation 1. A relationship between two variables wherein their ratio remains constant. **2.** An equation or function expressing such a relationship. { də 'rekt ˌver·ē'ā·shən }

Dirichlet conditions The requirement that a function be bounded, and have finitely many maxima, minima, and discontinuities on the closed interval $[-\pi, \pi]$. { ˌdē·rē'klā kənˌdish·ənz }

Dirichlet drawer principle *See* pigeonhole principle. { ˌdē·rē'klā 'drȯ·ər ˌprin·sə·pəl }

Dirichlet problem To determine a solution to Laplace's equation which satisfies certain conditions in a region and on its boundary. { ˌdē·rē'klā ˌpräb·ləm }

Dirichlet series A series whose nth term is a complex number divided by n to the zth power. { ˌdē·rē'klā ˌsir·ēz }

Dirichlet test for convergence If Σb_n is a series whose sequence of partial sums is bounded, and if $\{a_n\}$ is a monotone decreasing null sequence, then the series

$$\sum_{n=1}^{\infty} a_n b_n$$

converges. { ˌdē·rē'klā ˌtest fər kən'vər·jəns }

Dirichlet transform For a function $f(x)$, this is the integral of $f(x)\cdot\sin(kx)/x$; its convergence determines the convergence of the Fourier series of $f(x)$. { ˌdē·rē'klā ˌtranzˌfȯrm }

disconnected set A set in a topological space that is the union of two nonempty sets A and B for which both the intersection of the closure of A with B and the intersection of the closure of B with A are empty. { ˌdis·kə¦nek·təd 'set }

discontinuity A point at which a function is not continuous. { disˌkänt·ən'ü·əd·ē }

discrete set A set with no cluster points. { di'skrēt 'set }

discrete variable A variable for which the possible values form a discrete set. { di¦skrēt 'ver·ē·ə·bəl }

discretization error The error in the numerical calculation of an integral that results from using an approximate expression for the true mathematical function to be integrated. { ˌdis·krə·də'zā·shən ˌer·ər }

discriminant 1. The quantity $b^2 - 4ac$, where a,b,c are coefficients of a given quadratic polynomial: $ax^2 + bx + c$. **2.** More generally, for the polynomial equation

$a_0x^n + a_1x^{n-1} + \cdots + a_nx_0 = 0$, a_0^{2n-2} times the product of the squares of all the differences of the roots of the equation, taken in pairs. { di'skrim·ə·nənt }

disintegration of measure The representation of a measure as an integral of a family of positive measures. { dis‚in·tə'grā·shən əv 'mezh·ər }

disjoint sets Sets with no elements in common. { dis'jȯint 'sets }

disjunction The connection of two statements by the word "or." { dis'jəŋk·shən }

disk The region in the plane consisting of all points with norm less than 1 (sometimes less than or equal to 1). Also spelled disc. { disk }

displacement operator A difference operator, denoted E, defined by the equation $Ef(x) = f(x + h)$, where h is a constant denoting the difference between successive points of interpolation or calculation. Also known as forward shift operator. { dis'plās·mənt ‚äp·ə‚rād·ər }

dissimilar terms Terms that do not contain the same unknown factors or that do not contain the same powers of these factors. { di¦sim·ə·lər 'tərmz }

distance **1.** A nonnegative number associated with pairs of geometric objects. **2.** The spatial separation of two points, measured by the length of a hypothetical line joining them. **3.** For two parallel lines, two skew lines, or two parallel planes, the length of a line joining the two objects and perpendicular to both. **4.** For a point and a line or plane, the length of the perpendicular from the point to the line or plane. { 'dis·təns }

distribution An abstract object which generalizes the idea of function; used in applied mathematics, quantum theory, and probability theory; the delta function is an example. Also known as generalized function. { ‚dis·trə'byü·shən }

distributive lattice A lattice in which "greatest lower bound" obeys a distributive law with respect to "least upper bound," and vice versa. { di'strib·yəd·iv 'lad·əs }

distributive law A rule which stipulates how two binary operations on a set shall behave with respect to one another; in particular, if $+$, $\circ$ are two such operations then $\circ$ distributes over $+$ means $a \circ (b + c) = (a \circ b) + (a \circ c)$ for all a,b,c in the set. { di'strib·yəd·iv 'lȯ }

divergence For a vector-valued function, the sum of the diagonal entries of the Jacobian matrix; it is the scalar product of the del operator and the vector. { də'vər·jəns }

divergence theorem *See* Gauss' theorem. { də'vər·jəns ‚thir·əm }

divergent integral An improper integral which does not have a finite value. { də'vər·jənt 'in·tə·grəl }

divergent sequence A sequence which does not converge. { də'vər·jənt 'sē·kwəns }

divergent series An infinite series whose sequence of partial sums does not converge. { də'vər·jənt 'sir·ēz }

divide One object (integer, polynomial) divides another if their quotient is an object of the same type. { də'vīd }

divided differences Quantities which are used in the interpolation or numerical calculation or integration of a function when the function is known at a series of points which are not equally spaced, and which are formed by various operations on the difference between the values of the function at successive points. { də'vīd·əd 'dif·rən·səs }

dividend A quantity which is divided by another quantity in the operation of division. { 'div·ə‚dend }

division The inverse operation of multiplication; the number a divided by the number b is the number c such that b multiplied by c is equal to a. { də'vizh·ən }

division algebra A hypercomplex system that is also a skew field. { də'vizh·ən ‚al·jə·brə }

division modulo *p* Division in the finite field with p elements, where p is a prime number. { də'vizh·ən ¦mäj·ə·lō 'pē }

division ring 1. A ring in which the set of nonzero elements form a group under multiplication. 2. More generally, a nonassociative ring with nonzero elements in which, for any two elements a and b, there are elements x and y such that $ax = b$ and $ya = b$. { di'vizh·ən ‚riŋ }

division sign 1. The symbol ÷, used to indicate division. 2. The diagonal /, used to indicate a fraction. { di'vizh·ən ‚sīn }

divisor 1. The quantity by which another quantity is divided in the operation of division. 2. An element b in a commutative ring with identity is a divisor of an element a if there is an element c in the ring such that $a = bc$. { də'vīz·ər }

divisor of zero A nonzero element x of a commutative ring such that $xy = 0$ for some nonzero element y of the ring. Also known as zero divisor. { di¦vī·zər əv 'zir·ō }

dodecagon A 12-sided polygon. { dō'dek·ə‚gän }

dodecahedron A polyhedron with 12 faces. { dō‚dek·ə'hē‚drən }

domain 1. For a function, the set of values of the independent variable. 2. A nonempty open connected set in euclidean space. Also known as region. 3. *See* Abelian field. { dō'mān }

domain of dependence For an initial-value problem for a partial differential equation, a portion of the range such that the initial values on this portion determine the solution over the entire range. { ¦dō¦mān əv di'pen·dəns }

dominant strategy Relative to a given pure strategy for one player of a game, a second pure strategy for that player that has at least as great a payoff as the given strategy for any pure strategy of the opposing player. { 'däm·ə·nənt 'strad·ə·jē }

dominated convergence theorem If a sequence $\{f_n\}$ of Lebesgue measurable functions converges almost everywhere to f and if the absolute value of each f_n is dominated by the same integrable function, then f is integrable and $\lim \int f_n dm = \int f dm$. { 'däm·ə‚nād·əd kən'vər·jəns ‚thir·əm }

dominating integral An improper integral whose nonnegative, nonincreasing integrand function has the property that its value for all sufficiently large positive integers n is no smaller than the nth term of a given series of positive terms; used in the integral test for convergence. { 'däm·ə‚nad·iŋ 'in·tə·grəl }

dominating series A series, each term of which is larger than the respective term in some other given series; used in the comparison test for convergence of series. { 'däm·ə͵nad·iŋ 'sir·ēz }

double cusp A point on a curve through which two branches of the curve with the same tangent pass, and at which each branch extends in both directions of the tangent. Also known as point of osculation; tacnode. { 'dəb·əl ͵kəsp }

double integral The Riemann integral of functions of two variables. { ¦dəb·əl 'in·tə·grəl }

double law of the mean *See* second mean-value theorem. { ¦dəb·əl ¦lȯ əv thə 'mēn }

double minimal surface A minimal surface that is also a one-sided surface. { ¦dəb·əl ¦min·ə·məl 'sər·fəs }

double point A point on a curve at which a curve crosses or touches itself, or has a cusp; that is, a point at which the curve has two tangents (which may be coincident). { ¦dəb·əl 'pȯint }

double root A number a such that $(x - a)^2p(x) = 0$ where $p(x)$ is a polynomial of which a is not a root. { ¦dəb·əl 'rüt }

double tangent **1.** A line which is tangent to a curve at two distinct noncoincident points. Also known as bitangent. **2.** Two coincident tangents to branches of a curve at a given point, such as the tangents to a cusp. { ¦dəb·əl 'tan·jənt }

doubly ruled surface A ruled surface that can be generated by either of two distinct moving straight lines; quadric surfaces are the only surfaces of this type. { ¦dəb·lē ¦rüld 'sər·fəs }

doubly stochastic matrix A matrix of nonnegative real numbers such that every row sum and every column sum are equal to 1. { ¦dəb·lē stō¦kas·tik 'mā·triks }

dual coordinates Point coordinates and plane coordinates are dual in geometry since an equation about one determines an equation about the other. { ¦dü·əl kō'ȯrd·ən·əts }

dual graph A planar graph corresponding to a planar map obtained by replacing each country with its capital and each common boundary by an arc joining the two countries. { ¦dü·əl 'graf }

dual group The group of all homomorphisms of an Abelian group G into the cyclic group of order n, where n is the smallest integer such that g^n is the identity element of G. { ¦dü·əl 'grüp }

duality principle A principle that if a theorem is true, it remains true if each object and operation is replaced by its dual; important in projective geometry and Boolean algebra. Also known as principle of duality. { dü'al·əd·ē ͵prin·sə·pəl }

duality theorem **1.** A theorem which asserts that for a given space, the $(n - p)$ dimensional homology group is isomorphic to a p-dimensional cohomology group for each $p = 0, ..., n$, provided certain conditions are met. **2.** Let G be either a compact group or a discrete group, let X be its character group, and let G' be the character group of X; then there is an isomorphism of G onto G' so that the groups G and G' may be identified. **3.** If either of two dual linear-programming problems has a solution, then so does the other. { dü'al·əd·ē ͵thir·əm }

dual linear programming Linear programming in which the maximum and minimum number are the same number. { 'dü·əl ¦lin·ē·ər 'prō͵gram·iŋ }

dual operation In projective geometry, an operation that is obtained from a given operation by replacing points with lines, lines with points, the drawing of a line through a point with the marking of a point on a line, and so forth. { ¦dül äp·ə'rā·shən }

dual space The vector space consisting of all linear transformations from a given vector space into its scalar field. { 'dü·əl 'spās }

dual tensor The product of a given tensor, covariant in all its indices, with the contravariant form of the determinant tensor, contracting over the indices of the given tensor. { 'dü·əl 'ten·sər }

dual theorem In projective geometry, the theorem that is obtained from a given theorem by replacing points with lines, lines with points, and operations with their dual operations. Also known as reciprocal theorem. { ¦dül 'thir·əm }

dual variables Mutually dependent variables. { 'dü·əl 'ver·ē·ə·bəlz }

Duhamel's theorem If f and g are continuous functions, then

$$\lim_{|\Delta x| \to 0} \sum_{i=1}^{n} f(x_i')g(x_i'')\Delta x_i = \int_a^b f(x)g(x)dx$$

where x_i' and x_i'' are between x_{i-1} and x_i, $i = 1, ..., n$, and $|\Delta x| = \max\{x_i - x_{i-1}\}$ for a partition $a = x_0 < x_1 < \cdots < x_n = b$. { dyə'melz ˌthir·əm }

dummy suffix A suffix which has no true mathematical significance and is used only to facilitate notation; usually an index which is summed over. { ¦dəm·ē 'səf·iks }

dummy variable A variable which has no true mathematical significance and is used only to facilitate notation; usually a variable which is integrated over. { ¦dəm·ē 'ver·ē·ə·bəl }

duodecimal number system A representation system for real numbers using 12 as the base. { ˌdü·ə¦des·məl 'nəm·bər ˌsis·təm }

Dupin's theorem The proposition that, given three families of mutually orthogonal surfaces, the line of intersection of any two surfaces of different families is a line of curvature for both the surfaces. { dyü'paz ˌthir·əm }

Durer's conchoid A plane curve consisting of points that lie on a variable line passing through points Q and R and are a constant distance a from Q, where Q and R have cartesian coordinates $(q,0)$ and $(0,r)$ and q and r satisfy the equation $q + r = b$, where b is a constant. { 'dùr·ərz 'käŋˌkòid }

dyad An abstract object which is a pair of vectors **AB** in a given order on which certain operations are defined. { 'dīˌad }

dyadic expansion The representation of a number in the binary number system. { dī'ad·ik ik'span·chən }

dyadic operation An operation that has only two operands. { dī'ad·ik ˌäp·ə'rā·shən }

dyadic rational A fraction whose denominator is a power of 2. { dī'ad·ik 'rash·ən·əl }

dynamical system An abstraction of the concept of a family of solutions to an ordinary differential equation; namely, an action of the real numbers on a topological space satisfying certain "flow" properties. { dī¦nam·ə·kəl ′sis·təm }

dynamic programming A mathematical technique, more sophisticated than linear programming, for solving a multidimensional optimization problem, which transforms the problem into a sequence of single-stage problems having only one variable each. { dī¦nam·ik ′prō·grə·miŋ }

E

e The base of the natural logarithms; the number defined by the equation

$$\int_1^e \frac{1}{x}\,dx = 1;$$

approximately equal to 2.71828.

eccentric angle **1.** For an ellipse having semimajor and semiminor angles of lengths a and b respectively, lying along the x and y axes of a coordinate system respectively, and for a point (x,y) on the ellipse, the angle

$$\text{arc cos}\,\frac{x}{a} = \text{arc sin}\,\frac{y}{b}$$

2. For a hyperbola having semitransverse and semiconjugate axes of lengths a and b respectively, lying along the x and y axes of a coordinate system respectively, and for a point (x,y) on the hyperbola, the angle

$$\text{arc sec}\,\frac{x}{a} = \text{arc tan}\,\frac{y}{b}$$

{ ek¦sen·trik 'ang·əl }

eccentric circles **1.** For an ellipse, two circles whose centers are at the center of the ellipse and whose diameters are, respectively, the major and minor axes of the ellipse. **2.** For a hyperbola, two circles whose centers are at the center of symmetry of the hyperbola and whose diameters are, respectively, the transverse and conjugate axes of the hyperbola. { ek¦sen·trik 'sərk·əlz }

eccentricity The ratio of the distance of a point on a conic from the focus to the distance from the directrix. { ˌek·sən'tris·əd·ē }

edge **1.** A line along which two plane faces of a solid intersect. **2.** A line segment connecting nodes or vertices in a graph (a geometric representation of the relation among situations). Also known as arc. { ej }

edge of regression The curve swept out by the characteristic point of a one-parameter family of surfaces. { ¦ej əv rē'gresh·ən }

effectively computable function *See* computable function. { ə¦fek·tiv·lē kəm¦pyüd·ə·bəl 'fəŋk·shən }

effective transformation group A transformation group in which the identity element is the only element to leave all points fixed. { ə¦fek·tiv ˌtranz·fər'mā·shən ˌgrüp }

Egerov's theorem If a sequence of measurable functions converges almost everywhere on a set of finite measure to a real-valued function, then given any $\epsilon > 0$ there is a set of measure smaller than ϵ on whose complement the sequence converges uniformly. { 'eg·ə‚räfs ‚thir·əm }

eigenfunction Also known as characteristic function. **1.** An eigenvector for a linear operator on a vector space whose vectors are functions. Also known as proper function. **2.** A solution to the Sturm-Liouville partial differential equation. { 'ī·gən‚fəŋk·shən }

eigenfunction expansion By using spectral theory for linear operators defined on spaces composed of functions, in certain cases the operator equals an integral or series involving its eigenvectors; this is known as its eigenfunction expansion and is particularly useful in studying linear partial differential equations. { 'ī·gən‚fəŋk·shən ik'span·chən }

eigenmatrix Corresponding to a diagonalizable matrix or linear transformation, this is the matrix all of whose entries are 0 save those on the principal diagonal where appear the eigenvalues. { 'ī·gən‚mā·triks }

eigenvalue The one of the scalars λ such that $T(v) = \lambda v$, where T is a linear operator on a vector space, and v is an eigenvector. Also known as characteristic number; characteristic root; characteristic value; latent root; proper value. { 'ī·gən‚val·yü }

eigenvalue equation *See* characteristic equation. { 'ī·gən‚val·yü i‚kwā·zhən }

eigenvalue problem *See* Sturm-Liouville problem. { 'ī·gən‚val·yü ‚präb·ləm }

eigenvector A nonzero vector v whose direction is not changed by a given linear transformation T; that is, $T(v) = \lambda v$ for some scalar λ. Also known as characteristic vector. { 'ī·gən‚vek·tər }

eight curve A plane curve whose equation in cartesian coordinates x and y is $x^4 = a^2(x^2 - y^2)$, where a is a constant. Also known as lemniscate of Gerono. { 'āt ‚kərv }

Einstein space A Riemannian space in which the contracted curvature tensor is proportional to the metric tensor. { ¦īn¦stīn 'spās }

Einstein's summation convention A notational convenience used in tensor analysis whereupon it is agreed that any term in which an index appears twice will stand for the sum of all such terms as the index assumes all of a preassigned range of values. { 'īn‚stīnz sə'mā·shən kən‚ven·chən }

element **1.** In an array such as a matrix or determinant, a quantity identified by the intersection of a given row or column. **2.** In network topology, an edge. **3.** The generatrix of a ruled surface at any one fixed position. { 'el·ə·mənt }

elementary function Any function which can be formed from algebraic functions and the exponential, logarithmic, and trigonometric functions by a finite number of operations consisting of addition, subtraction, multiplication, division, and composition of functions. { ‚el·ə'men·trē 'fəŋk·shən }

elementary symmetric functions For a set of n variables, a set of n functions, $\sigma_1, \sigma_2, \ldots, \sigma_n$, where σ_k is the sum of all products of k of the n variables. { el·ə'men·trē si¦me·trik 'fəŋk·shənz }

elimination A process of deriving from a system of equations a new system with fewer variables, but with precisely the same solutions. { ə‚lim·ə'nā·shən }

ellipse The locus of all points in the plane at which the sum of the distances from a fixed pair of points, the foci, is a given constant. { ə'lips }

ellipsoid A surface whose intersection with every plane is an ellipse (or circle). { ə'lip‚sȯid }

ellipsoidal coordinates Coordinates in space determined by confocal quadrics. { ə‚lip'sȯid·əl kō'ȯrd·ən·əts }

ellipsoidal harmonics Lamé functions that play a role in potential problems on an ellipsoid analogous to that played by spherical harmonics in potential problems on a sphere. { ə¦lip‚sȯid·əl ‚här'män·iks }

ellipsoidal wave functions *See* Lamé wave functions. { ə¦lip‚sȯid·əl 'wāv ‚fəŋk·shənz }

ellipsoid of revolution An ellipsoid generated by rotation of an ellipse about one of its axes. Also known as spheroid. { ə'lip‚sȯid əv ‚rev·ə'lü·shən }

elliptic cone A cone whose base is an ellipse. { ə¦lip·tik 'kōn }

elliptic conical surface A conical surface whose directrix is an ellipse. { ə¦lip·tik ¦kän·ə·kəl 'sər·fəs }

elliptic coordinates The coordinates of a point in the plane determined by confocal ellipses and hyperbolas. { ə'lip·tik kō'ȯrd·ən·əts }

elliptic cylinder A cylinder whose directrix is an ellipse. { ə¦lip·tik 'sil·ən·dər }

elliptic differential equation A general type of second-order partial differential equation which includes Laplace's equation and has the form

$$\sum_{i,j=1}^{n} A_{ij}\,(\partial^2 u/\partial x_i \partial x_j) + \sum_{i=1}^{n} B_i\,(\partial u/\partial x_i) + Cu + F = 0$$

where A_{ij}, B_i, C, and F are suitably differentiable real functions of $x_1, x_2, \ldots, x_n$, and there exists at each point $(x_1, x_2, \ldots, x_n)$ a real linear elliptic partial differential equation transformation on the x_i which reduces the quadratic form

$$\sum_{i,j=1}^{n} A_{ij}\, x_i\, x_j$$

to a sum of n squares, all of the same sign. Also known as elliptic partial differential equation. { ə'lip·tik dif·ə¦ren·chəl i'kwā·zhən }

elliptic function An inverse function of an elliptic integral; alternatively, a doubly periodic, meromorphic function of a complex variable. { ə'lip·tik 'fəŋk·shən }

elliptic geometry The geometry obtained from euclidean geometry by replacing the parallel line postulate with the postulate that no line may be drawn through a given point, parallel to a given line. Also known as Riemannian geometry. { ə'lip·tik jē'äm·ə·trē }

elliptic integral An integral over x whose integrand is a rational function of x and the square root of $p(x)$, where $p(x)$ is a third- or fourth-degree polynomial without multiple roots. { ə'lip·tik 'int·ə·grəl }

elliptic integral of the first kind Any elliptic integral which is finite for all values of the limits of integration and which approaches a finite limit when one of the limits of integration approaches infinity. { ə¦lip·tik ¦int·ə·grəl əv thə ¦fərst ‚kīnd }

elliptic integral of the second kind Any elliptic integral which approaches infinity as one of the limits of integration y approaches infinity, or which is infinite for some value of y, but which has no logarithmic singularities in y. { ə¦lip·tik ¦int·ə·grəl əv thə ¦sek·ənd ˌkīnd }

elliptic integral of the third kind Any elliptic integral which has logarithmic singularities when considered as a function of one of its limits of integration. { ə¦lip·tik ¦int·ə·grəl əv thə ¦thərd ˌkīnd }

ellipticity Also known as oblateness. **1.** For an ellipse, the difference between the semimajor and semiminor axes of the ellipse, divided by the semimajor axis. **2.** For an oblate spheroid, the difference between the equatorial diameter and the axis of revolution, divided by the equatorial diameter. { ēˌlip'tis·əd·ē }

elliptic paraboloid A surface which can be so situated that sections parallel to one coordinate plane are parabolas while those parallel to the other plane are ellipses. { ə¦lip·tik pə'rab·əˌlȯid }

elliptic partial differential equation *See* elliptic differential equation. { ə'lip·tik ¦pär·shəl dif·ə¦ren·chəl i'kwā·zhən }

elliptic point A point on a surface at which the total curvature is strictly positive. { ə'lip·tik 'pȯint }

elliptic type A type of simply connected Riemann surface that can be mapped conformally on the closed complex plane, including the point at infinity. { ə¦lip·tik 'tīp }

elliptic wedge The surface generated by a moving straight line that remains parallel to a given plane and intersects both a given straight line and an ellipse whose plane is parallel to the given line but does not contain it. { ə¦lip·tik 'wej }

embedding An injective homomorphism between two algebraic systems of the same type. { em'bed·iŋ }

empirical curve A smooth curve drawn through or close to points representing measured values of two variables on a graph. { em'pir·ə·kəl 'kərv }

empty set The set with no elements. { 'em·tē 'set }

Encke roots For any two numbers a_1 and a_2, the numbers $-x_1$ and $-x_2$, where x_1 and x_2 are the roots of the equation $x^2 + a_1x + a_2 = 0$, with $x_1 < x_2$. { 'eŋ·kə ˌrüts }

endogenous variables In a mathematical model, the dependent variables; their values are to be determined by the solution of the model equations. { en'däj·ə·nəs 'ver·ē·ə·bəlz }

endomorphism A function from a set with some structure (such as a group, ring, vector space, or topological space) to itself which preserves this structure. { 'en·də'mȯrˌfiz·əm }

end point Either of two values or points that mark the ends of an interval or line segment. { 'end ˌpȯint }

entire function A function of a complex variable which is analytic throughout the entire complex plane. Also known as integral function. { en¦tīr ¦fəŋk·shən }

entire ring *See* integral domain. { en¦tīr 'riŋ }

entire series A power series which converges for all values of its variable; a power series with an infinite radius of convergence. { en¦tīr 'sir·ēz }

entire surd A surd that does not contain a rational factor or term. { en¦tīr 'sərd }

entropy In a mathematical context, this concept is attached to dynamical systems, transformations between measure spaces, or systems of events with probabilities; it expresses the amount of disorder inherent or produced. { 'en·trə·pē }

entropy of a partition If ξ is a finite partition of a probability space, the entropy of ξ is the negative of the sum of the products of the probabilities of elements in ξ with the logarithm of the probability of the element. { 'en·trə·pē əv ə pär'tish·ən }

entropy of a transformation *See* Kolmogorov-Sinai invariant. { 'en·trə·pē əv ə tranz·fər'mā·shən }

entropy of a transformation given a partition If T is a measure preserving transformation on a probability space and ξ is a finite partition of the space, the entropy of T given ξ is the limit as $n \rightarrow \infty$ of $1/n$ times the entropy of the partition which is the common refinement of ξ, $T^{-1}\xi$, ..., $T^{-n+1}\xi$. { 'en·trə·pē əv ə tranz·fər'mā·shən 'giv·ən ə pär'tish·ən }

enumerable *See* countable. { ē'nüm·rə·bəl }

envelope **1.** The envelope of a one-parameter family of curves is a curve which has a common tangent with each member of the family. **2.** The envelope of a one-parameter family of surfaces is the surface swept out by the characteristic curves of the family. { 'en·və,lōp }

epicenter The center of a circle that generates an epicycloid or hypocycloid. { 'ep·ə,sen·tər }

epicycle The circle which generates an epicycloid or hypocycloid. { 'ep·ə,sī·kəl }

epicycloid The curve traced by a point on a circle as it rolls along the outside of a fixed circle. { ,ep·ə'sī,klȯid }

epi spiral A plane curve whose equation in polar coordinates (r, θ) is $r \cos n\theta = a$, where a is a constant and n is an integer. { 'ep·ē ,spī·rəl }

epitrochoid A curve traced by a point rigidly attached to a circle at a point other than the center when the circle rolls without slipping on the outside of a fixed circle. { ¦ep·ə'trō,kȯid }

epsilon chain A finite sequence of points such that the distance between any two successive points is less than the positive real number epsilon (ϵ). { 'ep·sə,län ,chān }

epsilon neighborhood The set of all points in a metric space whose distance from a given point is less than some number; this number is designated ϵ. { 'ep·sə,län 'nā·bər,hu̇d }

epsilon symbols The symbols

$$\epsilon^{i_1 i_2 \ldots i_n} \text{ and } \epsilon_{i_1 i_2 \ldots i_n}$$

which are $+1$ if $i_1, i_2, \ldots, i_n$ is an even permutation of $1, 2, \ldots, n$; -1 if it is an odd permutation; and zero otherwise. { 'ep·sə,län ,sim·bəlz }

equal Being the same in some sense determined by context. { 'ē·kwəl }

equality The state of being equal. { ē′kwal·əd·ē }

equal ripple property For any continuous function $f(x)$ on the interval −1,1, and for any positive integer n, a property of the polynomial $pn(x)$, which is the best possible approximation to $f(x)$ in the sense that the maximum absolute value of $en(x) = f(x) - pn(x)$ is as small as possible; namely, that $en(x)$ assumes its extreme values at least $n + 2$ times, with the consecutive extrema having opposite signs. { ¦ē·kwəl ′rip·əl ‚präp·ərd·ē }

equal sets Sets with precisely the same elements. { ¦ē·kwəl ′sets }

equate To state algebraically that two expressions are equal to one another. { ē′kwāt }

equation A statement that each of two expressions is equal to the other. { i′kwā·zhən }

equation of mixed type A partial differential equation which is of hyperbolic, parabolic, or elliptic type in different parts of a region. { i¦kwā·zhən əv ¦mikst ′tīp }

equiangular polygon A polygon all of whose interior angles are equal. { ¦ē·kwē¦aŋ·gyə·lər ′päl·ə‚gän }

equiangular spiral *See* logarithmic spiral. { ¦ē·kwē¦aŋ·gyə·lər ′spī·rəl }

equicontinuous family of functions A family of functions with the property that given any $\epsilon > 0$ there is a $\delta > 0$ such that whenever $x - y < \delta$, $f(x) - f(y) < \epsilon$ for every function $f(x)$ in the family. { ¦ē·kwē·kən′tiŋ·yə·wəs ′fam·lē əv ′fəŋk·shənz }

equidistant Being the same distance from some given object. { ¦ē·kwə¦dis·tənt }

equidistant system A system of parametric curves on a surface obtained by setting surface coordinates u and v equal to various constants, where the coordinates are chosen so that an element of length ds on the surface is given by $ds^2 = du^2 + F\,dudv + dv^2$, where F is a function of u and v. { ¦ē·kwə¦dis·tənt ′sis·təm }

equilateral polygon A polygon all of whose sides are the same length. { ¦ē·kwə¦lad·ə·rəl ′päl·ə‚gän }

equilateral polyhedron A polyhedron all of whose faces are identical. { ¦ē·kwə¦lad·ə·rəl ‚päl·ə′hē·drən }

equitangential curve *See* tractrix. { ‚ē·kwə·tan′jen·chəl ′kərv }

equivalence A logic operator having the property that if P, Q, R, etc., are statements, then the equivalence of P, Q, R, etc., is true if and only if all statements are true or all statements are false. { i′kwiv·ə·ləns }

equivalence classes The collection of pairwise disjoint subsets determined by an equivalence relation on a set; two elements are in the same equivalence class if and only if they are equivalent under the given relation. { i′kwiv·ə·ləns ‚klas·əs }

equivalence relation A relation which is reflexive, symmetric, and transitive. { i′kwiv·ə·lənsri′lā·shən }

equivalence transformation A mapping which associates with each square matrix A the matrix $B = SAT$, where S and T are nonsingular matrices. Also known as equivalent transformation. { i′kwiv·ə·ləns ‚tranz·fər‚mā·shən }

equivalent continued fractions Continued fractions whose values to n terms are the same for $n = 1, 2, 3, \ldots$. { i¦kwiv·ə·lənt kən¦tin·yüd 'frak·shənz }

equivalent elements *See* associates. { ə¦kwiv·ə·lənt 'el·ə·məns }

equivalent equations Equations that have the same set of solutions. { i¦kwiv·ə·lənt i'kwā·zhənz }

ergodic theory The study of measure-preserving transformations. { ər'gäd·ik 'thē·ə·rē }

ergodic transformation A measure-preserving transformation on X with the property that whenever X is written as a union of two disjoint invariant subsets, one of these must have measure zero. { ər'gäd·ik tranz·fər'mā·shən }

error function The real function defined as the integral from 0 to x of $e^{-t^2}\,dt$ or $e^{-t^2}\,dt$, or the integral from x to ∞ of $e^{-t^2}\,dt$. { 'er·ər ˌfəŋk·shən }

escribed circle For a triangle, a circle that lies outside of the triangle and is tangent to one side of the triangle and to the extensions of the other two sides. Also known as excircle. { ə¦skrībd 'sər·kəl }

essential bound For a function f, a number A such that the set of points x for which the absolute value of $f(x)$ is greater than A is of measure zero. { i¦sen·chəl 'bau̇nd }

essential constants A set of constants in an equation that cannot be replaced by a smaller number of constants in another equation that has the same solutions. { i¦sen·chəl 'kän·stəns }

essentially bounded function A function that has an essential bound. { i¦sen·chə·lē ¦bau̇nd·əd 'fəŋk·shən }

essential mapping A mapping between topological spaces that is not homotopic to a mapping whose range is a single point. { i¦sen·chəl 'map·iŋ }

essential singularity An isolated singularity of a complex function which is neither removable nor a pole. { i'sen·chəl siŋ·gyə'lar·əd·ē }

essential supremum For an essentially bounded function, the greatest lower bound of the essential bounds. { i¦sen·chəl sə'prēm·əm }

euclidean algorithm A method of finding the greatest common divisor of a pair of integers. { yü'klid·ē·ən 'al·gəˌrith·əm }

euclidean geometry The study of the properties preserved by isometries of two- and three-dimensional euclidean space. { yü'klid·ē·ən jē'äm·ə·trē }

euclidean space A space consisting of all ordered sets $(x_1, \ldots, x_n)$ of n numbers with the distance between $(x_1, \ldots, x_n)$ and $(y_1, \ldots, y_n)$ being given by

$$\left[\sum_{i=1}^{n} (x_i - y_i)^2\right]^{1/2};$$

the number n is called the dimension of the space. { yü'klid·ē·ən 'spās }

Euler characteristic of a topological space *X* The number $\chi(X) = \Sigma(-1)^q\beta_q$, where β_q is the qth Betti number of X. { 'ȯi·lər ˌkar·ik·tə'ris·tik əv ə ˌtäp·ə¦läj·i·kəl ¦spās 'eks }

Euler diagram A diagram consisting of closed curves, used to represent relations between logical propositions or sets; similar to a Venn diagram. { 'ȯi·lər ˌdī·əˌgram }

Eulerian path A path that traverses each of the lines in a graph exactly once. { ȯi'ler·ē·ən 'path }

Euler-Lagrange equation A partial differential equation arising in the calculus of variations, which provides a necessary condition that $y(x)$ minimize the integral over some finite interval of $f(x,y,y')dx$, where $y' = dy/dx$; the equation is $(\delta f(x,y,y')/\delta y) - (d/dx)(\delta f(x,y,y')/\delta y') = 0$. Also known as Euler's equation. { ¦ȯi·lər lə'grānj iˌkwā·zhən }

Euler-Maclaurin formula A formula used in the numerical evaluation of integrals, which states that the value of an integral is equal to the sum of the value given by the trapezoidal rule and a series of terms involving the odd-numbered derivatives of the function at the end points of the interval over which the integral is evaluated. { ¦ȯi·lər mə'klȯr·ən ˌfȯr·myə·lə }

Euler's constant The limit as n approaches infinity, of $1 + 1/2 + 1/3 + \cdots + 1/n - \ln n$, equal to approximately 0.5772. Denoted γ. Also known as Mascheroni's constant. { 'ȯi·lərz ¦kän·stənt }

Euler's criterion A criterion for the congruence $x^n \equiv a \pmod{m}$ to have a solution, namely that $a^{\phi/d} \equiv 1 \pmod{m}$, where $\phi = \phi(m)$ is Euler's phi function evaluated at m, and d is the greatest common divisor of ϕ and n. { 'ȯi·lərz krī'tir·ē·ən }

Euler's equation *See* Euler-Lagrange equation. { 'ȯi·lərz i¦kwā·zhən }

Euler's formula The formula $e^{ix} = \cos x + i \sin x$, where $i = \sqrt{-1}$. { 'ȯi·lərz ˌfȯr·myə·lə }

Euler's numbers The numbers E_{2n} defined by the equation

$$\frac{1}{\cos z} = \sum_{n=0}^{\infty} (-1)^n \frac{E_{2n}}{(2n)!} z^{2n}$$

{ 'ȯi·lərz ˌnəm·bərz }

Euler's phi function A function ϕ, defined on the positive integers, whose value $\phi(n)$ is the number of integers equal to or less than n and relatively prime to n. Also known as indicator; phi function; totient. { 'ȯi·lərz 'fī ˌfəŋk·shən }

Euler's spiral *See* Cornu's spiral. { 'ȯi·lərz ¦spī·rəl }

Euler's theorem For any polyhedron, $V - E + F = 2$, where V, E, F represent the number of vertices, edges, and faces respectively. { 'ȯi·lərz ˌthir·əm }

Euler transformation A method of obtaining from a given convergent series a new series which converges faster to the same limit, and for defining sums of certain divergent series; the transformation carries the series $a_0 - a_1 + a_2 - a_3 + \cdots$ into a series whose nth term is

$$\sum_{r=0}^{n-1} (-1)^r \binom{n-1}{r} a_r / 2^n$$

{ 'ȯi·lər ˌtranz·fər'mā·shən }

even function A function with the property that $f(x) = f(-x)$ for each number x. { 'ē·vən ˌfəŋk·shən }

even number An integer which is a multiple of 2. { 'ē·vən ˌnəm·bər }

even permutation A permutation which may be represented as a result of an even number of transpositions. { ¦ē·vən pər·myə'tā·shən }

Everett's interpolation formula A formula for estimating the value of a function at an intermediate value of the independent variable, when its value is known at a series of equally spaced points (such as those that appear in a table), in terms of the central differences of the function of even order only and coefficients which are polynomial functions of the independent variable. { ¦ev·rəts ˌin·tər·pə'lā·shən ˌfȯr·myə·lə }

evolute **1.** The locus of the centers of curvature of a curve. **2.** The two surfaces of center of a given surface. { 'ev·əˌlüt }

exact differential equation A differential equation obtained by setting the exact differential of some function equal to zero. { ig'zakt dif·ə'ren·chəl iˌkwā·zhən }

exact differential form A differential form which is the differential of some other form. { ig'zakt dif·ə'ren·chəl ˌfȯrm }

exact division Division wherein the remainder is zero. { ig'zakt di'vizh·ən }

exact divisor A divisor that leaves a remainder of zero. { ig'zakt di'vī·zər }

exact sequence A sequence of homomorphisms with the property that the kernel of each homomorphism is precisely the image of the previous homomorphism. { ig'zakt 'sē·kwəns }

excenter The center of the escribed circle of a given triangle. { ¦ek'sen·tər }

except A logical operator which has the property that if P and Q are two statements, then the statement "P except Q" is true only when P alone is true; it is false for the other three combinations (P false Q false, P false Q true, and P true Q true). { ek'sept }

exceptional group One of five Lie groups which leave invariant certain forms constructed out of the Cayley numbers; they are Lie groups with maximum symmetry in the sense that, compared with other simple groups with the same rank (number of independent invariant operators), they have maximum dimension (number of generators). { ek¦sep·shən·əl ¦grüp }

exceptional Jordan algebra A Jordan algebra that cannot be written as a symmetrized product over a matrix algebra; used in formulating a generalization of quantum mechanics. { ek'sep·shən·əl ¦jȯrd·ən'al·jə·brə }

excircle *See* escribed circle. { ¦ek'sər·kəl }

exclusive or A logic operator which has the property that if P is a statement and Q is a statement, then P exclusive or Q is true if either but not both statements are true, false if both are true or both are false. { ik¦sklü·siv 'ȯr }

existential quantifier A logical relation, often symbolized ∃, that may be expressed by the phrase "there is a" or "there exists"; if P is a predicate, the statement $(\exists x)P(x)$ is true if there exists at least one value of x in the domain of P for which $P(x)$ is true, and is false otherwise. { ˌeg·zə¦sten·chəl 'kwän·təˌfī·ər }

exogenous variables In a mathematical model, the independent variables, which are predetermined and given outside the model. { ˌek'säj·ə·nəs 'ver·ē·ə·bəlz }

expansion The expression of a quantity as the sum of a finite or infinite series of terms, as a finite or infinite product of factors, or, in general, in any extended form. { ik'span·shən }

expectation *See* expected value. { ˌekˌspek'tā·shən }

expected value 1. For a random variable x with probability density function $f(x)$, this is the integral from $-\infty$ to ∞ of $xf(x)dx$. Also known as expectation. **2.** For a random variable x on a probability space (Ω,P), the integral of x with respect to the probability measure P. { ek'spek·təd 'val·yü }

explementary angles *See* conjugate angles. { ˌek·splə¦men·tə·rē 'aŋ·gəlz }

exponent A number or symbol placed to the right and above some given mathematical expression. { ik'spō·nənt }

exponential For a bounded linear operator A on a Banach space, the sum of a series which is formally the exponential series in A. { ˌek·spə'nen·chəl }

exponential curve A graph of the function $y = a^x$, where a is a positive constant. { ˌek·spə'nen·chəl 'kərv }

exponential density function A probability density function obtained by integrating a function of the form $\exp(-x - m/\sigma)$, where m is the mean and σ the standard deviation. { ˌek·spə'nen·chəl den·səd·ē ˌfəŋk·shən }

exponential equation An equation containing e^x (the Naperian base raised to a power) as a term. { ˌek·spə'nen·chəl i'kwā·zhən }

exponential function The function $f(x) = e^x$, written $f(x) = \exp(x)$. { ˌek·spə'nen·chəl 'fəŋk·shən }

exponential integral The function defined to be the integral from x to ∞ of $(e^{-t}/t)\,dt$ for x positive. { ˌek·spə'nen·chəl 'int·ə·grəl }

exponential law *See* law of exponents. { ˌek·spə'nen·chəl 'lȯ }

exponential series The Maclaurin series expansion of e^x, namely,

$$e^x = 1 + \sum_{n=1}^{\infty} \frac{x^n}{n!}$$

{ ˌek·spə'nen·chəl 'sir·ēz }

exsecant The trigonometric function defined by subtracting unity from the secant, that is $\operatorname{exsec}\theta = \sec\theta - 1$. { ˌek'sē·kant }

extended mean-value theorem *See* second mean-value theorem. { ik¦sten·dəd ˌmēn ˌval·yü 'thir·əm }

extension *See* extension fields. { ik'sten·chən }

extension field An extension field of a given field E is a field F such that E is a subfield of F. Also known as extension. { ik'sten·chən ˌfēld }

extension map An extension map of a map f from a set A to a set L is a map g from a set B to L such that A is a subset of B and the restriction of g to A equals f. { ik'sten·chən ˌmap }

exterior For a set A in a topological space, the largest open set contained in the complement of A. { ek′stir·ē·ər }

exterior algebra An algebra whose structure is analogous to that of the collection of differential forms on a Riemannian manifold. Also known as Grassmann algebra. { ek′stir·ē·ər ′al·jə·brə }

exterior angle **1.** An angle between one side of a polygon and the prolongation of an adjacent side. **2.** An angle made by a line (the transversal) that intersects two other lines, and either of the latter on the outside. { ek′stir·ē·ər ′aŋ·gəl }

exterior content *See* exterior Jordan content. { ek′stir·ē·ər ′kän‚tent }

exterior Jordan content For a set of points on a line, the largest number C such that the sum of the lengths of a finite number of closed intervals that includes every point in the set is always equal to or greater than C. Also known as exterior content. { ek¦stir·ē·ər ¦jȯrd·ən ′kän‚tent }

exterior measure *See* Lebesgue exterior measure. { ek¦stir·ē·ər ′mezh·ər }

external angle The angle defined by an arc around the boundaries of an internal angle or included angle. { ek′stərn·əl ′aŋ·gəl }

extract a root To determine a root of a given number, usually a positive real root, or a negative real odd root of a negative number. { ik′strakt ə ′rüt }

extraneous root A root that is introduced into an equation in the process of solving another equation, but is not a solution of the equation to be solved. { ik¦strān·ē·əs ′rüt }

extrapolation Estimating a function at a point which is larger than (or smaller than) all the points at which the value of the function is known. { ik‚strap·ə′lā·shən }

extremals For a variational problem in the calculus of variaitons entailing use of the Euler-Lagrange equation, the extremals are the solutions of this equation. { ek′strem·əlz }

extreme *See* extremum. { ek′strēm }

extreme and mean ratio *See* golden section. { ek′strēm ən ′mēn ‚rā·shō }

extreme point **1.** A maximum or minimum value of a function. **2.** A point in a convex subset K of a vector space is called extreme if it does not lie on the interior of any line segment contained in K. { ek′strēm ′pȯint }

extreme terms The first and last terms in a proportion. { ek¦strēm ′tərmz }

extreme value problem A set of mathematical conditions which may be met by values that are less than or greater than an upper or a lower bound, that is, an extreme value. { ek¦strēm ′val·yü präb·ləm }

extremum A maximum or minimum value of a function. Also known as extreme. { ek′strēm·əm }

F

face 1. One of the plane polygons bounding a polyhedron. **2.** A face of a simplex is the subset obtained by setting one or more of the coordinates a_i, defining the simplex, equal to 0; for example, the faces of a triangle are its sides and vertices. { fās }

face angle An angle between two successive edges of a polyhedral angle. { 'fās ˌaŋ·gəl }

factor 1. For an integer n, any integer which gives n when multiplied by another integer. **2.** For a polynomial p, any polynomial which gives p when multiplied by another polynomial. { 'fak·tər }

factorable integer An integer that has factors other than unity and itself. { 'fak·trə·bəl 'int·ə·jər }

factorable polynomial A polynomial which has polynomial factors other than itself. { 'fak·tə·rə·bəl ˌpäl·ə'nō·mē· əl }

factor analysis Given sets of variables which are related linearly, factor analysis studies techniques of approximating each set relative to the others; usually the variables denote numbers. { 'fak·tər əˌnal·ə·səs }

factor group *See* quotient group. { 'fak·tər ˌgrüp }

factorial The product of all positive integers less than or equal to n; written $n!$; by convention $0! = 1$. { fak'tȯr·ē·əl }

factorial series The series $1 + (1/1!) + (1/2!) + (1/3!) + \cdots$, whose $(n + 1)$st term is $1/n!$ for $n = 1, 2, \ldots$; its sum is the number e. { fak¦tȯr·ē·əl 'sirˌēz }

factoring Finding the factors of an integer or polynomial. { 'fak·tə·riŋ }

factoring of the secular equation Factoring the polynomial that results from expanding the secular determinant of a matrix, in order to find the roots of this polynomial, which are the eigenvalues of the matrix. { 'fak·tə·riŋ əv thə ¦sek·yə· lər i'kwā·zhən }

factor module The factor module of a module M over a ring R by a submodule N is the quotient group M/N, where the product of a coset $x + N$ by an element a in R is defined to be the coset $ax + N$. { 'fak·tər ˌmä·jül }

factor of proportionality Two quantities A and B are related by a factor of proportionality μ if either $A = \mu B$ or $B = \mu A$. { 'fak·tər əv prəˌpȯrsh·ən'al·əd·ē }

factor ring *See* quotient ring. { 'fak·tər ˌriŋ }

factor space *See* quotient space. { 'fak·tər ˌspās }

factor theorem of algebra A polynomial $f(x)$ has $(x - a)$ as a factor if and only if $f(a) = 0$. { 'fak·tər ˌthir·əm əv 'al·jə·brə }

fair game A game in which all of the participants have equal expectation of gain. { ¦fer 'gām }

faithful module A module M over a commutative ring R such that if a is an element in R for which $am = 0$ for all m in M, then $a = 0$. { ¦fāth·fu̇l 'mä·jül }

faithful representation A homomorphism h of a group onto some group of matrices or linear operators such that h is an injection. { ¦fāth·fu̇l ˌrep·rə·zen'tā·shən }

faltung A family of functions where the convolution of any two members of the family is also a member of the family. Also known as convolution family. { 'fälˌtu̇ŋ }

Fano plane A projective plane in which the points of intersection of the three possible pairs of opposite sides of a quadrilateral are collinear. { 'fä·nō ˌplān }

Fano's axiom The postulate that the points of intersection of the three possible pairs of opposite sides of any quadrilateral in a given projective plane are noncollinear; thus a projective plane satisfying Fano's axiom is not a Fano plane, and a Fano plane does not satisfy Fano's axiom. { ¦fä·nōz 'ak·sē·əm }

Farey sequence The Farey sequence of order n is the increasing sequence, from 0 to 1, of fractions whose denominator is equal to or less that n, with each fraction expressed in lowest terms. { 'far·ē ˌsē·kwəns }

fast Fourier transform A Fourier transform employing the Cooley-Tukey algorithm to reduce the number of operations. { ¦fast ˌfu̇r·ēˌā 'tranzˌfo̊rm }

Fatou-Lebesgue lemma Given a sequence f_n of positive measurable functions on a measure space (X, μ), then

$$\int_x (\liminf_{n\to\infty} f_n) d\mu \le \lim_{n\to\infty} \int_x f_n d\mu$$

{ ˌfä'tü lə'beg ¦lem·ə }

Feit-Thompson theorem The proposition that every group of odd order is solvable. { ¦fīt ¦täm·sən ˌthir·əm }

Fermat numbers The numbers of the form $F_n = (2^{(2n)}) + 1$ for $n = 0, 1, 2, \ldots$. { 'fer·mä ˌnəm·bərz }

Fermat's last theorem The conjecture that there are no positive integer solutions of the equation $x^n + y^n = z^n$ for $n \ge 3$. { fer'mäz ¦last 'thir·əm }

Fermat's spiral A plane curve whose equation in polar coordinates (r,θ) is $r^2 = a^2\theta$, where a is a constant. { fer'mäz ˌspī·rəl }

Fermat's theorem The proposition that, if p is a prime number and a is a positive integer which is not divisible by p, then $a^{p-1} - 1$ is divisible by p. { 'ferˌmäz ˌthir·əm }

Ferrers diagram An array of dots associated with an integer partition $n = a_1 + \cdots + a_k$, whose ith row contains a_i dots. { 'fer·ərz ˌdi·əˌgram }

fiber The set of points in the total space of a bundle which are sent into the same element of the base of the bundle by the projection map. { 'fī·bər }

fiber bundle A bundle whose total space is a *G*-space *X* and whose base is the homomorphic image of the orbit space of *X* and whose fibers are isomorphic to the orbits of points in the base space under the action of *G*. { 'fī·bər ˌbən·dəl }

Fibonacci number A number in the Fibonacci sequence whose first two terms are $f_1 = f_2 = 1$. { ¦fib·ə¦nä·chē 'nəm·bər }

Fibonacci sequence The sequence 1,1,2,3,5,8,13,21, ..., or any sequence where each entry is the sum of the two previous entries. { ˌfē·bə'näch·ē ˌsē·kwəns }

field An algebraic system possessing two operations which have all the properties that addition and multiplication of real numbers have. { fēld }

field of planes on a manifold A continuous assignment of a vector subspace of tangent vectors to each point in the manifold. Also known as plane field. { 'fēld əv 'plānz ȯn ə 'man·əˌfōld }

field of vectors on a manifold A continuous assignment of a tangent vector to each point in the manifold. Also known as vector field. { 'fēld əv 'vek·tərz ȯn ə 'man·əˌfōld }

field theory The study of fields and their extensions. { 'fēld ˌthē·ə·rē }

filter A family of subsets of a set *S*: it does not include the empty set, the intersection of any two members of the family is also a member, and any subset of *S* containing a member is also a member. { 'fil·tər }

filter base A family of subsets of a given set with the property that it does not include the empty set, and the intersection of any finite number of members of the family includes another member. { 'fil·tər ˌbās }

final-value theorem The theorem that if $f(t)$ is a function which has a Laplace transform $F(s)$, and if the derivative of $f(t)$ with respect to t is also Laplace transformable, and if the limit of $f(t)$ as t approaches infinity exists, then this limit is equal to the limit of $sF(s)$ as s approaches zero. { ¦fīn·əl ¦val·yü ¦thir·əm }

finite character **1.** A property of a family *C* of sets such that any finite subset of a member of *C* belongs to *C*, and *C* includes any set all of whose finite subsets belong to *C*. **2.** A characteristic of a property of subsets of a set such that a subset *S* has the property if and only if all the nonempty finite subsets of *S* have the property. { 'fīˌnīt 'kar·ik·tər }

finite difference The difference between the values of a function at two discrete points, used to approximate the derivative of the function. { ¦fīˌnīt 'dif·rəns }

finite-difference equations Equations arising from differential equations by substituting difference quotients for derivatives, and then using these equations to approximate a solution. { ¦fīˌnīt ¦dif·rəns iˌkwā·zhənz }

finite discontinuity A discontinuity of a function that lies at the center of an interval on which the function is bounded. { 'fīˌnīt ˌdis·kän·tə'nü·əd·ē }

finite extension An extension field *F* of a given field *E* such that *F*, viewed as a vector space over *E*, has finite dimension. { ¦fīˌnīt ik'sten·chən }

finite group A group which contains a finite number of distinct elements. { ¦fī‚nīt 'grüp }

finite intersection property of a family of sets If the intersection of any finite number of them is nonempty, then the intersection of all the members of the family is nonempty. { ¦fī‚nīt ‚in·tər'sek·shən ‚präp·ərd·ē əv ə 'fam·lē əv 'sets }

finitely generated extension A finitely generated extension of a field k is the smallest field which contains k and some finite set of elements. { ¦fī‚nīt· ¦gen·ə‚rād·əd ik'sten·chən }

finite mathematics **1.** Those parts of mathematics which deal with finite sets. **2.** Those fields of mathematics which make no use of the concept of limit. { ¦fī‚nīt ‚math·ə'mad·iks }

finite matrix A matrix with a finite number of rows and columns. { ¦fī‚nīt 'mā·triks }

finite measure space A measure space in which the measure of the entire space is a finite number. { 'fī‚nīt ¦mezh·ər ‚spās }

finite moment theorem The theorem that if $f(x)$ is a continuous function, and if the integral of $f(x)x^n$ over a finite interval is zero for all positive integers n, then $f(x)$ is identically zero in that interval. { ¦fī‚nīt 'mō·mənt ‚thir·əm }

finite plane In projective geometry, a plane with a finite number of points and lines. { ¦fī‚nīt 'plān }

finite quantity Any bounded quantity. { ¦fī‚nīt 'kwän·əd·ē }

finite series A series that has a limited number of terms. { 'fī‚nīt 'sir‚ēz }

finite set A set whose elements can be indexed by integers 1,2,3, ...,n inclusive. { ¦fī‚nīt 'set }

Finsler geometry The study of the geometry of a manifold in terms of the various possible metrics on it by means of Finsler structures. { 'fin·slər jē'äm·ə·trē }

Finsler structure on a manifold A family of metrics varying continuously from point to point. { 'fin·slər ‚strək·chər ȯn ə 'man·ə‚fōld }

first category **1.** A set is of first category if it is a countable union of nowhere dense sets. **2.** A set S is of first category at a point x if there is a neighborhood of x whose intersection with S is of first category. { 'fərst 'kad·ə‚gȯr·ē }

first countable topological space A topological space in which every point has a countable number of open neighborhoods so that any neighborhood of this point contains one of these. { ¦fərst 'kau̇nt·ə·bəl ‚täp·ə¦läj·ə·kəl 'spās }

first derived curve *See* derived curve. { ¦fərst də¦rīvd 'kərv }

first law of the mean *See* mean value theorem. { 'fərst ‚lȯ əv thə 'mēn }

first law of the mean for integrals The proposition that the definite integral of a continuous function over an interval equals the length of the interval multiplied by the value of the function at some point in the interval. { ¦fərst ¦lȯ əv thə ¦mēn fȯr 'int·ə·grəlz }

first negative pedal *See* negative pedal. { 'fərst 'neg·əd·iv 'ped·əl }

first-order difference A member of a sequence that is formed from a given sequence by subtracting each term of the original sequence from the next succeeding term. { ¦fərst ¦ȯrd·ər 'dif·rəns }

first-order theory A logical theory in which predicates are not allowed to have other functions or predicates as arguments and in which predicate quantifiers and function quantifiers are not permitted. { ¦fərst ˌȯrd·ər 'thē·ə·rē }

first pedal curve *See* pedal curve. { 'fərst 'ped·əl ˌkərv }

first positive pedal curve *See* pedal curve. { 'fərst 'päz·əd·iv ˌped·əl ˌkərv }

first quadrant **1.** The range of angles from 0 to 90°. **2.** In a plane with a system of cartesian coordinates, the region in which the x and y coordinates are both positive. { ¦fərst 'kwäd·rənt }

first species The class of sets G_0 such that one of the sets G_n is the null set, where, in general, G_n is the derived set of G_{n-1}. { 'fərst 'spēˌshēz }

Fisher's inequality The inequality whereby the number b of blocks in a balanced incomplete block design is equal to or greater than the number v of elements arranged among the blocks. { ¦fish·ərz ˌin·i'kwäl·əd·ē }

five-dimensional space A vector space whose basis has five vectors. { 'fīv dəˌmen·chən·əl 'spās }

fixed point For a function f mapping a set S to itself, any element of S which f sends to itself. { ¦fikst 'pȯint }

fixed-point theorem Any theorem, such as the Brouwer theorem or Schauder's fixed-point theorem, which states that a certain type of mapping of a set into itself has at least one fixed point. { ¦fikst 'pȯint ˌthir·əm }

fixed radix notation A form of positional notation in which successive digits are interpreted as coefficients of successive powers of an integer called the base or radix. { ¦fikst 'rāˌdiks nōˌtā·shən }

flat space A Riemannian space for which a coordinate system exists such that the components of the metric tensor are constants throughout the space; equivalently, a space in which the Riemann-Christoffel tensor vanishes throughout the space. { ¦flat ˌspās }

flecnode A node that is also a point of inflection of one of the two branches of the curve that cross at the node. { 'flekˌnōd }

floating arithmetic *See* floating-point arithmetic. { ¦flōd·iŋ ə'rith·mə·tik }

floating-decimal arithmetic *See* floating-point arithmetic. { ¦flōd·iŋ ¦des·məl ə'rith·mə·tik }

floating-point arithmetic A method of performing arithmetical operations, used especially by automatic computers, in which numbers are expressed as integers multiplied by the radix raised to an integral power, as 87×10^{-4} instead of 0.0087. Also known as floating arithmetic; floating-decimal arithmetic. { ¦flōd·iŋ ¦pȯint ə'rith·mə·tik }

Floquet theorem A second-order linear differential equation whose coefficients are periodic single-valued functions of an independent variable x, has a solution of the form $e\mu x P(x)$ where μ is a constant and $P(x)$ a periodic function. { flō'kā ˌthir·əm }

F martingale A stochastic process $\{X_t, t > 0\}$ such that the conditional expectation of X_t given F_s equals X_s whenever $s\ t$, where $F = \{F_t, t \geq 0\}$ is an increasing family of sigma algebras that represents the amount of information increasing with time. { ¦ef 'mart·ənˌgāl }

focal chord For a conic, a chord that passes through a focus of the conic. { 'fō·kəl ¦kȯrd }

focal property **1.** The property of an ellipse or hyperbola whereby lines drawn from the foci to any point on the conic make equal angles with the tangent to the conic at that point. **2.** The property of a parabola whereby a line from the focus to any point on the parabola, and a line through this point parallel to the axis of the parabola, make equal angles with the tangent to the parabola at this point. { 'fō·kəl 'präp·ər·dē }

focal radius For a conic, a line segment from a focus to any point on the conic. { 'fō·kəl 'rād·ē·əs }

focus A point in the plane which together with a line (directrix) defines a conic section. { 'fō·kəs }

folium A plane curve that is a pedal curve (first positive pedal) of the deltoid. { 'fō·lē·əm }

folium of Descartes A plane cubic curve whose equation in cartesian coordinates x and y is $x^3 + y^3 = 3axy$, where a is some constant. Also known as leaf of Descartes. { 'fō·lē·əm əv dā'kärt }

formal logic The study of the permissible relationships between propositions, a study that concerns the form rather than the content. { ¦fȯr·məl 'läj·ik }

formula An equation or rule relating mathematical objects or quantities. { 'fȯr·myə·lə }

forward difference One of a series of quantities obtained from a function whose values are known at a series of equally spaced points by repeatedly applying the forward difference operator to these values; used in interpolation or numerical calculation and integration of functions. { ¦fȯr·wərd 'dif·rəns }

forward difference operator A difference operator, denoted Δ, defined by the equation $\Delta f(x) = f(x + h) - f(x)$, where h is a constant indicating the difference between successive points of interpolation or calculation. { ¦fȯr·wərd ¦dif·rəns 'äp·əˌrād·ər }

forward shift operator *See* displacement operator. { ¦fȯr·wərd ¦shift 'äp·əˌrād·ər }

four-color problem The problem of proving the statement that, given any map in the plane, it is possible to color the regions with four colors so that any two regions with a common boundary have different colors. { ¦fȯr 'kəl·ər ˌpräb·ləm }

Fourier analysis The study of convergence of Fourier series and when and how a function is approximated by its Fourier series or transform. { ˌfu̇r·ēˌā əˌnal·ə·səs }

Fourier-Bessel integrals Given a function $F(r,\theta)$ independent of θ where r,θ are the polar coordinates in the plane, these integrals have the form

$$\int_0^\infty udu \int_0^\infty F(r)J_m(ur)rdr$$

where J_m is a Bessel function order m. { ˌfu̇r·ēˌā ¦bes·əl ′int·ə·grəlz }

Fourier-Bessel series For a function $f(x)$, the series whose mth term is $a_mJ_0(j_mx)$, where $j_1, j_2, ...$ are positive zeros of the Bessel function J_0 arranged in ascending order, and a_m is the product of $2/J_1^2(j_m)$ and the integral over t from 0 to 1 of $tf(t)J_0(j_mt)$; J_1 is a Bessel function. { ˌfu̇r·ēˌā ¦bes·əl ˌsir·ēz }

Fourier-Bessel transform *See* Hankel transform. { ˌfu̇r·ēˌā ¦bes·əl ′tranzˌfȯrm }

Fourier expansion *See* Fourier series. { ˌfu̇r·ēˌā ik′span·chən }

Fourier integrals For a function $f(x)$ the Fourier integrals are

$$\frac{1}{\pi}\int_0^\infty du \int_{-\infty}^\infty f(t) \cos u(x - t)\, dt$$

$$\frac{1}{\pi}\int_0^\infty du \int_{-\infty}^\infty f(t) \sin u(x - t)\, dt$$

{ ˌfu̇r·ēˌā ′int·ə·grəlz }

Fourier kernel Any kernel $K(x,y)$ of an integral transform which may be written in the form $K(x,y) = k(xy)$ and which is identical with the kernel of the inverse transform. { ′fȯr·ē·ā ˌkər·nəl }

Fourier-Legendre series Given a function $f(x)$, the series from $n = 0$ to infinity of $a_nP_n(x)$, where $P_n(x)$, $n = 0, 1, 2, ...$, are the Legendre polynomials, and a_n is the product of $(2n + 1)/2$ and the integral over x from -1 to 1 of $f(x)P_n(x)$. { ˌfu̇r·ēˌā lə′zhän·drə ˌsir·ēz }

Fourier series The Fourier series of a function $f(x)$ is

$$\tfrac{1}{2} a_0 + \sum_{n=1}^{\infty} (a_n \cos nx + b_n \sin nx)$$

$$a_n = \frac{1}{\pi}\int_{-\pi}^{\pi} f(x) \cos nx dx$$

$$b_n = \frac{1}{\pi}\int_{-\infty}^{\pi} f(x) \sin nx dx$$

Also known as Fourier expansion. { ˌfu̇r·ēˌā ˌsir·ēz }

Fourier's half-range series A Fourier series that either contains only terms that are even in the independent variable (the cosine series) or contains only terms that are odd (the sine series). { ′fȯr·ēˌāz ¦haf ¦rānj ˌsirˌēz }

Fourier space The space in which the Fourier transform of a function is defined. { ˌfu̇r·ēˌā ˌspās }

Fourier's theorem If $f(x)$ satisfies the Dirichlet conditions on the interval $-\pi < x < \pi$, then its Fourier series converges to $f(x)$ for all values of x in this interval at which $f(x)$ is continuous, and approaches $1/2[f(x + 0) + f(x - 0)$ at points at which $f(x)$ is discontinuous, where $f(x - 0)$ is the limit on the left of f at x and $f(x + 0)$ is the limit on the right of f at x. { ˌfu̇r·ēˌāz ˌthir·əm }

Fourier-Stieltjes series For a function $f(x)$ of bounded variation on the interval $[0,2\pi]$, the series from $n = 0$ to infinity of $c_n \exp(inx)$, where c_n is $1/2\pi$ times the integral from $x = 0$ to $x = 2\pi$ of $\exp(-inx)df(x)$. { ˌfu̇r·ēˌā 'stēl·yes ˌsir·ēz }

Fourier-Stieltjes transform For a function $f(y)$ of bounded variation on the interval $(-\infty, \infty)$, the function $F(x)$ equal to $1/\sqrt{2\pi}$ times the integral from $y = -\infty$ to $y = \infty$ of $\exp(-ixy)df(y)$. { ˌfu̇r·ēˌā 'stēl·yes ˌtranzˌfȯrm }

Fourier synthesis The determination of a periodic function from its Fourier components. { ˌfu̇r·ēˌā 'sin·thə·səs }

Fourier transform For a function $f(t)$, the function $F(x)$ equal to $1/\sqrt{2\pi}$ times the integral over t from $-\infty$ to ∞ of $f(t) \exp(itx)$. { ˌfu̇r·ēˌā 'tranzˌfȯrm }

four-point A set of four points in a plane, no three of which are collinear. Also known as complete four-point. { 'fȯr ˌpȯint }

fourth proportional For numbers a, b, and c, a number x such that $a/b = c/x$. { 'fȯrth prə'pȯr·shən·əl }

fourth quadrant **1.** The range of angles from 270 to 360°. **2.** In a plane with a system of cartesian coordinates, the region in which the x coordinate is positive and the y coordinate is negative. { ¦fȯrth 'kwäd·rənt }

F process A stochastic process $\{X_t, t > 0\}$ whose value at time t is determined by the information up to time t; more precisely, the events $\{X_t \leq a\}$ belong to F_t for every t and a, where $F = \{F_t, t \geq 0\}$ is an increasing family of sigma algebras that represents the amount of information increasing with time. { 'ef ˌpräs·əs }

fractal A geometrical shape whose structure is such that magnification by a given factor reproduces the original object. { 'frakt·əl }

fractal dimensionality A number D associated with a fractal which satisfies the equation $N = b^D$, where b is the factor by which the length scale changes under a magnification in each step of a recursive procedure defining the object, and N is the factor by which the number of basic units increases in each such step. { 'frak·təl diˌmen·shə'nal·əd·ē }

fraction An expression which is the product of a real number or complex number with the multiplicative inverse of a real or complex number. { 'frak·shən }

fractional equation **1.** Any equation that contains fractions. **2.** An equation in which the unknown variable appears in the denominator of one or more terms. { ¦frak·shən·əl i'kwā·zhən }

fractional ideal A submodule of the quotient field of an integral domain. { ¦frak·shən·əl i'dēl }

fraction in lowest terms A fraction from which all common factors have been divided out of the numerator and denominator. { 'frak·shən in ¦lō·əst 'tərmz }

Fréchet space **1.** A quasi-normed linear space in which every Cauchy sequence converges to a point in the space. **2.** A quasi-normed linear space which is locally convex under the topology generated by the norm, and in which every Cauchy sequence converges to a point in the space. **3.** A complete metrizable locally convex topological space. { frā'shā ˌspās }

Fredholm determinant A power series obtained from the function $K(x,y)$ of the Fredholm equation which provides solutions to the equation under certain conditions. { 'fredˌhōm di¦tər·mə·nənt }

Fredholm integral equations Given functions $f(x)$ and $K(x,y)$, the Fredholm integral equations with unknown function y are

$$\text{type 1: } f(x) = \int_a^b K(x, t)y(t)dt$$

$$\text{type 2: } y(x) = f(x) + \lambda \int_a^b K(x, t)y(t)dt$$

{ 'fredˌhōm ¦int·ə·grəl i'kwā·zhənz }

Fredholm operator A linear operator between Banach spaces which has closed range, and both the Fredholm operator and its adjoint have finite dimensional null space. { 'fredˌhōm ˌäp·əˌrād·ər }

Fredholm theorem A Fredholm equation of type 2 with continuous $f(x)$ has a unique continuous solution, or else the corresponding equation of type 1 has a positive number of linearly independent solutions. { 'fredˌhōm ˌthir·əm }

Fredholm theory The study of the solutions of the Fredholm equations. { 'fredˌhōm ˌthē·ə·rē }

free group A group whose generators satisfy the equation $x \cdot y = e$ (e is the identity element in the group) only when $x = y^{-1}$ or $y = x^{-1}$. { 'frē ˌgrüp }

free module A module which is a free group with respect to its additive group. { ¦frē ¦mäj·yül }

Freeth's nephroid The strophoid of a circle with respect to a pole located at the center and a fixed point located on the circumference. Also known as nephroid of Freeth. { 'frāths 'nefˌrȯid }

free tree A tree graph in which there is no node which is distinguished as the root. { 'frē ˌtrē }

free variable In logic, a variable that has an occurrence which is not within the scope of a quantifier and thus can be replaced by a constant. { ¦frē 'ver·ē·ə·bəl }

Frenet-Serret formulas Formulas in the theory of space curves, which give the directional derivatives of the unit vectors along the tangent, principal normal and binormal of a space curve in the direction tangent to the curve. Also known as Serret-Frenet formulas. { fre'nā sə'rā ˌfōr·myə·ləz }

frequency distribution A function which measures the relative frequency or probability that a variable can take on a set of values. { ¦frē·kwən·sē ˌdis·trə'byü·shən }

Fresnel integrals Given a parameter x, the integrals over t from 0 to x of $\sin t^2$ and of $\cos t^2$ or from x to ∞ of $(\cos t)/t^{1/2}$ and of $(\sin t)/t^{1/2}$. { frā'nel 'int·ə·grəlz }

friendship theorem The proposition that, among a finite set of people, if every pair of people has exactly one common friend, then there is someone who knows everyone else. { 'fren,ship ,präb·ləm }

Frobenius method A method of finding a series solution near a point for a linear homogeneous ordinary differential equation. { frō'ben·yu̇s ,meth·əd }

frontier For a set in a topological space, all points in the closure of the set but not in its interior. Also known as boundary. { frən'tir əv ə 'set }

frustum The part of a solid between two cutting parallel planes. { 'frəs·təm }

Fubini's theorem The theorem stating conditions under which

$$\int\int f(u,v)dudv = \int du \int f(u,v)dv = \int dv \int f(u,v)du$$

{ fü'bē·nēz ,thir·əm }

Fuchsian differential equation A homogeneous, linear differential equation whose coefficients are analytic functions whose only singularities, if any, are poles of order one. { ¦fyük·sē·ən ,dif·ə¦ren·chəl i'kwā·zhən }

Fuchsian group A Kleinian group G for which there is a region D in the complex plane, consisting of either the interior of a circle or the portion of the plane on one side of a straight line, such that D is mapped onto itself by every element of G. { 'fyük·sē·ən ,grüp }

full linear group The group of all nonsingular linear transformations of a complex vector space whose group operation is composition. { ¦fu̇l 'lin·ē·ər ,grüp }

fully parenthesized notation A method of writing arithmetic expressions in which parentheses are placed around each pair of operands and its associated operator. { 'fu̇l·ē pə,ren·thə,sīzd nō'tā·shən }

function A mathematical rule between two sets which assigns to each member of the first, exactly one member of the second. { 'fəŋk·shən }

functional Any function from a vector space into its scalar field. { 'fəŋk·shən·əl }

functional analysis A branch of analysis which studies the properties of mappings of classes of functions from one topological vector space to another. { ¦'fəŋk·shən·əl ə'nal·ə·səs }

functional constraint A mathematical equation which must be satisfied by the independent parameters in an optimization problem, representing some physical principle which governs the relationship among these parameters. { 'fəŋk·shən·əl kən'strānt }

function space A metric space whose elements are functions. { 'fəŋk·shən ,spās }

function table A table that lists the values of a function for various values of the variable. { 'fəŋk·shən ,tā·bəl }

functor A function between categories which associates objects with objects and morphisms with morphisms. { 'fəŋk·tər }

fundamental affine connection An affine connection whose coefficients arise from the covariant and contravariant metric tensors of a space. { ¦fən·də¦ment·əl ə'fīn kə'nek·shən }

fundamental forms of a surface Differential forms which express the area and curvature of the surface. { ¦fən·də¦ment·əl 'fȯrmz əv ə 'sər·fəs }

fundamental group For a topological space, the group of homotopy classes of all closed paths about a point in the space; this group yields information about the number and type of "holes" in a surface. { ¦fən·də¦ment·əl 'grüp }

fundamental region Any region in the complex plane that can be mapped conformally onto all of the complex plane. { ¦fən·də¦ment·əl 'rē·jən }

fundamental sequence *See* Cauchy sequence. { ¦fən·də¦ment·əl 'sē·kwəns }

fundamental tensor *See* metric tensor. { ¦fən·də¦ment·əl 'ten·sər }

fundamental theorem of algebra Every polynomial of degree n with complex coefficients has exactly n roots counted according to multiplicity. { ¦fən·də¦ment·əl ¦thir·əm əv 'al·jə·brə }

fundamental theorem of arithmetic Every positive integer greater than 1 can be factored uniquely into the form $P_1^{n_1} \ldots P_i^{n_i} \ldots P_k^{n_k}$, where the P_i are primes, the n_i positive integers. { ¦fən·də¦ment·əl ¦thir·əm əv ə'rith·mə·tik }

fundamental theorem of calculus Given a continuous function $f(x)$ on the closed interval $[a,b]$ the functional

$$F(x) = \int_a^x f(t)\,dt$$

is differentiable on $[a,b]$ and $F'(x) = f(x)$ for every x in $[a,b]$, and if G is any function on $[a,b]$ such that $G'(x) = f(x)$ for all x in $[a,b]$, then

$$\int_a^b f(t)\,dt = G(b) - G(a)$$

{ ¦fən·də¦ment·əl ¦thir·əm əv 'kal·kyə·ləs }

fuzzy logic The logic of approximate reasoning, bearing the same relation to approximate reasoning that two-valued logic does to precise reasoning. { ¦fəz·ē 'läj·ik }

fuzzy model A finite set of fuzzy relations that form an algorithm for determining the outputs of a process from some finite number of past inputs and outputs. { ¦fəz·ē 'mäd·əl }

fuzzy relation A fuzzy subset of the cartesian product $X \times Y$, denoted as a relation from a set X to a set Y. { ¦fəz·ē ri'lā·shən }

fuzzy relational equation An equation of the form $A \circ R = B$, where A and B are fuzzy sets, R is a fuzzy relation, and $A \circ R$ stands for the composition of A with R. { ¦fəz·ē ri¦lā·shən·əl i'kwā·zhən }

fuzzy set An extension of the concept of a set, in which the characteristic function which determines membership of an object in the set is not limited to the two values 1 (for membership in the set) and 0 (for nonmembership), but can take on any value between 0 and 1 as well. { 'fəz·ē 'set }

fuzzy value A membership function of a fuzzy set that serves as the value assigned to a variable. { ¦fəz·ē 'val,yü }

G

Galois field A type of field extension obtained from considering the coefficients and roots of a given polynomial. Also known as root field. { 'gal‚wä ‚fēld }

Galois group A group of isomorphisms of a particular field extension associated with a polynomial's roots. { 'gal‚wä ‚grüp }

Galois theory The study of the Galois field and Galois group corresponding to a polynomial. { 'gal‚wä ‚thē·ə·rē }

game A mathematical model expressing a contest between two or more players under specified rules. { gām }

game theory The mathematical study of games or abstract models of conflict situations from the viewpoint of determining an optimal policy or strategy. Also known as theory of games. { 'gām ‚thē·ə·rē }

game tree A tree graph used in the analysis of strategies for a game, in which the vertices of the graph represent positions in the game, and a given vertex has as its successors all vertices that can be reached in one move from the given position. Also known as lookahead tree. { 'gām ‚trē }

gamma function The complex function given by the integral with respect to t from 0 to ∞ of $e^{-t}t^{z-1}$; this function helps determine the general solution of Gauss' hypergeometric equation. { 'gam·ə ‚fəŋk·shən }

Gaskin's theorem A theorem in projective geometry which states that if a circle circumscribes a triangle which is identical with its conjugate triangle with respect to a given conic, then the tangent to the circle at either of its intersections with the director circle of the conic is perpendicular to the tangent to the director circle at the same intersection. { 'gas·kinz ‚thir·əm }

Gauss-Bonnet theorem The theorem that the Euler characteristic of a compact Riemannian surface is $1/(2\pi)$ times the integral over the surface of the Gaussian curvature. { ¦gau̇s bə'nā ‚thir·əm }

Gauss-Codazzi equations Equations dealing with the components of the fundamental tensor and Riemann-Christoffel tensor of a surface. { ¦gau̇s kō'dat·sē i‚kwā·zhənz }

Gauss formulas Formulas dealing with the sine and cosine of angles in a spherical triangle. Also known as Delambre analogies. { 'gau̇s ‚fȯr·myə·ləz }

Gauss' hypergeometric equation The differential equation, arising in many physical contexts, $x(1 - x)y'' + [c - (a + b + 1)x]y' - aby = 0$. { 'gau̇s ˌhī·pərˌjē·ə'me·trik i'kwā·zhən }

Gaussian complex integers Complex numbers whose real and imaginary parts are both integers. { ¦gau̇·sē·ən ¦kämˌpleks 'int·ə·jərz }

Gaussian curvature The invariant of a surface specified by Gauss' theorem. Also known as total curvature. { ¦gau̇·sē·ən 'kər·və·chər }

Gaussian elimination A method of solving a system of n linear equations in n unknowns, in which there are first $n - 1$ steps, the mth step of which consists of subtracting a multiple of the mth equation from each of the following ones so as to eliminate one variable, resulting in a triangular set of equations which can be solved by back substitution, computing the nth variable from the nth equation, the $(n - 1)$st variable from the $(n - 1)$st equation, and so forth. { ¦gau̇·sē·ən əˌlim·ə'nā·shən }

Gaussian integer A complex number whose real and imaginary parts are both ordinary (real) integers. Also known as complex integer. { ¦gäu̇s·ē·ən 'int·ə·jər }

Gaussian noise *See* Wiener process. { ¦gau̇·sē·ən 'nȯiz }

Gaussian reduction A procedure of simplification of the rows of a matrix which is based upon the notion of solving a system of simultaneous equations. Also known as Gauss-Jordan elimination. { ¦gau̇·sē·ən ri'dək·shən }

Gauss-Jordan elimination *See* Gaussian reduction. { ¦gau̇s ¦jȯrd·ən əˌlim·ə'nā·shən }

Gauss' law of the arithmetic mean The law that a harmonic function can attain its maximum value only on the boundary of its domain of definition, unless it is a constant. { 'gau̇s ˌlȯ əv thə ˌa·rith¦med·ik 'mēn }

Gauss-Legendre rule An approximation technique of definite integrals by a finite series which uses the zeros and derivatives of the Legendre polynomials. { ¦gau̇s lə'zhän·drə ˌrül }

Gauss' mean value theorem The value of a harmonic function at a point in a planar region is equal to its integral about a circle centered at the point. { 'gau̇s 'mēn ˌval·yü ˌthir·əm }

Gauss-Seidel method *See* Seidel method. { ¦gau̇s 'zīd·əl ˌmeth·əd }

Gauss test In an infinite series with general term a_n, if $a_{n+1}/a_n = 1 - (x/n) - [f(n)/n\lambda]$ where x and λ are greater than 1, and $f(n)$ is a particular integer function, then the series converges. { 'gau̇s ˌtest }

Gauss' theorem **1.** The assertion, under certain light restrictions, that the volume integral through a volume V of the divergence of a vector function is equal to the surface integral of the exterior normal component of the vector function over the boundary surface of V. Also known as divergence theorem. **2.** At a point on a surface the product of the principal curvatures is an invariant of the surface, called the Gaussian curvature. { 'gau̇s ˌthir·əm }

gcd *See* greatest common divisor.

Gegenbauer polynomials A family of polynomials solving a special case of the Gauss hypergeometric equation. Also known as ultraspherical polynomials. { 'gāg·ənˌbau̇r ˌpäl·i'nō·mē·əlz }

Gelfond-Schneider theorem The theorem that if a and b are algebraic numbers, where a is not equal to 0 or 1, and b is not a rational number, then a^b is a transcendental number. { ¦gel‚fänd 'shnīd·ər ‚thir·əm }

general integral *See* general solution. { ¦gen·rəl 'int·ə·grəl }

generalized function *See* distribution. { 'jen·rə‚līzd 'fəŋk·shən }

generalized mean-value theorem *See* second mean-value theorem. { ¦jen·rə‚līzd 'mēn ¦val·yü ‚thir·əm }

generalized Poincaré conjecture The question as to whether every closed n-manifold which has the homotopy type of the n-sphere is homeomorphic to the n-sphere. { 'jen·rə‚līzd ¦Pwän·ka‚rā kən'jek·chər }

generalized power For a positive number a and an irrational number x, the number a^x defined by the equation $a^x = e^{x \log a}$, where e is the base of the natural logarithms and log a is taken to that base. { 'jen·rə‚līzd 'pau·ər }

general solution For an nth-order differential equation, a function of the independent variables of the equation and of n parameters such that assignment of any numerical values to the parameters yields a solution to the equation. Also known as general integral. { ¦jen·rəl sə'lü·shən }

general term The general term of a sequence or series is an expression subscripted by an integer which determines any desired entry. { ¦jen·rəl 'tərm }

general topology The branch of topology that studies the relationships between the basic topological properties that spaces may possess. Also known as point-set topology. { ¦jen·rəl tə'päl·ə·jē }

generating function A function $g(x,y)$ corresponding to a family of orthogonal polynomials $f_0(x), f_1(x),...$, where a Taylor series expansion of $g(x,y)$ in powers of y will have the polynomial $f_n(x)$ as the coefficient for the term y^n. { 'jen·ə‚rād·iŋ ‚fəŋk·shən }

generator **1.** One of the set of elements of an algebraic system such as a group, ring, or module which determine all other elements when all admissible operations are performed upon them. **2.** *See* generatrix. { 'jen·ə‚rād·ər }

generatrix The straight line generating a ruled surface. Also known as generator. { ¦jen·ə¦rā·triks }

Genocchi number An integer of the form $G_n = 2(2^{2n} - 1)B_n$, where B_n is the nth Bernoulli number. { gə'näk·ē‚nəm·bər }

genus An integer associated to a surface which measures the number of holes in the surface. { 'jē·nəs }

geodesic A curve joining two points in a Riemannian manifold which has minimum length. { ¦jē·ə¦des·ik }

geodesic circle The locus of all points on a given surface whose geodesic distance from a given point on the surface (called the center of the circle) is a given constant. { ¦jē·ə¦des·ik 'sər·kəl }

geodesic curvature For a point on a curve lying on a surface, the curvature of the orthogonal projection of the curve onto the tangent plane to the surface at the point; it measures the departure of the curve from a geodesic. Also known as tangential curvature. { ¦jē·ə¦des·ik 'kərv·ə·chər }

geodesic distance For two points in a Riemannian manifold, the length of a geodesic connecting them. { ¦jē·ə¦des·ik 'di·stəns }

geodesic ellipse The locus of all points on a given surface at which the sum of geodesic distances from a fixed pair of points is a constant. { ¦jē·ə¦des·ik i'lips }

geodesic hyperbola The locus of all points on a given surface at which the difference between the geodesic distances to two fixed points is a constant. { ¦jē·ə¦des·ik hī'pər·bə·lə }

geodesic line The shortest line between two points on a mathematically derived surface. { ¦jē·ə¦des·ik 'līn }

geodesic parallels Two curves on a given surface such that the lengths of geodesics between the curves that intersect both curves orthogonally is a constant. { ¦jē·ə¦des·ik 'par·ə‚lelz }

geodesic parameters Coordinates u and v of a surface such that the curves obtained by setting u equal to various constants form a family of geodesic parallels, while the curves obtained by setting v equal to various constants form the corresponding orthogonal family, of length $u_2 - u_1$ between the points (u_1, v) and (u_2, v). { ¦jē·ə¦des·ik pə'ram·əd·ərz }

geodesic polar coordinates Coordinates u and v of a surface such that the curves obtained by setting u equal to various constants are geodesic circles with a common center P and geodesic radius u, and the curves obtained by setting v equal to various constants are geodesics passing through P such that v_0 is the angle between the tangents at P to the lines $v = 0$ and $v = v_0$. { ¦jē·ə¦des·ik ¦pōl·ər kō'ȯrd·ə·nəts }

geodesic radius For a geodesic circle on a surface, the geodesic distance from the center of a circle to the points on the circle. { ¦jē·ə¦des·ik 'rād·ē·əs }

geodesic torsion **1.** For a given point on a surface and a given direction, the torsion of the geodesic on the surface through the point and in the given direction. **2.** For a given curve on a surface at a given point, the torsion of the geodesic through the point in the same direction as the given curve. { ¦jē·ə¦des·ik 'tȯr·shən }

geodesic triangle The figure formed by three geodesics joining three points on a given surface. { ¦jē·ə¦des·ik 'trī‚aŋ·gəl }

geodetic triangle *See* spheroidal triangle. { ¦jē·ə¦ded·ik 'trī‚aŋ·gəl }

geometric average *See* geometric mean. { ¦jē·ə¦me·trik 'av·rij }

geometric mean The geometric mean of n given quantities is the nth root of their product. Also known as geometric average. { ¦jē·ə¦me·trik 'mēn }

geometric moment of inertia The geometric moment of inertia of a plane figure about an axis in or perpendicular to the plane is the integral over the area of the figure of the square of the distance from the axis. Also known as second moment of area. { ¦jē·ə¦me·trik ¦mō·mənt əv i'nər·shə }

geometric number theory The branch of number theory studying relationships among numbers by examining the geometric properties of ordered pair sets of such numbers. { ¦jē·ə¦me·trik 'nəm·bər ‚thē·ə·rē }

geometric progression A sequence which has the form a, ar, ar^2, ar^3, ... { ¦jē·ə ¦me·trik prə′gresh·ən }

geometric series An infinite series of the form $a + ar + ar^2 + ar^3 + \cdots$. { ¦jē·ə ¦me·trik ′sir·ēz }

geometry The qualitative study of shape and size. { jē′äm·ə·trē }

Gershgorin's method A method of obtaining bounds on the eigenvalue of a matrix, based on the fact that the absolute value of any eigenvalue is equal to or less than the maximum over the rows of the matrix of the sum of the absolute values of the entries in a row, and is also equal to or less than the maximum over the columns of the matrix of the sum of the absolute values of the entries in a column. { gərsh′gȯr·ənz ˌmeth·əd }

gibbous Bounded by convex curves. { ′jib·əs }

Gibbs' phenomenon A convergence phenomenon occurring when a function with a discontinuity is approximated by a finite number of terms from a Fourier series. { ′gibz fəˌnäm·əˌnän }

Gibrat's distribution The distribution of a variable whose logarithm has a normal distribution. { zhē′bräz di·strə′byü·shən }

give-and-take lines Straight lines which are used to approximate the boundary of an irregular, curvilinear figure for the purpose of approximating its area; they are placed so that small portions excluded from the area under consideration are balanced by other small portions outside the boundary. { ¦giv ən ′tāk ˌlīnz }

Givens's method A transformation method for finding the eigenvalues of a matrix, in which each of the orthogonal transformations that reduce the original matrix to a triple-diagonal matrix makes one pair of elements, a_{ij} and a_{ji}, lying off the principal diagonal and the diagonals immediately above and below it, equal to zero, without affecting zeros obtained earlier. { ′giv·ən·zəz ˌmeth·əd }

glb *See* greatest lower bound.

glisette A curve, such as Watt's curve, traced out by a point attached to a curve which moves so that it always touches two fixed curves, or the envelope of any line or curve attached to the moving curve. { gli′set }

Glivenko-Cantelli lemma The empirical distribution functions of a random variable converge uniformly in probability to the distribution function of the random variable. { gli′veŋ·kō kan′tel·ē ′lem·ə }

gnomon A geometric figure formed by removing from a parallelogram a similar parallelogram that contains one of its corners. { ′nō·mən }

Gödel numbering *See* arithmetization. { ′gərd·əl ˌnəm·bə·riŋ }

Gödel's proof Any formal arithmetical system is incomplete in the sense that, given any consistent set of arithmetical axioms, there are true statements in the resulting arithmetical system that cannot be derived from these axioms. { ′gərd·əlz ′prüf }

Goldbach conjecture The unestablished conjecture that every even number except the number 2 is the sum of two primes. { ′gōlˌbäk kəŋˌjek·chər }

golden section The division of a line so that the ratio of the whole line to the larger interval equals the ratio of the larger interval to the smaller. Also known as extreme and mean ratio. { 'gōl·dən 'sek·shən }

gon *See* grade. { gän }

googol A name for 10 to the power 100. { 'gü׳gȯl }

googolplex A name for 10 to the power googol. { 'gü׳gȯl׳pleks }

grade A unit of plane angle, equal to 0.01 right angle, or $\pi/200$ radians, or 0.9°. Also known as gon. { grād }

graded Lie algebra A generalization of a Lie algebra in which both commutators and anticommutators occur. { ¦grād·əd ¦lē 'al·jə·brə }

gradient A vector obtained from a real function $f(x_1, x_2,...,x_n)$ whose components are the partial derivatives of f; this measures the maximum rate of change of f in a given direction. { 'grād·ē·ənt }

gradient method A finite iterative procedure for solving a system of n equations in n unknowns. { 'grād·ē·ənt ׳meth·əd }

gradient projection method Computational method used in nonlinear programming when constraint functions are linear. { 'grād·ē·ənt prə'jek·shən ׳meth·əd }

Graeffe's method A method of solving algebraic equations by means of squaring the exponents and making appropriate substitutions. { 'gref·əz ׳meth·əd }

Gram determinant The Gram determinant of vectors v_1, ..., v_n from an inner product space is the determinant of the $n \times n$ matrix with the inner product of v_i and v_j as entry in the ith column and jth row; its vanishing is a necessary and sufficient condition for linear dependence. { 'gram di'tərm·ə·nənt }

Gram-Schmidt orthogonalization process A process by which an orthogonal set of vectors is obtained from a linearly independent set of vectors in an inner product space. { ¦gram 'shmit ׳ȯr¦thäg·ən·əl·ə'zā·shən ׳präs·əs }

Gram's theorem A set of vectors are linearly dependent if and only if their Gram determinant vanishes. { 'gramz ׳thir·əm }

graph **1.** The planar object, formed from points and line segments between them, used in the study of circuits and networks. **2.** The graph of a function f is the set of all ordered pairs $[x,f(x)]$, where x is in the domain of f. **3.** *See* graphical representation. { graf }

graph component A particular type of maximal connected subgraph of a graph. { 'graf kəm'pō·nənt }

graphical analysis The study of interdependent phenomena by analyzing graphical representations. { ¦graf·ə·kəl ə'nal·ə·səs }

graphical representation The plot of the points in the plane which constitute the graph of a given real function or a pictorial diagram depicting interdependence of variables. Also known as graph. { ¦graf·ə·kəl ׳rep·rə·zen'tā·shən }

graph theory **1.** The mathematical study of the structure of graphs and networks. **2.** The body of techniques used in graphing functions in the plane. { 'graf ׳thē·ə·rē }

Grassmann algebra *See* exterior algebra. { 'gräs·mən ˌal·jə·brə }

Grassmannian *See* Grassmann manifold. { ¦gräs¦man·ē·ən }

Grassmann manifold The differentiable manifold whose points are all k-dimensional planes passing through the origin in n-dimensional euclidean space. Also known as Grassmannian. { 'gräs·mən 'man·əˌfōld }

great circle The circle on the two-sphere produced by a plane passing through the center of the sphere. { 'grāt ¦sər·kəl }

greatest common divisor The greatest common divisor of integers $n_1, n_2, \ldots, n_k$ is the largest of all integers that divide each n_i. Abbreviated gcd. Also known as highest common factor (hcf). { 'grād·əst ¦käm·ən di'vīz·ər }

greatest lower bound The greatest lower bound of a set of numbers S is the largest number among the lower bounds of S. Abbreviated glb. Also known as infimum (inf). { 'grād·əst ¦lō·ər 'bau̇nd }

Green's dyadic A vector operator which plays a role analogous to a Green's function in a partial differential equation expressed in terms of vectors. { 'grēnz dī'ad·ik }

Green's function A function, associated with a given boundary value problem, which appears as an integrand for an integral representation of the solution to the problem. { 'grēnz ˌfəŋk·shən }

Green's identities Formulas, obtained from Green's theorem, which relate the volume integral of a function and its gradient to a surface integral of the function and its partial derivatives. { 'grēnz i'den·əˌdēz }

Green's theorem Under certain general conditions, an integral along a closed curve C involving the sum of functions $P(x,y)$ and $Q(x,y)$ is equal to a surface integral, over the region D enclosed by C, of the partial derivatives of P and Q; namely,

$$\int_C P\,dx + Q\,dy = \int\!\!\int_D \left(\frac{\partial Q}{\partial x} - \frac{\partial P}{\partial y}\right) dx\,dy$$

{ 'grēnz ˌthir·əm }

Gregory formula A formula used in the numerical evaluation of integrals derived from the Newton formula. { 'greg·ə·rē ˌfȯr·myə·lə }

group A set G with an associative binary operation where $g_1 \cdot g_2$ always exists and is an element of G; each g has an inverse element g^{-1}, and G contains an identity element. { grüp }

groupoid A set having a binary relation everywhere defined. { 'grüˌpȯid }

group theory The study of the structure of groups which especially deals with the classification of finite groups. { 'grüp ˌthē·ə·rē }

group without small subgroups A topological group in which there is a neighborhood of the identity element that contains no subgroup other than the subgroup consisting of the identity element alone. { ¦grüp withˌau̇t ˌsmȯl 'səbˌgrüps }

growth index For a function of bounded growth f, the smallest real number a such that for some positive real constant M the quantity Me^{ax} is greater than the absolute value of $f(x)$ for all positive x; for a function that is not of bounded growth, the quantity $+\infty$. { 'grōth ˌinˌdeks }

G space A topological space X together with a topological group G and a continuous function on the cartesian product of X and G to X such that if the values of this function at (x,g) are denoted by xg, then $x(g_1g_2) = (xg_1)g_2$ and $xe = x$ where e is the identity in G and g_1,g_2 are elements in G. { 'jē ˌspās }

Gudermannian The function y of the variable x satisfying $\tan y = \sinh x$ or $\sin y = \tanh x$; written gd$x$. { 'güd·ərˌmän·ē·ən }

Gutschoven's curve *See* kappa curve. { 'gütˌshō·fənz ˌkərv }

H

Haar measure A measure on the Borel subsets of a locally compact topological group whose value on a Borel subset U is unchanged if every member of U is multiplied by a fixed element of the group. { 'här ˌmezh·ər }

Hadamard's conjecture The conjecture that any partial differential equation that is essentially different from the wave equation fails to satisfy Huygens' principle. { 'had·əˌmärdz kən'jek·chər }

Hadamard's inequality An inequality that gives an upper bound for the square of the absolute value of the determinant of a matrix in terms of the squares of the matrix entries; the upper bound is the product, over the rows of the matrix, of the sum of the squares of the absolute values of the entries in a row. { 'had·əˌmärdz ˌin·ə¦kwäl·əd·ē }

Hadamard's three-circle theorem The theorem that if the complex function $f(z)$ is analytic in the ring $a\ z\ b$, and if $m(r)$ denotes the maximum value of $f(z)$ on the circle $z = r$ with arb, then log $m(r)$ is a convex function of log r. { 'had·əˌmärdz ¦thrē ¦sər·kəl 'thir·əm }

Hahn-Banach extension theorem The theorem that every continuous linear functional defined on a subspace or linear manifold in a normed linear space X may be extended to a continuous linear functional defined on all of X. { ¦hän ¦bän·äk ek'sten·chən ˌthir·əm }

Hahn decomposition The Hahn decomposition of a measurable space X with signed measure m consists of two disjoint subsets A and B of X such that the union of A and B equals X, A is positive with respect to m, and B is negative with respect to m. { ¦hän dēˌkäm·pə'zish·ən }

half-angle formulas In trigonometry, formulas that express the trigonometric functions of half an angle in terms of trigonometric functions of the angle. { 'haf ˌaŋ·gəl ˌfȯr·myə·ləz }

half line *See* ray. { 'haf ¦līn }

half plane The portion of a plane lying on one side of some line in the plane; in particular, all points of the complex plane either above or below the real axis. { 'haf ¦plān }

half-side formulas In trigonometry, formulas that express the tangents of one-half of each of the sides of a spherical triangle in terms of its angles. { 'haf ˌsīd ˌfōr·myə·ləz }

half space A space bounded only by an infinite plane. { 'haf ˌspās }

half-width For a function which has a maximum and falls off rapidly on either side of the maximum, the difference between the two values of the independent variable for which the dependent variable has one-half its maximum value. { 'haf ¦width }

Hall's theorem *See* marriage theorem. { 'hȯlz ˌthir·əm }

Hamel basis For a normed space, a collection of vectors with every finite subset linearly independent, while any vector of the space is a linear combination of at most countably many vectors from this subset. { 'ham·əl ¦bā·səs }

Hamilton-Cayley theorem *See* Cayley-Hamilton theorem. { 'ham·əl·tən 'kā·lē ˌthir·əm }

Hamiltonian cycle *See* Hamiltonian path. { ˌham·əl'tō·nē·ən ˌsī·kəl }

Hamiltonian graph A graph which has a Hamiltonian path. { ˌham·əl'tō·nē·ən ˌgraf }

Hamiltonian path A path along the edges of a graph that traverses every vertex exactly once and terminates at its starting point. Also known as Hamiltonian cycle. { ˌham·əl'tō·nē·ən ˌpath }

Hamilton-Jacobi equation A particular partial differential equation useful in studying certain systems of ordinary equations arising in the calculus of variations, dynamics, and optics: $H(q_1, \ldots, q_n, \partial\phi/\partial q_1, \ldots, \partial\phi/\partial q_n, t) + \partial\phi/\partial t = 0$, where $q_1, \ldots, q_n$ are generalized coordinates, t is the time coordinate, H is the Hamiltonian function, and ϕ is a function that generates a transformation by means of which the generalized coordinates and momenta may be expressed in terms of new generalized coordinates and momenta which are constants of motion. { 'ham·əl·tən jə'kō·bē iˌkwā·zhən }

Hamilton-Jacobi theory The study of the solutions of the Hamilton-Jacobi equation and the information they provide concerning solutions of the related systems of ordinary differential equations. { 'ham·əl·tən jə'kō·bē ˌthē·ə·rē }

Hankel functions The Bessel functions of the third kind, occurring frequently in physical studies. { 'häŋk·əl ˌfəŋk·shənz }

Hankel transform The Hankel transform of order m of a real function $f(t)$ is the function $F(s)$ given by the integral from 0 to ∞ of $f(t)tJ^m(st)dt$, where J^m denotes the mth-order Bessel function. Also known as Bessel transform; Fourier-Bessel transform. { 'häŋk·əl ˌtranzˌfȯrm }

harmonic A solution of Laplace's equation which is separable in a specified coordinate system. { här'män·ik }

harmonic analysis A study of functions by attempting to represent them as infinite series or integrals which involve functions from some particular well-understood family; it subsumes studying a function via its Fourier series. { här'man·ik ə'nal·ə·səs }

harmonic conjugates **1.** Two points, P_3 and P_4, that are collinear with two given points, P_1 and P_2, such that P_3 lies in the line segment P_1P_2 while P_4 lies outside it, and, if x_1, x_2, x_3, and x_4 are the abscissas of the points, $(x_3 - x_1)/(x_3 - x_2) = -(x_4 - x_1)/(x_4 - x_2)$. **2.** A pair of harmonic functions, u and v, such that $u + iv$ is an analytic function, or, equivalently, u and v satisfy the Cauchy-Riemann equations. { här'män·ik 'kän·jə·gəts }

harmonic division The division of a line segment externally and internally in the same ratio; that is, the division of a line segment by the harmonic conjugates of its end points. { här'män·ik di'vizh·ən }

harmonic function 1. A function of two real variables which is a solution of Laplace's equation in two variables. **2.** A function of three real variables which is a solution of Laplace's equation in three variables. { här'män·ik 'fəŋk·shən }

harmonic mean For n positive numbers $x_1, x_2, \ldots, x_n$ their harmonic mean is the number $n/(1/x_1 + 1/x_2 + \cdots + 1/x_n)$. { här'män·ik 'mēn }

harmonic measure Let D be a domain in the complex plane bounded by a finite number of Jordan curves Γ, and let Γ be the disjoint union of α and β, where α and β are Jordan arcs; the harmonic measure of α with respect to D is the harmonic function on D which assumes the value 1 on α and the value 0 on β. { här¦män·ik 'mezh·ər }

harmonic pencil The configuration of four lines, passing through a single point, such that any line that is not parallel to one of the four cuts the four lines at points which are harmonic conjugates. { här¦män·ik 'pen·səl }

harmonic progression A sequence of numbers whose reciprocals form an arithmetic progression. { här'män·ik prə'gresh·ən }

harmonic range The configuration of four collinear points which are harmonic conjugates. { här¦män·ik 'rānj }

harmonic ratio A cross ratio that is equal to -1. { här¦män·ik 'rā·sho }

harmonic series A series whose terms form a harmonic progression. { här¦män·ik 'sir,ēz }

Harnack's first convergence theorem The theorem that if a sequence of functions harmonic in a common domain of three-dimensional space and continuous on the boundary of the domain converges uniformly on the boundary, then it converges uniformly in the domain to a function which is itself harmonic; the sequence of any partial derivative of the functions in the original sequence converges uniformly to the corresponding partial derivative of the limit function in every closed subregion of the domain. { 'här·naks ¦fərst kən'vər·jəns ,thir·əm }

Harnack's second convergence theorem The theorem that if a sequence of functions is harmonic in a common domain of three-dimensional space and their values are monotonically decreasing at any point in the domain, then convergence of the sequence at any point in the domain implies uniform convergence of the sequence in every closed subregion of the domain to a function which is itself harmonic. { 'här·naks ¦sek·ənd kən'vər·jəns ,thir·əm }

Hausdorff maximal principle The principle that every partially ordered set has a linearly ordered subset S which is maximal in the sense that S is not a proper subset of another linearly ordered subset. { 'haủs·dȯrf 'mak·sə·məl ,prin·sə·pəl }

Hausdorff space A topological space where each pair of distinct points can be enclosed in disjoint open neighborhoods. Also known as T_2 space. { 'haủs·dȯrf ,spās }

hav *See* haversine.

haversine The haversine of an angle A is half of the versine of A, or $\frac{1}{2}(1 - \cos A)$. Abbreviated hav. { 'ha·vər,sīn }

hcf *See* greatest common divisor.

Heaviside calculus A type of operational calculus that is used to completely analyze a linear dynamical system which represents some vibrating physical system. { ′hev·ē‚sīd ‚kal·kyə·ləs }

Heaviside's expansion theorem A theorem providing an infinite series representation for the inverse Laplace transforms of functions of a particular type. { ′hev·ē‚sīdz ik′span·chən ‚thir·əm }

Heaviside unit function The real function $f(x)$ whose value is 0 if x is negative and whose value is 1 otherwise. { ′hev·ē‚sīd ′yü·nət ¦fəŋk·shən }

height **1.** The perpendicular distance between horizontal lines or planes passing through the top and bottom of an object. **2.** The height of a rational number q is the maximum of $|m|$ and $|n|$, where m and n are relatively prime integers such that $q = m/n$. { hīt }

Heine-Borel theorem The theorem that the only compact subsets of the real line are those which are closed and bounded. { ′hī·nə bȯ′rel ‚thir·əm }

helical Pertaining to a cylindrical spiral, for example, a screw thread. { ′hel·ə·kəl }

helicoid A surface generated by a curve which is rotated about a straight line and also is translated in the direction of the line at a rate that is a constant multiple of its rate of rotation. { ′hel·ə‚kȯid }

helix A curve traced on a cylindrical or conical surface where all points of the surface are cut at the same angle. { ′hē‚liks }

helix angle The constant angle between the tangent to a helix and a generator of the cylinder upon which the helix lies. { ′hē‚liks ‚aŋ·gəl }

Helmholtz equation A partial differential equation obtained by setting the Laplacian of a function equal to the function multiplied by a negative constant. { ′helm‚hōlts i‚kwā·zhən }

Helmholtz's theorem The theorem determining a general class of vector fields as being everywhere expressible as the sum of an irrotational vector with a divergence-free vector. { ′helm‚hōlt·səz ‚thir·əm }

hemicycle A curve in the form of a semicircle. { ′he·mē‚sī·kəl }

hemisphere One of the two pieces of a sphere divided by a great circle. { ′he·mē ‚sfir }

hemispheroid One of the halves into which a spheroid is divided by a plane of symmetry. { ‚he·mē′sfir‚ȯid }

heptagon A seven-sided polygon. { ′hep·tə‚gän }

heptakaidecagon A polygon with 17 sides. { ¦hep·tə‚kī′dek·ə‚gän }

Hermite polynomials A family of orthogonal polynomials which arise as solutions to Hermite's differential equation, a particular case of the hypergeometric differential equation. { er′mēt ‚päl·ə′nō·mē·əlz }

Hermite's differential equation A particular case of the hypergeometric equation; it has the form $w'' - 2zw' + 2nw = 0$, where n is an integer. { er′mēts dif·ə¦ren·chəl i′kwā·zhən }

Hermitian conjugate For a matrix A, the transpose of the complex conjugate of A. Also known as adjoint; associate matrix. { er′mish·ən ′kän·jə·gət }

Hermitian conjugate operator *See* adjoint operator. { er′mish·ən ′kän·jə·gət ′äp·ə‚rād·ər }

Hermitian form 1. A polynomial in n real or complex variables where the matrix constructed from its coefficients is Hermitian. **2.** More generally, a sesquilinear form g such that $g(x,y) = \overline{g(y,x)}$ for all values of the independent variables x and y, where $\overline{g(x,y)}$ is the image of $g(x,y)$ under the automorphism of the underlying ring. { er′mish·ən ′fȯrm }

Hermitian inner product *See* inner product. { er′mish·ən ′in·ər ¦präd·əkt }

Hermitian kernel A kernel $K(x,t)$ of an integral transformation or integral equation is Hermitian if $K(x,t)$ equals its adjoint kernel, $K^*(t,x)$. { er′mish·ən ′kər·nəl }

Hermitian matrix A matrix which equals its conjugate transpose matrix, that is, is self-adjoint. { er′mish·ən ′mā·triks }

Hermitian operator A linear operator A on vectors in a Hilbert space, such that if x and y are in the range of A then the inner products (Ax,y) and (x,Ay) are equal. { er′mish·ən ′äp·ə‚rād·ər }

Hermitian scalar product *See* inner product. { er′mish·ən ′skāl·ər ¦präd·əkt }

Hermitian space *See* inner product space. { er′mish·ən ′spās }

hermit point *See* isolated point. { ′hər·mit ‚pȯint }

Heron's formula *See* Hero's formula. { ′her·ənz ‚fȯr·myə·lə }

Hero's formula A formula expressing the area of a triangle in terms of the sides a, b, and c as

$$\sqrt{s(s-a)(s-b)(s-c)}$$

$$\text{where } s = \sqrt{(1/2)(aa + b + c}$$

Also known as Heron's formula. { ′hir·ōz ‚fȯr·myə·lə }

Hesse's theorem A theorem in projective geometry which states that, from the three pairs of lines containing the two pairs of opposite sides and the diagonals of a quadrilateral, if any two pairs are conjugate lines with respect to a given conic, then so is the third. { ′hes·əz ‚thir·əm }

Hessian For a function $f(x_1, \ldots, x_n)$ of n real variables, the real-valued function of $(x_1, \ldots, x_n)$ given by the determinant of the matrix with entry $\partial^2 f/\partial x_i \partial x_j$ in the ith row and jth column; used for analyzing critical points. { ′hesh·ən }

heterogeneous Pertaining to quantities having different degrees or dimensions. { ‚hed·ə′räj·ə·nəs }

heuristic method A method of solving a problem in which one tries each of several approaches or methods and evaluates progress toward a solution after each attempt. { hyu̇′ris·tik ′meth·əd }

hexadecimal Pertaining to a number system using the base 16. Also known as sexadecimal. { ‚hek·sə′des·məl }

hexadecimal number system A digital system based on powers of 16, as compared with the use of powers of 10 in the decimal number system. Also known as sexadecimal number system. { ˌhek·səˈdes·məl ˈnəm·bər ˌsis·təm }

hexafoil A multifoil consisting of six congruent arcs of a circle arranged around a regular hexagon. { ˈhek·səˌfȯil }

hexagon A six-sided polygon. { ˈhek·səˌgän }

hexahedron A polyhedron with six faces. { ˌhek·səˈhē·drən }

higher plane curve Any algebraic curve whose degree exceeds 2. { ˈhī·ər ˌplān ¦kərv }

highest common factor *See* greatest common divisor. { ˈhī·əst ˈkäm·ən ˈfak·tər }

Hilbert cube The topological space which is the cartesian product of a countable number of copies of *I*, the unit interval. { ˈhil·bərt ˌkyüb }

Hilbert parallelotope **1.** A subset of an infinite-dimensional Hilbert space with coordinates $x_1, x_2, \ldots$, for which the absolute value of x_n is equal to or less than $(1/2)^n$ for each n. **2.** The subset of this space for which the absolute value of x_n is equal to or less than $1/n$ for each n. { ˈhil·bərt ˌpar·əˈlel·əˌtōp }

Hilbert-Schmidt theory A body of theorems which investigates the kernel of an integral equation via its eigenfunctions, and then applies these functions to help determine solutions of the equation. { ¦hil·bərt ˈshmit ˌthē·ə·rē }

Hilbert space A Banach space which also is an inner-product space with the inner product of a vector with itself being the same as the square of the norm of the vector. { ˈhil·bərt ¦spās }

Hilbert's theorem The proposition that the ring of polynomials with coefficients in a commutative Noetherian ring is itself a Noetherian ring. { ˈhilˌbərts ˌthir·əm }

Hilbert transform The transform of a function $f(x)$ realized by taking the integral of $f(x)[1 + \cot (y - x)/2]dx$. { ˈhil·bərt ¦tranzˌfȯrm }

hill-climbing Any numerical procedure for finding the maximum or maxima of a function. { ˈhil ˌklim·iŋ }

hippopede A plane curve whose equation in polar coordinates r and θ is $r^2 = 4b\ (a - b \sin^2 \theta)$, where a and b are positive constants. Also known as horse fetter. { ˈhip·əˌpēd }

Hjelmslev plane *See* affine Hjelmslev plane. { ˈhyelmˌslev ˌplān }

Hodge conjecture The $2p$-dimensional rational cohomology classes in an n-dimensional algebraic manifold M which are carried by algebraic cycles are those with dual cohomology classes representable by differential forms of bidegree $(n - p, n - p)$ on M. { ˈhäj kənˌjek·chər }

Hölder condition **1.** A function $f(x)$ satisfies the Hölder condition in a neighborhood of a point x_0 if $|f(x) - f(x_0)| \leq c\,|(x - x_0)|^n$, where c and n are constants. **2.** A function $f(x)$ satisfies a Hölder condition in an interval or in a region of the plane if $|f(x) - f(y)| \leq c\,|x - y|^n$ for all x and y in the interval or region, where c and n are constants. { ˈhel·dər kənˌdish·ən }

Hölder's inequality Generalization of the Schwarz inequality: for real functions $|\int f(x)g(x)dx| \leq (\int |f(x)|^p dx)^{1/p} \cdot (\int |g(x)|^q dx)^{1/q}$ where $1/p + 1/q = 1$. { 'hel·dərz ˌin·i'kwäl·əd·ē }

Hölder summation A method of attributing a sum to certain divergent series in which a new series is formed, each of whose partial sums is the average of the first n partial sums of the original series, and this process is repeated until a stage is reached where the limit of this average exists. { 'hel·dər səˌmā·shən }

holomorphic function *See* analytic function. { ¦häl·ō¦mȯr·fik 'fəŋk·shən }

homeomorphic spaces Two topological spaces with a homeomorphism existing between them; intuitively one can be obtained from the other by stretching, twisting, or shrinking. { ¦hō·mē·ə¦mȯr·fik 'spās·əz }

homeomorphism A continuous map between topological spaces which is one-to-one, onto, and its inverse function is continuous. Also known as bicontinuous function; topological mapping. { ¦hō·mē·ə¦mȯrˌfiz·əm }

homogeneous Pertaining to a group of mathematical symbols of uniform dimensions or degree. { ˌhä·mə'jē·nē·əs }

homogeneous coordinates To a point in the plane with cartesian coordinates (x,y) there corresponds the homogeneous coordinates (x_1,x_2,x_3), where $x_1/x_3 = x$, $x_2/x_3 = y$; any polynomial equation in cartesian coordinates becomes homogeneous if a change into these coordinates is made. { ˌhä·mə'jē·nē·əs kō'ȯrd·ən·əts }

homogeneous differential equation A differential equation where every scalar multiple of a solution is also a solution. { ˌhä·mə'jē·nē·əs ˌdif·ə'ren·chəl iˌkwā·zhən }

homogeneous equation An equation that can be rewritten into the form having zero on one side of the equal sign and a homogeneous function of all the variables on the other side. { ˌhä·mə'jē·nē·əs i'kwā·zhən }

homogeneous function A real function $f(x_1,x_2, ...,x_n)$ is homogeneous of degree r if $f(ax_1,ax_2,...,ax_n) = a^r f(x_1,x_2, ...,x_n)$ for every real number a. { ˌhä·mə'jē·nē·əs 'fəŋk·shən }

homogeneous integral equation An integral equation where every scalar multiple of a solution is also a solution. { ˌhä·mə'jē·nē·əs 'int·ə·grəl iˌkwā·zhən }

homogeneous polynomial A polynomial all of whose terms have the same total degree; equivalently it is a homogenous function of the variables involved. { ˌhä·mə'jē·nē·əs ˌpäl·ə'nō·mē·əl }

homogeneous space A topological space having a group of transformations acting upon it, that is, a transformation group, where for any two points x and y some transformation from the group will send x to y. { ˌhä·mə'jē·nē·əs 'spās }

homogeneous transformation *See* linear transformation. { ˌhä·mə'jē·nē·əs ˌtranz·fər'mā·shən }

homographic transformations *See* Möbius transformations. { ¦hä·mə¦graf·ik ˌtranz·fər'mā·shənz }

homological algebra The study of the structure of modules, particularly by means of exact sequences; it has application to the study of a topological space via its homology groups. { ¦hä·mə¦läj·ə·kəl 'al·jə·brə }

homology group Associated to a topological space X, one of a sequence of Abelian groups $H_n(X)$ that reflect how n-dimensional simplicial complexes can be used to fill up X and also help determine the presence of n-dimensional holes appearing in X. Also known as Betti group. { hə'mäl·ə·jē ˌgrüp }

homology theory Theory attempting to compare topological spaces and investigate their structures by determining the algebraic nature and interrelationships appearing in the various homology groups. { hə'mäl·ə·jē ˌthē·ə·rē }

homomorphism A function between two algebraic systems of the same type which preserves the algebraic operations. { ˌhä·mə'mȯrˌfiz·əm }

homothetic curves For a given point, a set of curves such that any straight line through the point intersects all the curves in the set at the same angle. { ¦häm·əˌthed·ik 'kərvz }

homothetic figures Similar figures which are placed so that lines joining corresponding points pass through a common point and are divided in a constant ratio by this point. { ¦häm·əˌthed·ik 'fig·yərz }

homothetic ratio *See* ratio of similitude. { ¦hō·mə¦thed·ik 'rā·shō }

homothetic transformation A transformation that leaves the origin of coordinates fixed and multiplies the distance between any two points by the same fixed constant. Also known as transformation of similitude. { ¦häm·əˌthed·ik ˌtranz·fər 'mā·shən }

homotopy Between two mappings of the same topological spaces, a continuous function representing how, in a step-by-step fashion, the image of one mapping can be continuously deformed onto the image of the other. { hō'mäd·ə·pē }

homotopy groups Associated to a topological space X, the groups appearing for each positive integer n, which reflect the number of different ways (up to homotopy) than an n-dimensional sphere may be mapped to X. { hō'mäd·ə·pē ˌgrüps }

homotopy theory The study of the topological structure of a space by examining the algebraic properties of its various homotopy groups. { hō'mäd·ə·pē ˌthē·ə·rē }

horn angle A geometric figure formed by two tangent plane curves that lie on the same side of their mutual tangent line in the neighborhood of the point of tangency. { 'hȯrn ˌaŋ·gəl }

Horner's method A technique for approximating the real roots of an algebraic equation; a root is located between consecutive integers, then a successive search is performed. { 'hȯrn·ərz ˌmeth·əd }

horse fetter *See* hippopede. { 'hȯrs ˌfed·ər }

Householder's method A transformation method for finding the eigenvalues of a symmetric matrix, in which each of the orthogonal transformations that reduce the original matrix to a triple-diagonal matrix reduces one complete row to the required form. { 'hau̇sˌhōl·dərz ˌmeth·əd }

Hughes plane A finite projective plane with nine points on each line that can be represented by a nonlinear ternary ring generated by a four-point in the plane. { 'hyüz ˌplān }

Hurwitz polynomial A polynomial whose zeros all have negative real parts. { 'hər·vitz ˌpäl·ə'nō·mē·əl }

Hurwitz's criterion A criterion that determines whether a polynomial is a Hurwitz polynomial, based on the signs of a set of determinants formed from the polynomial's coefficients. { 'hərˌwit·səz krīˌtir·ē·ən }

Huygens' approximation The length of a small circular arc is approximately $\frac{1}{3}(8c' - c)$, where c is the chord of the arc and c' is the chord of half the arc. { 'hī·gənz əˌpräk·səˌmā·shən }

hyperbola The plane curve obtained by intersecting a circular cone of two nappes with a plane parallel to the axis of the cone. { hī·pər·bə·lə }

hyperbolic cosecant A function whose value is equal to the reciprocal of the value of the hyperbolic sine. Abbreviated csch. { ¦hī·pər¦bäl·ik kō'sēˌkant }

hyperbolic cosine A function whose value for the complex number z is one-half the sum of the exponential of z and the exponential of $-z$. Abbreviated cosh. { ¦hī·pər¦bäl·ik 'kōˌsīn }

hyperbolic cotangent A function whose value is equal to the value of the hyperbolic cosine divided by the value of the hyperbolic sine. Abbreviated coth. { ¦hī·pər¦bäl·ik kō'tan·jənt }

hyperbolic cylinder A cylinder whose directrix is a hyperbola. { ¦hī·pər¦bäl·ik 'sil·ən·dər }

hyperbolic differential equation A general type of second-order partial differential equation which includes the wave equation and has the form

$$\sum_{i,j=1}^{n} A_{ij}\,(\partial^2 u/\partial x_i \partial x_j) + \sum_{i=1}^{n} B_i(\partial u/\partial x_i) + Cu + F = 0$$

where the A_{ij}, B_i, C, and F are suitably differentiable real functions of $x_1, x_2, \ldots, x_n$, and there exists at each point $(x_1, x_2, \ldots, x_n)$ a real linear transformation on the x_i which reduces the quadratic form

$$\sum_{ij=1}^{n} A_{ij} x_i x_j$$

to a sum of n squares not all of the same sign. { ¦hī·pər¦bäl·ik ˌdif·ə¦ren·chəl i'kwā·zhən }

hyperbolic form A nondegenerate, symmetric or alternating form on a vector space E such that E is a hyperbolic space under this form. { ¦hī·pərˌbäl·ik 'fȯrm }

hyperbolic functions The real or complex functions sinh (x), cosh (x), tanh (x), coth (x), sech (x), csch (x); they are related to the hyperbola in somewhat the same fashion as the trigonometric functions are related to the circle, and have properties analogous to those of the trigonometric functions. { ¦hī·pər¦bäl·ik 'fəŋk·shənz }

hyperbolic geometry *See* Lobachevski geometry. { ¦hī·pər¦bäl·ik jē'äm·ə·trē }

hyperbolic logarithm *See* logarithm. { ¦hī·pər¦bäl·ik 'läg·əˌrith·əm }

hyperbolic paraboloid A surface which can be so situated that sections parallel to one coordinate plane are parabolas while those parallel to the other plane are hyperbolas. { ¦hī·pər¦bäl·ik pə'rab·əˌlȯid }

hyperbolic plane A two-dimensional vector space E on which there is a nondegenerate, symmetric or alternating form $f(x,y)$ such that there exists a nonzero element w in E for which $f(w,w) = 0$. { ¦hī·pər‚bäl·ik 'plān }

hyperbolic point A point on a surface where the Gaussian curvature is strictly negative. { ¦hī·pər¦bäl·ik 'pȯint }

hyperbolic secant A function whose value is equal to the reciprocal of the value of the hyperbolic cosine. Abbreviated sech. { ¦hī·pər¦bäl·ik 'sē‚kant }

hyperbolic sine A function whose value for the complex number z is one-half the difference between the exponential of z and the exponential of $-z$. Abbreviated sinh. { ¦hī·pər¦bäl·ik 'sīn }

hyperbolic space A space described by hyperbolic rather than cartesian coordinates. { ¦hī·pər¦bäl·ik 'spās }

hyperbolic spiral A plane curve for which the radius vector is inversely proportional to the polar angle. Also known as reciprocal spiral. { ¦hī·pər¦bäl·ik 'spi·rəl }

hyperbolic tangent A function whose value is equal to the value of the hyperbolic sine divided by the value of the hyperbolic cosine. Abbreviated tanh. { ¦hī·pər ¦bäl·ik 'tan·jənt }

hyperbolic type A type of simply connected Riemann surface that can be mapped conformally on the interior of the unit circle. { ¦hī·pər¦bäl·ik 'tīp }

hyperboloid A quadric surface given by an equation of the form $(x^2/a^2) \pm (y^2/b^2) - (z^2/c^2) = 1$; in certain cases it is a hyperboloid of revolution, which can be realized by rotating the pieces of a hyperbola about an appropriate axis. { hī'pər·bə‚lȯid }

hyperboloid of one sheet A surface whose equation in stardard form is $(x^2/a^2) + (y^2/b^2) - (z^2/c^2) = 1$, so that it is in one piece, and cuts planes perpendicular to the x or y axes in hyperbolas and planes perpendicular to the z axis in ellipses. { hī'pər·bə‚lȯid əv 'wən ‚shēt }

hyperboloid of revolution A surface generated by rotating a hyperbola about one of its axes. { hī'pər·bə‚lȯid əv ‚rev·ə'lü·shən }

hyperboloid of two sheets A surface whose equation in standard form is $(x^2/a^2) - (y^2/b^2) - (z^2/c^2) = 1$, so that it is in two pieces, and cuts planes perpendicular to the y and z axes in hyperbolas and planes perpendicular to the x axis in ellipses, except for the interval $-axa$, where there is no intersection. { hī'pər·bə‚lȯid əv 'tü ‚shēts }

hypercircle method A geometric method of obtaining approximate solutions of linear boundary value problems of mathematical physics that cannot be solved exactly, in which a correspondence is made between physical variables and vectors in a function space. { ¦hī·pər¦sər·kəl ‚meth·əd }

hypercomplex number *See* quaternion. { ¦hī·pər¦käm‚pleks 'nəm·bər }

hypercomplex system *See* algebra. { ¦hī·pər¦käm‚pleks 'sis·təm }

hypercube The analog of a cube in n dimensions ($n = 2, 3, \ldots.$), with 2^n vertices, $n2^{n-1}$ edges, and $2n$ cells; for an object with edges of length $2a$, the coordinates of the vertices are $(+a, +a, \ldots., +a)$. { 'hī·pər ‚kyüb }

hypergeometric function A function which is a solution to the hypergeometric equation and obtained as an infinite series expansion. { ˌhī·pərˌjē·ə′me·trik ′fəŋk·shən }

hypergeometric series A particular infinite series which in certain cases is a solution to the hypergeometric equation, and having the form

$$1 + \frac{ab}{c} z + \frac{1}{2!} \frac{a(a+1)b(b+1)}{c(c+1)} z^2 + \cdots$$

{ ˌhī·pərˌjē·ə′me·trik ′sir·ēz }

hyperplane A hyperplane is an $(n - 1)$-dimensional subspace of an n-dimensional vector space. { ¦hī·pərˌplān }

hyperplane of support Relative to a convex body in a normed vector space, a hyperplane whose distance from the body is zero, and which separates the normed vector space into two halves, one of which contains no points of the convex body. { ′hī·pərˌplān əv sə′pȯrt }

hypocycloid The curve which is traced in the plane as a given point fixed on a circle moves while this circle rolls along the inside of another circle. { ¦hī·pō′sīˌklȯid }

hypotenuse On a right triangle, the side opposite the right angle. { hī′pät·ənˌüs }

hypotrochoid A curve traced by a point rigidly attached to a circle at a point other than the center when the circle rolls without slipping on the inside of a fixed circle. { ¦hī·pō′trōˌkȯid }

I

icosahedron A 20-sided polyhedron. { ī¦kä·sə¦hē·drən }

ideal A subset I of a ring R where $x - y$ is in I for every x,y in I and either rx is in I for every r in R and x in I or xr is in I for every r in R and x in I; in the first case I is called a left ideal, and in the second a right ideal; an ideal is two-sided if it is both a left and a right ideal. { ī'dēl }

ideal line The collection of all ideal points, each corresponding to a given family of parallel lines. Also known as line at infinity. { ī'dēl 'līn }

ideal point In projective geometry, all lines parallel to a given line are hypothesized to meet at a point at infinity, called an ideal point. Also known as point at infinity. { ī'dēl 'pȯint }

ideal theory The branch of algebra studying the properties of ideals. { ī'dēl 'thē·ə·rē }

idem factor The dyadic $I = ii + jj + kk$ such that scalar multiplication of I by any vector yields that vector. { 'ī‚dem ‚fak·tər }

idempotent **1.** An element x of an algebraic system satisfying the equation $x^2 = x$. **2.** An algebraic system in which every element x satisfies $x^2 = x$. { ¦ī‚dem¦pōt·ənt }

idempotent law A law which states that an element x of an algebraic system satisfies $x^2 = x$. { ¦ī‚dem¦pōt·ənt 'lȯ }

idempotent matrix A matrix E satisfying the equation $E^2 = E$. { ¦ī‚dem¦pōt·ənt 'mā·triks }

identity **1.** An equation satisfied for all possible choices of values for the variables involved. **2.** *See* identity element. { ī'den·ə‚dē }

identity element The unique element e of a group where $g \cdot e = e \cdot g = g$ for every element g of the group. Also known as identity. { ī'den·ə‚dē ‚el·ə·mənt }

identity function The function of a set to itself which assigns to each element the same element. Also known as identity operator. { ī'den·ə‚dē ‚fəŋk·shən }

identity matrix The square matrix all of whose entries are zero except along the principal diagonal where they all are 1. { ī'den·ə‚dē ‚mā·triks }

identity operator *See* identity function. { ī'den·ə‚dē ‚äp·ə‚rād·ər }

if and only if operation *See* biconditional operation. { ¦if ən 'ōn·lē ¦if ‚äp·ə‚rā·shən }

if-then operation *See* implication. { ¦if then ˌäp·əˈrā·shən }

ill-posed problem A problem which may have more than one solution, or in which the solutions depend discontinuously upon the initial data. Also known as improperly posed problem. { ˈil ¦pōzd ˈpräb·ləm }

image **1.** For a point x in the domain of a function f, the point $f(x)$. **2.** For a subset A of the domain of a function f, the set of all points that are equal to $f(x)$ for some point x in A. { ˈim·ij }

imaginary axis All complex numbers $x + iy$ where $x = 0$; the vertical coordinate axis for the complex plane. { əˈmaj·əˌner·ē ˈak·səs }

imaginary circle The set of points in the x-y plane that satisfy the equation $x^2 + y^2 = -r^2$, or $(x - h)^2 + (y - k)^2 = -r^2$, where r is greater than zero, and x, y, h, and k are allowed to be complex numbers. { i¦maj·əˌner·ē ˈsər·kəl }

imaginary number A complex number of the form $a + bi$, with b not equal to zero, where a and b are real numbers, and $i = \sqrt{-1}$; some mathematicians require also that $a = 0$. Also known as imaginary quantity. { əˈmaj·əˌner·ē ˈnəm·bər }

imaginary part For a complex number $x + iy$ the imaginary part is the real number y. { əˈmaj·əˌner·ē ˌpärt }

imaginary quantity *See* imaginary number. { əˈmaj·əˌner·ē ˈkwän·əd·ē }

imbedding A homeomorphism of one topological space to a subspace of another topological space. { imˈbed·iŋ }

immersion A mapping f of a topological space X into a topological space Y such that for every $x \in X$ there exists a neighborhood N of x, such that f is a homeomorphism of N onto $f(N)$. { əˈmər·zhən }

implication **1.** The logical relation between two statements p and q, usually expressed as "if p then q." **2.** A logic operator having the characteristic that if p and q are statements, the implication of p and q is false if p is true and q is false, and is true otherwise. Also known as conditional implication; if-then operation; material implication. { ˌim·pləˈkā·shən }

implicit differentiation The process of finding the derivative of one of two variables with respect to the other by differentiating all the terms of a given equation in the two variables and solving the resulting equation for this derivative. { imˈplis·it ˌdif·əˌren·chēˈā·shən }

implicit enumeration A method of solving integer programming problems, in which tests that follow conceptually from using implied upper and lower bounds on variables are used to eliminate all but a tiny fraction of the possible values, with implicit treatment of all other possibilities. { imˈplis·ət iˌnü·məˈrā·shən }

implicit function A function defined by an equation $f(x,y) = 0$, when x is considered as an independent variable and y, called an implicit function of x, as a dependent variable. { imˈplis·ət ˈfəŋk·shən }

implicit function theorem A theorem that gives conditions under which an equation in variables x and y may be solved so as to express y directly as a function of x; it states that if $F(x,y)$ and $\partial F(x,y)/\partial y$ are continuous in a neighborhood of the point (x_0,y_0) and if $F(x,y) = 0$ and $\partial F(x,y)/\partial y \neq 0$, then there is a number $\epsilon > 0$ such that there is one and only one function $f(x)$ that is continuous and satisfies

$F[x,f(x)] = 0$ for $x - x_0 < \epsilon$, and satisfies $f(x_0) = y_0$. { im'plis·ət ¦fəŋk·shən ˌthir·əm }

improper divisor An improper divisor of an element x in a commutative ring with identity is any unit of the ring or any associate of x. { im'präp·ər di'vī·zər }

improper fraction 1. In arithmetic, the quotient of two integers in which the numerator is greater than or equal to the denominator. 2. In algebra, the quotient of two polynomials in which the degree of the numerator is greater than or equal to that of the denominator. { im'präp·ər 'frak·shən }

improper integral Any integral in which either the integrand becomes unbounded on the domain of integration, or the domain of integration is itself unbounded. { im'präp·ər 'int·ə·grəl }

improperly posed problem *See* ill-posed problem. { im'präp·ər·lē ¦pōzd 'präb·ləm }

improper orthogonal transformation An orthogonal transformation such that the determinant of its matrix is -1. { im¦präp·ər ȯr¦thäg·ə·nəl ˌtranz·fər'mā·shən }

impulse function An idealized or generalized function defined not by its values but by its behavior under integration, such as the (Dirac) delta function. { 'im ˌpəls ˌfəŋk·shən }

incenter The center of the inscribed circle of a given triangle. { ¦in¦sen·tər }

incidence matrix In a graph the $p \times q$ matrix (b_{ij}) for which $b_{ij} = 1$ if the ith vertex is an end point of the jth edge, and $b_{ij} = 0$ otherwise. { 'in·səd·əns ˌmā·triks }

incircle *See* inscribed circle. { ¦in¦sər·kəl }

inclusive or *See* or. { in'klü·siv 'ȯr }

inclination 1. The inclination of a line in a plane is the angle made with the positive x axis. 2. The inclination of a line in space with respect to a plane is the smaller angle the line makes with its orthogonal projection in the plane. 3. The inclination of a plane with respect to a given plane is the smaller of the dihedral angles which it makes with the given plane. { ˌiŋ·klə'nā·shən }

inclusion relation 1. A set theoretic relation, usually denoted by the symbol ⊂, such that, if A and B are two sets, A ⊂ B if and only if every element of A is an element of B. 2. Any relation on a Boolean algebra which is reflexive, antisymmetric, and transitive. { iŋ'klü·zhən riˌlā·shən }

incommensurable line segments Two line segments the ratio of whose lengths is irrational. { ˌin·kə¦mens·ə·rə·bəl 'līn ˌseg·məns }

incommensurable numbers Two numbers whose ratio is irrational. { ˌin·kə'mens·ə·rə·bəl 'nəm·bərz }

incompatible equations Two or more equations that are not satisfied by any set of values for the variables appearing. Also known as inconsistent equations. { ˌin·kəm'pad·ə·bəl i¦kwā·zhənz }

incompatible inequalities Two or more inequalities that are not satisfied by any set of values of the variables involved. Also known as inconsistent inequalities. { ˌin·kəm'pad·ə·bəl ˌin·ə'kwäl·əd·ēz }

incomplete beta function The function $\beta_x(p,q)$ defined by

$$\beta_x(p,q) = \int_0^x t^{p-1}(1-t)^{q-1}\,dt$$

where $0 \leq x \leq 1$, $p > 0$, and $q > 0$. { ˌin·kəm'plēt 'bād·ə ˌfəŋk·shən }

incomplete gamma function Either of the functions $\gamma(a,x)$ and $\Gamma(a,x)$ defined by

$$\gamma(a,x) = \int_0^x t^{a-1} e^{-t}\,dt$$

$$\Gamma(a,x) = \int_x^\infty t^{a-1} e^{-t}\,dt$$

where $0 \leq x \leq \infty$ and $a > 0$. { ˌin·kəm'plēt 'gam·ə ˌfəŋk·shən }

inconsistent equations *See* incompatible equations. { ˌin·kən'sis·tənt i'kwā·zhənz }

inconsistent inequalities *See* incompatible inequalities. { ˌin·kən'sis·tənt ˌin·ə 'kwäl·əd·ēz }

increasing function *See* monotone nondecreasing function. { in'krēs·iŋ ˌfəŋk·shən }

increment A change in the argument or values of a function, usually restricted to being a small positive or negative quantity. { 'iŋ·krə·mənt }

indefinite integral An indefinite integral of a function $f(x)$ is a function $F(x)$ whose derivative equals $f(x)$. Also known as antiderivative; integral. { in'def·ə·nət 'int·ə·grəl }

independent axioms A list of axioms such that no axiom can be deduced as a theorem from the others. { ˌin·də'pen·dənt 'ak·sē·əmz }

independent equations A system of equations such that no one of them is necessarily satisfied by a solution to the rest. { ˌin·də'pen·dənt i'kwā·zhənz }

independent functions A set of functions such that knowledge of the values obtained by all but one of them at a point is insufficient to determine the value of the remaining function. { ˌin·də'pen·dənt 'fəŋk·shənz }

independent variable In an equation $y = f(x)$, the input variable x. Also known as argument. { ˌin·də'pen·dənt 'ver·ē·ə·bəl }

indeterminate equations A set of equations possessing an infinite number of solutions. { ˌin·də'tərm·ə·nət i'kwā·zhənz }

indeterminate forms Products, quotients, differences, or powers of functions which are undefined when the argument of the function has a certain value, because one or both of the functions are zero or infinite; however, the limit of the product, quotient, and so on as the argument approaches this value is well defined. { ˌin·də'tərm·ə·nət 'fȯrmz }

index **1.** Unity of a logarithmic scale, as the C scale of a slide rule. **2.** A subscript or superscript used to indicate a specific element of a set or sequence. **3.** The number above and to the left of a radical sign, indicating the root to be extracted. **4.** For a subgroup of a finite group, the order of the group divided by the order of the subgroup. **5.** For a continuous complex-valued function defined on a closed plane curve, the change in the amplitude of the function when traversing the curve in a counterclockwise direction, divided by 2π. { 'inˌdeks }

index line *See* isopleth. { ˈinˌdeks ˌlīn }

indicator *See* Euler's phi function. { ˈin·dəˌkād·ər }

indicator function *See* characteristic function. { ˈin·dəˌkād·ər ˌfəŋk·shən }

induced orientation An orientation of a face of a simplex S opposite a vertex p_i obtained by deleting p_i from the ordering defining the orientation of S. { in¦düsd ˌȯr·ē·ənˈtā·shən }

inequality A statement that one quantity is less than, less than or equal to, greater than, or greater than or equal to another quantity. { ˌin·iˈkwäl·əd·ē }

inessential mapping A mapping between topological spaces that is homotopic to a mapping whose range is a single point. { ˌin·ə¦sen·chəl ˈmap·iŋ }

inf *See* greatest lower bound.

infimum *See* greatest lower bound. { ˈin·fə·məm }

infinite Larger than any fixed number. { ˈin·fə·nət }

infinite discontinuity A discontinuity of a function for which the absolute value of the function can have arbitrarily large values arbitrarily close to the discontinuity. { ˈin·fə·nit ˌdisˌkänt·ənˈü·əd·ē }

infinite extension An extension field F of a given field E such that F, viewed as a vector space over E, has infinite dimension. { ˈin·fi·nit ikˈsten·chən }

infinite group A group that contains an infinite number of distinct elements. { ˈin·fə·nit ˈgrüp }

infinite integral An integral at least one of whose limits of integration is infinite. { ˈin·fə·nət ˈint·ə·grəl }

infinite sequence *See* sequence. { ˈin·fə·nət ˈsē·kwəns }

infinite series An indicated sum of an infinite sequence of quantities, written $a_1 + a_2 + a_3 + \cdots$, or

$$\sum_{k=1}^{\infty} a_k$$

{ ˈin·fə·nət ˈsir·ēz }

infinite set A set with more elements than any fixed integer; such a set can be put into a one to one correspondence with a proper subset of itself. { ˈin·fə·nət ˈset }

infinitesimal A function whose value approaches 0 as its argument approaches some specified limit.

infinitesimal generator A closed linear operator defined relative to some semigroup of operators and which uniquely determines that semigroup. { ¦inˌfin·ə¦tes·ə·məl ˈjen·əˌrād·ər }

infinity The concept of a value larger than any finite value. { inˈfin·əd·ē }

infix notation A method of forming mathematical or logical expressions in which operators are written between the operands on which they act. { ′in‚fiks nō‚tā·shən }

inflectional tangent A tangent to a curve at a point of inflection. { in¦flek·shə·nəl ′tan·jənt }

inflection point *See* point of inflection. { in′flek·shən ‚pȯint }

information function of a partition If ξ is a finite partition of a probability space, the information function of ξ is a step function whose sets of constancy are the elements of ξ and whose value on an element of ξ is the negative of the logarithm of the probability of this element. { ‚in·fər′mā·shən ¦fəŋk·shən əv ə pär′tish·ən }

information theory The branch of probability theory concerned with the likelihood of the transmission of messages, accurate to within specified limits, when the bits of information composing the message are subject to possible distortion. { ‚in·fər′mā·shən ‚thē·ə·rē }

initial line One of the two rays that form an angle and that may be regarded as remaining stationary while the other ray (the terminal line) is rotated about a fixed point on it to form the angle. { i¦nish·əl ′līn }

initial-value problem An nth-order ordinary or partial differential equation in which the solution and its first $(n - 1)$ derivatives are required to take on specified values at a particular value of a given independent variable. { i′nish·əl ¦val·yü ‚präb·ləm }

initial-value theorem The theorem that, if a function $f(t)$ and its first derivative have Laplace transforms, and if $g(s)$ is the Laplace transform of $f(t)$, and if the limit of $sg(s)$ as s approaches infinity exists, then this limit equals the limit of $f(t)$ as t approaches zero. { i′nish·əl ¦val·yü ‚thir·əm }

injection A mapping f from a set A into a set B which has the property that for any element b of B there is at most one element a of A for which $f(a) = b$. Also known as injective mapping; one-to-one mapping; univalent function. { in′jek·shən }

injective mapping *See* injection. { in′jek·tiv ′map·iŋ }

inner automorphism An automorphism h of a group where $h(g) = g_0^{-1} \cdot g \cdot g_0$, for every g in the group with g_0 some fixed group element. { ¦in·ər ‚ȯd·ō′mȯr·fiz·əm }

inner function A continuous open mapping of a topological space X into a topological space Y where the inverse image of each point in Y is zero dimensional. { ¦in·ər ′fəŋk·shən }

inner product **1.** A scalar valued function of pairs of vectors from a vector space, denoted by (x,y) where x and y are vectors, and with the properties that (x,x) is always positive and is zero only if $x = 0$, that $(ax + by,z) = a(x,z) + b(y,z)$ for any scalars a and b, and that $(x,y) = (y,x)$ if the scalars are real numbers, $(x,y) = \overline{(y,x)}$ if the scalars are complex numbers. Also known as Hermitian inner product; Hermitian scalar product. **2.** The inner product of vectors $(x_1, ..., x_n)$ and $(y_1, ..., y_n)$ from n-dimensional euclidean space is the sum of x_iy_i as i ranges from 1 to n. Also known as dot product; scalar product. **3.** The inner product of two functions f and g of a real or complex variable is the integral of $f(x)\overline{g(x)}dx$, where $\overline{g(x)}$ denotes the conjugate of $g(x)$. **4.** The inner product of two tensors is the contracted tensor obtained from their product by means of pairing contravariant indices of one with covariant indices of the other. { ¦in·ər ′präd·əkt }

inner product space A vector space that has an inner product defined on it. Also known as Hermitian space; unitary space. { ¦in·ər 'präd·əkt ˌspās }

inradius The radius of a circle or sphere inscribed in a given geometric figure. { 'inˌrād·ē·əs }

inscribed circle A circle that lies within a given triangle and is tangent to each of its sides. Also known as incircle. { in¦skrībd 'sər·kəl }

inscribed polygon A polygon that lies within a given circle or curve and whose vertices all lie on the circle or curve. { in¦skrībd 'päl·əˌgän }

inseparable degree Let E be a finite extension of a field F; the inseparable degree of E over F is the dimension of E viewed as a vector space over F divided by the separable degree of E over F. { in¦sep·rə·bəl di'grē }

integer Any positive or negative counting number or zero. { 'int·ə·jər }

integer partition For a positive integer n, a nonincreasing sequence of positive integers whose sum equals n. { 'int·ə·jər pär'tish·ən }

integral **1.** A solution of a differential equation is sometimes called an integral of the equation. **2.** An element a of a ring B is said to be integral over a ring A contained in B if it is the root of a polynomial with coefficients in A and with leading coefficient 1. **3.** *See* definite Riemann integral; indefinite integral. { 'int·ə·grəl }

integral calculus The study of integration and its applications to finding areas, volumes, or solutions of differential equations. { 'int·ə·grəl 'kal·kyəl·ləs }

integral closure The integral closure of a subring A of a ring B is the set of all elements in B that are integral over A. { 'int·ə·grəl 'klō·zhər }

integral curvature For a given region on a surface, the integral of the Gaussian curvature over the region. { 'int·ə·grəl 'kərv·ə·chər }

integral curves A family of curves that satisfy a particular differential equation. { 'int·ə·grəl 'kərvz }

integral domain A commutative ring with identity where the product of nonzero elements is never zero. Also known as entire ring. { 'int·ə·grəl dō'mān }

integral equation An equation where the unknown function occurs under an integral sign. { 'int·ə·grəl i'kwā·zhən }

integral extension An integral extension of a commutative ring A is a commutative ring B containing A such that every element of B is integral over A. { 'int·ə·grəl ik'sten·chən }

integral function **1.** A function taking on integer values. **2.** *See* entire function. { 'int·ə·grəl ˌfəŋk·shən }

integrally closed ring An integral domain which is equal to its integral closure in its quotient field. { in¦teg·rə·lē ¦klōzd 'riŋ }

integral map A homomorphism from a commutative ring A into a commutative ring B such that B is an integral extension of $f(A)$. { 'int·ə·grəl ˌmap }

integral operator A rule for transforming one function into another function by means of an integral; this often is in context a linear transformation on some vector space of functions. { 'int·ə·grəl 'äp·ə‚rād·ər }

integral test If $f(x)$ is a function that is positive and decreasing for positive x, then the infinite series with nth term $f(n)$ and the integral of $f(x)$ from 1 to ∞ are either both convergent (finite) or both infinite. { 'int·ə·grəl ‚test }

integral transform *See* integral transformation. { 'int·ə·grəl 'tranz‚fȯrm }

integral transformation A transform of a function $F(x)$ given by the function

$$f(y) = \int_a^b K(x,y)F(x)\,dx$$

where $K(x,y)$ is some function. Also known as integral transform. { 'int·ə·grəl ‚tranz·fər'mā·shən }

integrand The function which is being integrated in a given integral. { 'int·ə‚grand }

integrating factor A factor which when multiplied into a differential equation makes the portion involving derivatives an exact differential. { 'int·ə‚grād·iŋ 'fak·tər }

integration The act of taking a definite or indefinite integral. { ‚int·ə'grā·shən }

integration constant *See* constant of integration. { ‚int·ə'grā·shən ¦kän·stənt }

integration by parts A technique used to find the integral of the product of two functions by means of an identity involving another simpler integral; for functions of one variable the identity is

$$\int_a^b fg'\,dx + \int_a^b gf'\,dx = f(b)g(b) - f(a)g(a);$$

for functions of several variables the technique is tantamount to using Stokes' theorem or the divergence theorem. { ‚int·ə'grā·shən bī 'parts }

integrodifferential equation An equation relating a function, its derivatives, and its integrals. { in¦teg·rō‚dif·ə¦ren·chəl i'kwā·zhən }

intensification An operation that increases the value of the membership function of a fuzzy set if the value is equal to or greater than 0.5, and decreases it if it is less than 0.5. { in‚tens·ə·fə'kā·shən }

intercept The point where a straight line crosses one of the axes of a cartesian coordinate system. { ¦in·tər¦sept }

interior For a set A in a topological space, the set of all interior points of A. { in'tir·ē·ər }

interior angle **1.** An angle between two adjacent sides of a polygon that lies within the polygon. **2.** For a line (called the transversal) that intersects two other lines, an angle between the transversal and one of the two lines that lies within the space between the two lines. { in'tir·ē·ər 'aŋ·gəl }

interior content *See* interior Jordan content. { in¦tir·ē·ər 'kän‚tent }

interior Jordan content For a set a points on a line, the smallest number C such that the sum of the lengths of a finite number of open, nonoverlapping intervals that are completely contained in the set is always equal to or less than C. Also known as interior content. { in¦tir·ē·ər 'jȯrd·ən ¦kän‚tent }

interior measure *See* Lebesgue interior measure. { in¦tir·ē·ər 'mezh·ər }

interior point A point p in a topological space is an interior point of a set S if there is some open neighborhood of p which is contained in S. { in'tir·ē·ər 'pȯint }

intermediate value theorem If $f(x)$ is a continuous real-valued function on the closed interval from a to b, then, for any y between the least upper bound and the greatest lower bound of the values of f, there is an x between a and b with $f(x) = y$. { ˌin·tər'mēd·ē·ət ¦val·yü 'thir·əm }

interpolation A process used to estimate an intermediate value of one (dependent) variable which is a function of a second (independent) variable when values of the dependent variable corresponding to several discrete values of the independent variable are known. { inˌtər·pə'lā·shən }

intersection **1.** The point, or set of points, that is common to two or more geometric configurations. **2.** For two sets, the set consisting of all elements common to both of the sets. **3.** For two fuzzy sets A and B, the fuzzy set whose membership function has a value at any element x that is the minimum of the values of the membership functions of A and B at x. **4.** The intersection of two Boolean matrices A and B, with the same number of rows and columns, is the Boolean matrix whose element c_{ij} in row i and column j is the intersection of corresponding elements a_{ij} in A and b_{ij} in B. { ˌin·tər'sek·shən }

interval A set of numbers which consists of those numbers that are greater than one fixed number and less than another, and that may also include one or both of the end numbers. { 'in·tər·vəl }

interval of convergence The interval consisting of the real numbers for which a specified power series possesses a limit. { 'in·tər·vəl əv kən'vər·jəns }

intrinsic equations of a curve The equations describing the radius of curvature and torsion of a curve as a function of arc length; these equations determine the curve up to its position in space. { in'trin·sik i¦kwā·zhənz əv ə 'kərv }

intrinsic geometry of a surface The description of the intrinsic properties of a surface. { in'trin·sik jē'äm·ə·trē əv ə 'sərfəs }

intrinsic properties of a surface A property of a surface which can be described without reference to the surrounding space. { in'trin·sik ¦präp·ərd·ēz əv ə 'sər·fəs }

invariance *See* invariant property. { in'ver·ē·əns }

invariant **1.** An element x of a set E is said to be invariant with respect to a group G of mappings acting on E if $g(x) = x$ for all g in G. **2.** A subset F of a set E is said to be invariant with respect to a group G of mappings acting on E if $g(x)$ is in F for all x in F and all g in G. **3.** For an algebraic equation, an expression involving the coefficients that remains unchanged under a rotation or translation of the coordinate axes in the cartesian space whose coordinates are the unknown quantities. { in'ver·ē·ənt }

invariant function A function f on a set S is said to be invariant under a transformation T of S into itself if $f(Tx) = f(x)$ for all x in S. { in'ver·ē·ənt 'fəŋk·shən }

invariant measure A Borel measure m on a topological space X is invariant for a transformation group (G,X,π) if for all Borel sets A in X and all elements g in G, $m(A_g) = m(A)$, where A_g is the set of elements equal to $\pi(g,x)$ for some x in A. { in'ver·ē·ənt 'mezh·ər }

invariant property A mathematical property of some space unchanged after the application of any member from some given family of transformations. Also known as invariance. { in'ver·ē·ənt 'präp·ərd·ē }

invariant subgroup *See* normal subgroup. { in'ver·ē·ənt 'səb,grüp }

inverse **1.** The additive inverse of a real or complex number a is the number which when added to a gives 0; the multiplicative inverse of a is the number which when multiplied with a gives 1. **2.** The inverse of a fractional ideal I of an integral domain R is the set of all elements x in the quotient field K of R such that xy is in I for all y in I. { 'in,vərs }

inverse curves A pair of curves such that every point on one curve is the inverse point of some point on the other curve, with respect to a fixed circle. { 'in,vərs 'kərvz }

inverse element In a group G the inverse of an element g is the unique element g^{-1} such that $g \cdot g^{-1} = g^{-1} \cdot g = e$, where $\cdot$ denotes the group operation and e is the identity element. { 'in,vərs 'el·ə·mənt }

inverse function An inverse function for a function f is a function g whose domain is the range of f and whose range is the domain of f with the property that both f composed with g and g composed with f give the identity function. { 'in,vərs 'fənk·shən }

inverse function theorem If f is a continuously differentiable function of euclidean n-space to itself and at a point x_0 the matrix with the entry $(\partial f_i/\partial x_j)_{x_0}$ in the ith row and jth column is nonsingular, then there is a continuously differentiable function $g(y)$ defined in a neighborhood of $f(x_0)$ which is an inverse function for $f(x)$ at all points near x_0. { 'in,vərs 'fənk·shən ,thir·əm }

inverse implication The implication that results from replacing both the antecedent and the consequent of a given implication with their negations. { 'in,vərs ,im·plə'kā·shən }

inverse logarithm *See* antilogarithm. { 'in,vərs 'läg·ə,rith·əm }

inversely proportional quantities Two variable quantities whose product remains constant. { in¦vərs·lē prə¦pȯr·shən·əl 'kwän·əd·ēz }

inverse matrix The inverse of a nonsingular matrix A is the matrix A^{-1} where $A \cdot A^{-1} = A^{-1} \cdot A = I$, the identity matrix. { 'in,vərs 'mā·triks }

inverse operator The inverse of an operator L is the operator which is the inverse function of L. { 'in,vərs 'äp·ə,rād·ər }

inverse points A pair of points lying on a diameter of a circle or sphere such that the product of the distances of the points from the center equals the square of the radius. { 'in,vərs 'pȯins }

inverse probability principle *See* Bayes' theorem. { ¦in,vərs ,präb·ə'bil·əd·ē ,prin·sə·pəl }

inverse variation **1.** A relationship between two variables wherein their product is equal to a constant. **2.** An equation or function expressing such a relationship. { 'in,vərs ,ver·e'ā·shən }

inversion **1.** Given a point O lying in a plane or in space, a mapping of the plane or of space, excluding the point O, into itself in which every point is mapped into its inverse point with respect to a circle or sphere centered at O. **2.** The interchange of two adjacent members of a sequence. { in′vər·zhən }

invertible ideal A fractional ideal I of an integral domain R such that R is equal to the set of elements of the form xy, where x is in I and y is in the inverse of I. { in¦vərd·ə·bəl i′dēl }

involute **1.** A curve produced by any point of a perfectly flexible inextensible thread that is kept taut as it is wound upon or unwound from another curve. **2.** A curve that lies on the tangent surface of a given space curve and is orthogonal to the tangents to the given curve. **3.** A surface for which a given surface is one of the two surfaces of center. { ¦in·və¦lüt }

involution **1.** Any transformation that is its own inverse. **2.** In particular, a correspondence between the points on a line that is its own inverse, given algebraically by $x' = (ax + b)/(cx - a)$, where $a^2 + bc \neq 0$. **3.** A correspondence between the lines passing through a given point on a plane such that corresponding lines pass through corresponding points of an involution of points on a line. { ˌin·və′lü·shən }

irrational equation An equation having an unknown raised to some fractional power. Also known as radical equation. { i′rash·ən·əl i′kwā·zhən }

irrational number A number which is not the quotient of two integers. { i′rash·ən·əl ′nəm·bər }

irrational radical A radical that is not equivalent to a rational number or expression. { i′rāsh·ən·əl ′rad·ə·kəl }

irreducible element An element x of a ring which is not a unit and such that every divisor of x is improper. { ˌir·ə′düs·ə·bəl ′el·ə·mənt }

irreducible equation An equation that is equivalent to one formed by setting an irreducible polynomial equal to zero. { ˌir·ə′dü·sə·bəl i′kwā·zhən }

irreducible function *See* irreducible polynomial. { ˌir·ə′dü·sə·bəl ′fəŋk·shən }

irreducible lambda expression A lambda expression that cannot be converted to a reduced form by a sequence of applications of the renaming and reduction rules. { ˌir·ə′dü·sə·bəl ′lam·də ikˌspresh·ən }

irreducible polynomial A polynomial is irreducible over a field K if it cannot be written as the product of two polynomials of lesser degree whose coefficients come from K. Also known as irreducible function. { ˌir·ə′dü·sə·bəl ˌpäl·ə′nō·mē·əl }

irreducible representation of a group A representation of a group as a family of linear operators of a vector space V where there is no proper closed subspace of V invariant under these operators. { ˌir·ə′dü·sə·bəl ˌrep·rə·zən′tā·shən əv ə ′grüp }

irreducible tensor A tensor that cannot be written as the inner product of two tensors of lower degree. { ˌir·ə′dü·sə·bəl ′ten·sər }

irrotational vector field A vector field whose curl is identically zero; every such field is the gradient of a scalar function. Also known as lamellar vector field. { ¦ir·ə′tā·shən·əl ˌfēld }

isochrone *See* semicubical parabola. { ′ī·səˌkrōn }

isogonal conjugates *See* isogonal lines. { ī¦säg·ən·əl 'kän·jə·gəts }

isogonal lines Lines that pass through the vertex of an angle and make equal angles with the bisector of the angle. Also known as isogonal conjugates. { ī¦säg·ən·əl 'līnz }

isogonal transformation A mapping of the plane into itself which leaves the magnitudes of angles between intersecting lines unchanged but may reverse their sense. { ī¦säg·ən·əl ˌtranz·fər'mā·shən }

isolated point **1.** A point p in a topological space is an isolated point of a set if p is in the set and there is a neighborhood of p which contains no other points of the set. **2.** A point that satisfies the equation for a plane curve C but has a neighborhood that includes no other point of C. Also known as acnode; hermit point. { 'ī·səˌlād·əd 'point }

isolated set A set consisting entirely of isolated points. { 'ī·səˌlād·əd 'set }

isolated subgroup An isolated subgroup of a totally ordered Abelian group G is a subgroup of G which is also a segment of G. { ¦īs·əˌlād·əd 'səbˌgrüp }

isometric forms Two bilinear forms f and g on vector spaces E and F for which there exists a linear isomorphism of E onto F such that $f(x,y) = g(\sigma x,\sigma y)$ for all x and y in E. { ¦ī·sə'me·trik 'fȯrmz }

isometric spaces Two spaces between which an isometry exists. { ¦ī·sə'me·trik 'spā·səs }

isometry **1.** A mapping f from a metric space X to a metric space Y where the distance between any two points of X equals the distance between their images under f in Y. **2.** A linear isomorphism σ of a vector space E onto itself such that, for a given bilinear form g, $g(\sigma x,\sigma y) = g(x,y)$ for all x and y in E. { ī'säm·ə·trē }

isometry class A set consisting of all bilinear forms (on vector spaces over a given field) which are isometric to a given form. { ī'säm·ə·trē ˌklas }

isomorphic systems Two algebraic structures between which an isomorphism exists. { ˌī·sə¦mȯr·fik 'sis·təmz }

isomorphism A one to one function of an algebraic structure (for example, group, ring, module, vector space) onto another of the same type, preserving all algebraic relations; its inverse function behaves likewise. { ¦ī·sə¦mȯrˌfiz·əm }

isoperimetric figures Figures whose perimeters are equal. { ¦ī·sōˌper·ə'me·trik 'fig·yərz }

isoperimetric inequality The statement that the area enclosed by a plane curve is equal to or less than the square of its perimeter divided by 4π. { ˌī·səˌper·ə¦me·trik ˌin·i'kwäl·əd·ē }

isoperimetric problem In the calculus of variations this problem deals with finding a closed curve in the plane which encloses the greatest area given its length as fixed. { ¦ī·sōˌper·ə'me·trik 'präb·ləm }

isopleth The straight line which cuts the three scales of a nomograph at values satisfying some equation. Also known as index line. { 'ī·səˌpleth }

isoptic The locus of the intersection of tangents to a given curve that meet at a specified constant angle. { ī'säp·tik }

isosceles spherical triangle A spherical triangle that has two equal sides. { ī¦säs·ə‚lēz ¦sfer·ə·kəl 'trī‚aŋ·gəl }

isosceles triangle A triangle with two sides of equal length. { ī'säs·ə‚lēz 'trī‚aŋ·gəl }

isotropy group For an operation of a group G on a set S, the isotropy group of an element s of S is the set of elements g in G such that $gs = s$. { 'ī·sə‚trō·pē ‚grüp }

isthmus *See* bridge. { 'is·məs }

iterated integral An integral over an area or volume designated to be performed by successive integrals over line segments. { 'īd·ə‚rād·əd 'int·ə·grəl }

iteration *See* iterative method. { ‚īd·ə'rā·shən }

iterative method Any process of successive approximation used in such problems as numerical solution of algebraic equations, differential equations, or the interpolation of the values of a function. Also known as iteration. { 'īd·ə‚rād·iv 'meth·əd }

iterative process A process for calculating a desired result by means of a repeated cycle of operations, which comes closer and closer to the desired result; for example, the arithmetical square root of N may be approximated by an iterative process using additions, subtractions, and divisions only. { 'īd·ə‚rād·iv 'prä·səs }

Itô's formula *See* stochastic chain rule. { 'ē‚tōz ‚fȯr·myə·lə }

Itô's integral *See* stochastic integral. { 'ē‚tōz ‚int·ə·grəl }

J

Jacobian The Jacobian of functions $f_i(x_1, x_2, ..., x_n)$, $i = 1, 2, ..., n$, of real variables x_i is the determinant of the matrix whose ith row lists all the first-order partial derivatives of the function $f_i(x_1, x_2, ..., x_n)$. Also known as Jacobian determinant. { jə′kō·bē·ən }

Jacobian determinant *See* Jacobian. { jə′kō·bē·ən di′tər·mə·nənt }

Jacobian elliptic function For m a real number between 0 and 1, and u a real number, let ϕ be that number such that

$$\int_0^{\phi} d\theta/(1 - m \sin^2 \theta)^{1/2} = u;$$

the 12 Jacobian elliptic functions of u with parameter m are sn $(u|m) = \sin \phi$, cn $(u|m) = \cos \phi$, dn $(u|m) = (1 - m \sin^2 \phi)^{1/2}$, the reciprocals of these three functions, and the quotients of any two of them. { jə′kō·bē·ən ə¦lip·tik ′fəŋk·shən }

Jacobian matrix The matrix used to form the Jacobian. { jə′kō·bē·ən ′mā·triks }

Jacobi canonical matrix A form to which any matrix can be reduced by a collineatory transformation, with zeros below the principal diagonal and characteristic roots as elements of the principal diagonal. { jə¦kōb·ē kə¦nän·ə·kəl ′mā·triks }

Jacobi condition In the calculus of variations, a differential equation used to study the extremals in a variational problem. { jə′kō·bē kən‚dish·ən }

Jacobi polynomials Polynomials that are constructed from the hypergeometric function and satisfy the differential equation

$$(1 - x^2)y'' + [\beta - \alpha - (\alpha + \beta + 2)x]y' + n(\alpha + \beta + n + 1)y = 0$$

where n is an integer and α and β are constants greater than -1; in certain cases these generate the Legendre and Chebyshev polynomials. { jə′kō·bē ‚päl·ə′nō·mē·əlz }

Jacobi's method **1.** A method of determining the eigenvalues of a Hermitian matrix. **2.** A method for finding a complete integral of the general first-order partial differential equation in two independent variables; it involves solving a set of six ordinary differential equations. { jə′kō·bēz ‚meth·əd }

Jacobi's theorem The proposition that a periodic, analytic function of a complex variable is simply periodic or doubly periodic. { jə′kō·bēz ‚thir·əm }

Jacobi's transformations Transformations of Jacobian elliptic functions to other Jacobian elliptic functions given by change of parameter and variable. { jə'kō·bēz ˌtranz·fər'mā·shənz }

Jensen's inequality 1. A general inequality satisfied by a convex function

$$f\left(\sum_{i=1}^{n} a_i x_i\right) \leq \sum_{i=1}^{n} a_i f(x_i)$$

where the x_i are any numbers in the region where f is convex and the a_i are non-negative numbers whose sum is equal to 1. **2.** If $a_1, a_2, \ldots, a_n$ are positive numbers and $s > t > 0$, then $(a_1^s + a_2^s + \cdots + a_n^s)^{1/s}$ is less than or equal to $(a_1^t + a_2^t + \cdots + a_n^t)^{1/t}$. { 'jen·sənz ˌin·i'kwäl·ədē }

joint variation The relation of a variable x to two other variables y and z wherein x is proportional to the product of y and z. { ¦jȯint ˌver·ē'ā·shən }

Jordan algebra A nonassociative algebra over a field in which the products satisfy the Jordan identity $(xy)x^2 = x(yx^2)$. { zhȯr'dän ˌal·jə·brə }

Jordan arc *See* simple arc. { zhȯr'dän ˌärk }

Jordan content For a set whose exterior Jordan content and interior Jordan content are equal, the common value of these two quantities. Also known as content. { zhȯr'dän ˌkänˌtent }

Jordan curve A simple closed curve in the plane, that is, a curve that is closed, connected, and does not cross itself. { zhȯr'dän ˌkərv }

Jordan curve theorem The theorem that in the plane every simple closed curve separates the plane into two parts. { zhȯr'dän ˌkərv ˌthir·əm }

Jordan-Hölder theorem The theorem that for a group any two composition series have the same number of subgroups listed, and both series produce the same quotient groups. { zhȯr'dän 'hül·dər ˌthir·əm }

Jordan matrix A matrix whose elements are equal and nonzero on the principal diagonal, equal to 1 on the diagonal immediately above, and equal to 0 everywhere else. { zhȯr'dän ˌmā·triks }

jump discontinuity A point a where for a real-valued function $f(x)$ the limit on the left of $f(x)$ as x approaches a and the limit on the right both exist but are distinct. { 'jəmp disˌkänt·ən'ü·əd·ē }

jump function A function used to represent a sampled data sequence arising in the numerical study of linear difference equations. { 'jəmp ˌfəŋk·shən }

K

kampyle of Eudoxus A plane curve whose equation in cartesian coordinates x and y is $x^4 = a^2(x^2 + y^2)$, where a is a constant. { kam'pīl əv yü'däk·səs }

kappa curve A plane curve whose equation in cartesian coordinates x and y is $(x^2 + y^2)\,y^2 = a^2x^2$, where a is a constant. Also known as Gutschoven's curve. { 'kap·ə ˌkərv }

Kármán swirling flow problem The problem of describing fluid motion above a rotating infinite plane disk when the fluid at infinity does not rotate. { 'kärˌmän ¦swir·liŋ ¦flō ˌpräb·ləm }

Karmarkar's algorithm A method for solving linear programming problems that has a polynomial time bound and appears to be faster than the simplex method for many complex problems. { ¦kär·məˌkärz 'al·gəˌrith·əm }

Karush-Kuhn-Tucker conditions A system of equations and inequalities which the solution of a nonlinear programming problem must satisfy when the objective function and the constraint functions are differentiable. { ¦kär·əsh ¦kyün 'tək·ər kənˌdish·ənz }

Kekeya needle problem The problem of finding the smallest area of a plane region in which a line segment of unit length can be continuously moved so that it returns to its original position after turning through 360°. { käˌkē·ə 'nēd·əl ˌpräb·ləm }

Kempe chain A subgraph of a graph whose vertices have been colored, consisting of vertices which have been assigned a given color or colors and arcs connecting pairs of such vertices. { 'kem·pə ˌchān }

keratoid A plane curve whose equation in cartesian coordinates x and y is $y^2 = x^2y + x^5$. { 'ker·əˌtȯid }

keratoid cusp A cusp of a curve which has one branch of the curve on each side of the common tangent. Also known as single cusp of the first kind. { 'ker·əˌtȯid ˌkəsp }

kernel **1.** For any mapping f from a group A to a group B, the kernel of f, denoted $\ker f$, is the set of all elements a of A such that $f(a)$ equals the identity element of B. **2.** For a homomorphism h from a group G to a group H, this consists of all elements of G which h sends to the identity element of H. **3.** For Fredholm and Volterra integral equations, this is the function $K(x,t)$. **4.** For an integral transform, the function $K(x,t)$ in the transformation which sends the function $f(x)$ to the function $\int K(x,t)f(t)dt = F(x)$. **5.** *See* null space. { 'kərn·əl }

Killing's equations The equations for an isometry-generating vector field in a geometry. { 'kil·iŋz iˌkwā·zhənz }

Killing vector An element of a vector field in a geometry that generates an isometry. { ¦kil·iŋ ¦vek·tər }

Kirkman triple system A resolvable balanced incomplete block design with block size k equal to 3. { ¦kərk·mən ¦trip·əl 'sis·təm }

Klein bottle The nonorientable surface having only one side with no inside or outside; it resembles a bottle pulled into itself. { 'klīn ˌbäd·əl }

Kleinian group A group of conformal mappings of a Riemann surface onto itself which is discontinuous at one or more points and is not discontinuous at more than two points. { 'klī·nē·ən ˌgrüp }

Klein's four-group The noncyclic group of order four. { ¦klīnz 'fȯr ˌgrüp }

knapsack problem The problem, given a set of integers $\{A_1, A_2, \ldots, A_n\}$ and a target integer B, of determining whether a subset of the A_i can be selected without repetition so that their sum is the target B. { 'napˌsak ˌpräb·ləm }

knot In the general case, a knot consists of an embedding of an n-dimensional sphere in an $(n + 2)$-dimensional sphere; classically, it is an interlaced closed curve, homeomorphic to a circle. { nät }

knot theory The topological and algebraic study of knots emphasizing their classification and how one may be continuously deformed into another. { 'nät ˌthē·ə·rē }

Kobayashi potential A solution of Laplace's equation in three dimensions constructed by superposition of the solutions obtained by separation of variables in cylindrical coordinates. { ˌkō·bī'yä·shē pəˌten·chəl }

Koch curve A fractal which can be constructed by a recursive procedure; at each step of this procedure every straight segment of the curve is divided into three equal parts and the central piece is then replaced by two similar pieces. { 'kōk ˌkərv }

Koebe function The analytic function $k(z) = z(1 - z)^{-2} = z + 2z^2 + 3z^3 + \cdots$ that maps the unit disk onto the entire complex plane minus the part of the negative real axis to the left of $-1/4$. { 'kā·bē ˌfəŋk·shən }

Kolmogorov consistency conditions For each finite subset F of the real numbers or integers, let P_F denote a probability measure defined on the Borel subsets of the cartesian product of $k(F)$ copies of the real line indexed by elements in F, where $k(F)$ denotes the number of elements in F; the family $\{P_F\}$ of measures satisfy the Kolmogorov consistency conditions if given any two finite sets F_1 and F_2 with F_1 contained in F_2, the restriction of P_{F2} to those sets which are independent of the coordinates in F_2 which are not in F_1 coincides with P_{F1}. { ˌkȯl·mə'gȯ·rȯf kən 'sis·tən·sē kənˌdish·ənz }

Kolmogorov inequalities For each integer K let X_k be a random variable with finite variance σ_k and suppose $\{X_k\}$ is an independent sequence which is uniformly bounded by some constant c; then for every $\epsilon > 0$, and integer n,

$$1 - (\epsilon + 2c)^2 \Big/ \sum_{k=1}^{n} \sigma_k^2 \leq \text{Prob}\,\{\max_{k \leq n} |S_k + ES_k| \geq \epsilon\}$$

and

$$\frac{1}{\epsilon^2} \sum_{k=1}^{n} \sigma_k^2 \geq \text{Prob}\,\{\max_{k \leq n} |S_k + ES_k| \geq \epsilon\};$$

here $$S_k = \sum_{i=1}^{k} X_i$$

and ES_k denotes the expected value of S_k. { ˌkȯl·mə′gȯ·rȯf ˌin·i′kwäl·əd·ēz }

Kolmogorov-Sinai invariant An isomorphism invariant of measure-preserving transformations; if T is a measure-preserving transformation on a probability space, the Kolmogorov-Sinai invariant is the least upper bound of the set of entropies of T given each finite partition of the probability space. Also known as entropy of a transformation. { ˌkȯl·mə′gȯ·rȯf ′sīˌnī in¦ver·ē·ənt }

Königsberg bridge problem The problem of walking across seven bridges connecting four landmasses in a specified manner exactly once and returning to the starting point; this is the original problem which gave rise to graph theory. { ¦kərn·iksˌbərg ′brij ˌpräb·ləm }

Konig's theorem In combinatorial theory, the theorem that the minimum number m of lines is the same as the maximum number m of independent points. { ′kō·nigz ˌthir·əm }

Krawtchouk polynomials Families of polynomials which are orthogonal with respect to binomial distributions. { ¦krävˌchək ˌpäl·ə′nō·mē·əlz }

Krein-Milman theorem The theorem that in a locally convex topological vector space, any compact convex set K is identical with the intersection of all convex sets containing the extreme points of K. { ′krīn ′mil·mən ˌthir·əm }

Kronecker delta The function or symbol δ_{ij} dependent upon the subscripts i and j which are usually integers; its value is 1 if $i = j$ and 0 if $i \neq j$. { ′krō·nek·ər ˌdel·tə }

Kronecker product Given two different representations of the same group, their Kronecker product is a representation of the group constructed by taking direct products of matrices from the respective representations. { ′krō·nek·ər ˌpräd·əkt }

K theory The study of the mathematical structure resulting from associating an abelian group $K(X)$ with every compact topological space X in a geometrically natural way, with the aid of complex vector bundles over X. Also known as topological K theory. { ′kā ˌthē·ə·rē }

Kuratowski graphs Two graphs which appear in Kuratowski's theorem, the complete graph K_5 with five vertices and the bipartite graph $K_{3,3}$. { ˌku̇r·ə′təvˌskē ˌgrafs }

Kuratowski's lemma Each linearly ordered subset of a partially ordered set is contained in a maximal linearly ordered subset. { ku̇r·ə′tȯv·skēz ′lem·ə }

Kuratowski's theorem The proposition that a graph is nonplanar if and only if it has a subgraph which is either a Kuratowski graph or a subdivision of a Kuratowski graph. { ˌku̇r·ə′təvˌskēz ˌthir·əm }

Kureppa number A number of the form $!n = 0! + 1! + \cdots + (n-1)!$, where n is a positive integer. { ku̇′rep·ə ˌnəm·bər }

L

labeled graph A graph whose vertices are distinguished by names. { 'lā·bəld ˌgraf }

Lagrange's formula *See* mean value theorem. { lə'grān·jəz ˌfȯr·myə·lə }

Lagrange's theorem In a group of finite order, the order of any subgroup must divide the order of the entire group. { lə'grän·jəz ˌthir·əm }

Lagrangian multipliers A technique whereby potential extrema of functions of several variables are obtained. Also known as undetermined multipliers. { lə'grän·jē·ən 'məl·təˌplī·ərz }

Laguerre polynomials A sequence of orthogonal polynomials which solve Laguerre's differential equation for positive integral values of the parameter. { lə'ger ˌpäl·ə'nō·mē·əlz }

Laguerre's differential equation The equation $xy'' + (1 - x)y' + \alpha y = 0$, where α is a constant. { lə'gerz ˌdif·əˈren·chəl i'kwā·zhən }

lambda calculus A mathematical formalism to model the mathematical notion of substitution of values for bound variables. { 'lam·də ˌkal·kyə·ləs }

lambda expression An expression used to define a function in the lambda calculus; for example, the function $f(x) = x + 1$ is defined by the expression $\lambda x(x + 1)$. { 'lam·də ikˌspresh·ən }

Lamé functions Functions that arise when Laplace's equation is separated in ellipsoidal coordinates. { lä'mā ˌfəŋk·shənz }

lamellar vector field *See* irrotational vector field. { lə'mel·ər 'vek·tər ˌfēld }

Lamé polynomials Polynomials which result when certain parameters of Lamé functions assume integral values, and which are used to express physical solutions of Laplace's equation in ellipsoidal coordinates. { lä'mā ˌpäl·ə'nō·mē·əlz }

Lamé's equations A general collection of second-order differential equations which have five regular singularities. { lä'māz iˌkwā·zhənz }

Lamé's relations Six independent relations which when satisfied by the covariant metric tensor of a three-dimensional space provide necessary and sufficient conditions for the space to be euclidean. { lä'māz riˌlā·shənz }

Lamé wave functions Functions which arise when the wave equation is separated in ellipsoidal coordinates. Also known as ellipsoidal wave functions. { lä'mā 'wāv ˌfəŋk·shənz }

Lanczos's method A transformation method for diagonalizing a matrix in which the matrix used to transform the original matrix to triple-diagonal form is formed from a set of column vectors that are determined by a recursive process. { 'län͵chōz·əs ͵meth·əd }

language theory A branch of automata theory which attempts to formulate the grammar of a language in mathematical terms; it has been applied to automatic language translation and to the construction of higher-level programming languages and systems such as the propositional calculus, nerve networks, sequential machines, and programming schemes. { 'laŋ·gwij ͵thē·ə·rē }

Laplace operator The linear operator defined on differentiable functions which gives for each function the sum of all its nonmixed second partial derivatives. Also known as Laplacian. { lə'pläs ͵äp·ə͵rād·ər }

Laplace's equation The partial differential equation which states that the sum of all the nonmixed second partial derivatives equals 0; the potential functions of many physical systems satisfy this equation. { lə'pläs·əz i͵kwā·zhən }

Laplace's expansion An expansion by means of which the determinant of a matrix may be computed in terms of the determinants of all possible smaller square matrices contained in the original. { lə'pläs·əz ik͵span·chən }

Laplace transform For a function $f(x)$ its Laplace transform is the function $F(y)$ defined as the integral over x from 0 to ∞ of the function $e^{-yx}f(x)$. { lə'pläs 'trans͵fòrm }

Laplacian *See* Laplace operator. { lə'pläs·ē·ən }

latent root *See* eigenvalue. { 'lāt·ənt 'rüt }

lateral area The area of a surface with the bases (if any) excluded. { 'lad·ə·rəl 'er·ē·ə }

lateral face The lateral face for a prism or pyramid is any edge or face which is not part of a base. { 'lad·ə·rəl 'fās }

Latin square An $n \times n$ square array of n different symbols, each symbol appearing once in each row and once in each column; these symbols prove useful in ordering the observations of an experiment. { 'lat·ən 'skwer }

lattice A partially ordered set in which each pair of elements has both a greatest lower bound and least upper bound. { 'lad·əs }

latus rectum The length of a chord through the focus and perpendicular to the axis of symmetry in a conic section. { ¦lad·əs 'rek·təm }

Laurent expansion An infinite series in which an analytic function $f(z)$ defined on an annulus about the point z_0 may be expanded, with nth term $a_n(z - z_0)^n$, n ranging from $-\infty$ to ∞, and $a_n = 1/(2\pi i)$ times the integral of $f(t)/(t - z_0)^{n+1}$ along a simple closed curve interior to the annulus. Also known as Laurent series. { lò'rän ik͵span·chən }

Laurent series *See* Laurent expansion. { lò'ränz ͵sir·ēz }

law of cosines Given a triangle with angles A, B, and C and sides a, b, c opposite these angles respectively: $a^2 = b^2 + c^2 - 2bc \cos A$. { 'lò əv 'kō͵sīnz }

law of exponents Any of the laws $a^m a^n = a^{m+n}$, $a^m/a^n = a^{m-n}$, $(a^m)^n = a^{mn}$, $(ab)^n = a^n b^n$, $(a/b)^n = a^n/b^n$; these laws are valid when m and n are any integers, or when a and b are positive and m and n are any real numbers. Also known as exponential law. { 'lȯ əv ik'spō·nəns }

law of quadrants **1.** The law that any angle of a right spherical triangle (except a right angle) and the side opposite it are in the same quadrant. **2.** The law that when two sides of a right spherical triangle are in the same quadrant the third side is in the first quadrant, and when two sides are in different quadrants the third side is in the second quadrant. { 'lȯ əv 'kwäd·rəns }

law of signs The product or quotient of two numbers is positive if the numbers have the same sign, negative if they have opposite signs. { 'lȯ əv 'sīnz }

law of sines Given a triangle with angles A, B, and C and sides a, b, c opposite these angles respectively: $\sin A/a = \sin B/b = \sin C/c$. { 'lȯ əv 'sīnz }

law of species The law that one-half the sum of two angles in a spherical triangle and one-half the sum of the two opposite sides are of the same species, in that they are both acute or both obtuse angles. { 'lȯ əv 'spē,shēz }

law of tangents Given a triangle with angles A, B, and C and sides a, b, c opposite these angles respectively: $(a - b)/(a + b) = [\tan \frac{1}{2}(A - B)]/[\tan \frac{1}{2}(A + B)]$. { 'lȯ əv 'tan·jəns }

law of the mean *See* mean value theorem. { ¦lȯ əv thə 'mēn }

leading zeros Zeros preceding the first nonzero integer of a number. { 'lēd·iŋ 'zir·ōz }

leaf of Descartes *See* folium of Descartes. { 'lēf əv dā'kärt }

least common denominator The least common multiple of the denominators of a collection of fractions. { 'lēst 'käm·ən di'näm·ə,nād·ər }

least common multiple The least common multiple of a set of quantities (for example, numbers or polynomials) is the smallest quantity divisible by each of them. Abbreviated lcm. { 'lēst 'käm·ən 'məl·tə·pəl }

least upper bound The least upper bound of a subset A of a set S with ordering $<$ is the smallest element of S which is greater than or equal to every element of A. Abbreviated lub. Also known as supremum (sup). { ¦lēst ¦əp·ər 'bau̇nd }

Lebesgue exterior measure A measure whose value on a set S is the greatest lower bound of the Lebesgue measures of open sets that contain S. Also known as exterior measure. { lə'beg ik¦stir·ē·ər 'mezh·ər }

Lebesgue integral The generalization of Riemann integration of real valued functions, which allows for integration over more complicated sets, existence of the integral even though the function has many points of discontinuity, and convergence properties which are not valid for Riemann integrals. { lə'beg ,int·ə·grəl }

Lebesgue interior measure A measure whose value on a set S is the least upper bound of the Lebesgue measures of the closed sets contained in S. Also known as interior measure. { lə'beg in¦tir·ē·ər 'mezh·ər }

Lebesgue measure A measure defined on subsets of euclidean space which expresses how one may approximate a set by coverings consisting of intervals. { lə'beg ˌmezh·ər }

Lebesgue number The Lebesgue number of an open cover of a compact metric space X is a positive real number so that any subset of X whose diameter is less than this number must be completely contained in a member of the cover. { lə'beg ˌnəm·bər }

Lebesgue-Stieltjes integral A Lebesgue integral of the form

$$\int_b^a f(x)\,d\phi(x)$$

where ϕ is of bounded variation; if $\phi(x) = x$, it reduces to the Lebesgue integral of $f(x)$; if $\phi(x)$ is differentiable, it reduces to the Lebesgue integral of $f(x)\phi'(x)$. { lə'beg 'stēlt·yəs ˌint·ə·grəl }

left-continuous function A function $f(x)$ of a real variable is left-continuous at a point c if $f(x)$ approaches $f(c)$ as x approaches c from the left, that is, $x < c$ only. { 'left kən¦tin·yə·wəs 'fəŋk·shən }

left coset A left coset of a subgroup H of a group G is a subset of G consisting of all elements of the form ah, where a is a fixed element of G and h is any element of H. { ¦left 'käs·ət }

left-hand derivative The limit of the difference quotient $[f(x) - f(c)]/[x - c]$ as x approaches c from the left, that is, $x < c$ only. { 'left ˌhand də'riv·əd·iv }

left-handed coordinate system 1. A three-dimensional rectangular coordinate system such that when the thumb of the left hand extends in the positive direction of the first (or x) axis, the fingers fold in the direction in which the second (or y) axis could be rotated about the first axis to coincide with the third (or z) axis. **2.** A coordinate system of a Riemannian space which has negative scalar density function. { 'left ¦hand·əd kō'ȯrd·ən·ət ˌsis·təm }

left-handed curve A space curve whose torsion is positive at a given point. Also known as sinistrorse curve; sinistrorsum. { 'left ¦hand·əd 'kərv }

left-hand limit *See* limit on the left. { 'left ¦hand 'lim·ət }

left identity In a set on which a binary operation $\circ$ is defined, an element e with the property that $e \circ a = a$ for every element a in the set. { ¦left i'den·ə·dē }

leg Either side adjacent to the right angle of a right triangle. { leg }

Legendre contact transformation *See* Legendre transformation. { lə'zhän·drə 'känˌtak ˌtranz·fərˌmā·shən }

Legendre equation The second-order linear homogeneous differential equation $(1 - x^2)y'' - 2xy' + \nu(\nu + 1)y = 0$, where ν is real and nonnegative. { lə'zhän·drə iˌkwā·zhən }

Legendre function Any solution of the Legendre equation. { lə'zhän·drə ˌfəŋk·shən }

Legendre polynomials A collection of orthogonal polynomials which provide solutions to the Legendre equation for nonnegative integral values of the parameter. { lə'zhän·drə ˌpäl·i'nō·mē·əlz }

Legendre's symbol The symbol $(c|p)$, where p is an odd prime number, and $(c|p)$ is equal to 1 if c is a quadratic residue of p, and is equal to -1 if c is not a quadratic residue of p. { lə'zhän·drəz ˌsim·bəl }

Legendre transformation A mathematical procedure in which one replaces a function of several variables with a new function which depends on partial derivatives of the original function with respect to some of the original independent variables. Also known as Legendre contact transformation. { lə'zhän·drə ˌtranz·fər'mā·shən }

Leibnitz's rule A formula to compute the nth derivative of the product of two functions f and g:

$$d^n(f \cdot g)/dx^n = \sum_{k=0}^{n} \binom{n}{k} d^{n-k} f/dx^{n-k} \cdot d^k g/dx^k$$

where $\binom{n}{k} = n!/(n-k)!\,k!$ { 'līb·nit·səz ˌrül }

Leibnitz's test If the sequence of positive numbers a_n approaches zero monotonically, then the series

$$\sum_{n=1}^{\infty} (-1)^n a_n$$

is convergent. { 'līb·nit·səz ˌtest }

lemma A mathematical fact germane to the proof of some theorem. { 'lem·ə }

lemma of duBois-Reymond A continuous function $f(x)$ is constant in the interval (a,b) if for certain functions g whose integral over (a,b) is zero, the integral over (a,b) of f times g is zero. { 'lem·ə əv dyu̇b'wä rā'mōn }

lemniscate of Bernoulli The locus of points (x,y) in the plane satisfying the equation $(x^2 + y^2)^2 = a^2 (x^2 - y^2)$ or, in polar coordinates (r,θ), the equation $r^2 = a^2 \cos 2\theta$. { lem'nis·kət əv ber'nü·ē }

lemniscate of Gerono *See* eight curve. { lem'nis·kət əvje'rän·ō }

length of a curve A curve represented by $x = x(t)$, $y = y(t)$ for $t_1 \le t \le t_2$, with $x(t_1) = x_1$, $x(t_2) = x_2$, $y(t_1) = y_1$, $y(t_2) = y_2$, has length from (x_1,y_1) to (x_2,y_2) given by the integral from t_1 to t_2 of the function $\sqrt{(dx/dt)^2 + (dy/dt)^2}$. { 'leŋkth əv ə 'kərv }

Levi-Civita symbol A symbol $\epsilon_{i,j,\ldots,s}$ where $i, j, \ldots, s$ are n indices, each running from 1 to n; the symbol equals zero if any two indices are identical, and 1 or -1 otherwise, depending on whether $i, j, \ldots, s$ form an even or an odd permutation of $1, 2, \ldots, n$. { 'lā·vē chē·vē'tä ˌsim·bəl }

lexicographic order Given sets A and B with a common ordering $<$, one defines an ordering between all sequences (finite or infinite) of elements of A and of elements of B by $(a_1,a_2,\ldots) < (b_1,b_2,\ldots)$ if either $a_i = b_i$ for every i, or $a_n < b_n$, where n is the first place in which they differ; this is the way words are ordered in a dictionary. { ¦lek·sə·kō¦graf·ik 'ȯr·dər }

l'Hôpital's cubic *See* Tschirnhausen's cubic. { lō·pē·tälz 'kyü·bik }

l'Hôpital's rule A rule useful in evaluating indeterminate forms: if both the functions $f(x)$ and $g(x)$ and all their derivatives up to order $(n - 1)$ vanish at $x = a$, but the nth derivatives both do not vanish or both become infinite at $x = a$, then

$$\lim_{x \to a} f(x)/g(x) = f^{(n)}(a)/g^{(n)}(a),$$

$f^{(n)}$ denoting the nth derivative. { lō·pē·tälz ˌrül }

l'Huilier's equation An equation used in the solution of a spherical triangle, involving tangents of various functions of its angles and sides. { lə·wē'yāz iˌkwā·zhən }

Liapunov function *See* Lyapunov function. { 'lyä·pu̇·nȯf ˌfəŋk·shən }

Lie algebra The algebra of vector fields on a manifold with additive operation given by pointwise sum and multiplication by the Lie bracket. { 'lē ˌal·jə·brə }

Lie bracket Given vector fields X,Y on a manifold M, their Lie bracket is the vector field whose value is the difference between the values of XY and YX. { 'lē ˌbrak·ət }

Lie group A topological group which is also a differentiable manifold in such a way that the group operations are themselves analytic functions. { 'lē ˌgrüp }

lifting **1.** Given a fiber bundle $(\bar{X},B,p)$ and a continuous map of a topological space $\bar{Y}$ to B, $g{:}\bar{Y} \to B$, lifting entails finding a continuous map $\bar{g}{:}\bar{Y} \to \bar{X}$ such that the function g is the composition $p - \bar{g}$. **2.** *See* translation. { 'lift·iŋ }

likelihood The likelihood of a sample of independent values of $x_1, x_2, \ldots, x_n$, with $f(x)$ the probability function, is the product $f(x_1) - f(x_2) - \cdots - f(x_n)$. { 'līk·lēˌhu̇d }

like terms *See* similar terms. { ¦līk ¦tərmz }

limaçon The locus of points of the plane which in polar coordinates (r,θ) satisfy the equation $r = a \cos \theta + b$. Also known as Pascal's limaçon. { 'lim·əˌsän *or* ˌlim·ə'sōn }

limit **1.** A function $f(x)$ has limit L as x tends to c if given any positive number ϵ (no matter how small) there is a positive number δ such that if x is in the domain of f, x is not c, and $|x - c| < \delta$, then $|f(x) - L| < \epsilon$; written

$$\lim_{x \to c} f(x) = L$$

2. A sequence $\{a_n{:}n = 1, 2,\ldots\}$ has limit L if given a positive number ϵ (no matter how small), there is a positive integer N such that for all integers n greater than N, $|a_n - L| < \epsilon$. { 'lim·ət }

limit cycle For a differential equation, a closed trajectory C in the plane (corresponding to a periodic solution of the equation) where every point of C has a neighborhood so that every trajectory through it spirals toward C. { 'lim·ət ˌsīk·əl }

limit inferior **1.** The limit inferior of a sequence whose nth term is a_n is the limit as N approaches infinity of the greatest lower bound of the terms a_n for which n is greater than N; denoted by

$$\liminf_{n \to \infty} a_n \text{ or } \varliminf_{n \to \infty} a_n$$

2. The limit inferior of a function f at a point c is the limit as ϵ approaches zero of the greatest lower bound of $f(x)$ for $x - c \quad \epsilon$ and $< x \neq c$; denoted by

$$\liminf_{x \to c} f(x) \text{ or } \varliminf_{x \to c} f(x)$$

{ 'lim·ət in¦fir·ē·ər }

limit on the left The limit on the left of the function f at a point c is the limit of f at c

which would be obtained if only values of x less than c were taken into account; more precisely, it is the number L which has the property that for any positive number ϵ, there is a positive number δ so that if x is the domain of f and $0 < (c - x) < \delta$ then $|f(x) - L| < \epsilon$; denoted by

$$\lim_{x \to c^-} f(x) = L \text{ or } f(c^-) = L$$

Also known as left-hand limit. { 'lim·ət ȯn thə 'left }

limit on the right The limit on the right of the function $f(x)$ at a point c is the limit of f at c which would be obtained if only values of x greater than c were taken into account; more precisely, it is the number L which has the property that for any positive number ϵ there is a positive number δ so that if x is in the domain of f and $0 < (x - c) < \delta$, then $|f(x) - L| < \epsilon$; denoted by

$$\lim_{x \to c^+} f(x) = L \text{ or } f(c^+) = L$$

Also known as right-hand limit. { 'lim·ət ȯn thə 'rīt }

limit point *See* cluster point. { 'lim·ət ˌpȯint }

limits of integration The end points of the interval over which a function is being integrated. { 'lim·əts əv ˌint·ə'grā·shən }

limit superior 1. The limit superior of a sequence whose nth term is a_n is the limit as N approaches infinity of the least upper bound of the terms a_n for which n is greater than N; denoted by

$$\limsup_{h \to \infty} a_n \text{ or } \overline{\lim_{h \to \infty}}\, a_n$$

2. The limit superior of a function f at a point c is the limit as ϵ approaches zero of the least upper bound of $f(x)$ for $|x - c| < \epsilon$ and $x \neq c$; denoted by

$$\limsup_{x \to c} f(x) \text{ or } \overline{\lim_{x \to c}}\, f(x)$$

{ 'lim·ət sə'pir·ē·ər }

Lindelöf space A topological space where if a family of open sets covers the space, then a countable number of these sets also covers the space. { 'lin·dəˌlȯf ˌspās }

Lindelöf theorem The proposition that there is a countable subcover of each open cover of a subset of a space whose topology has a countable base. { 'lin·dəˌlef ˌthir·əm }

line The set of points $(x_1,...,x_n)$ in euclidean space, each of whose coordinates is a linear function of a single parameter t; $x_i = f_i(t)$. { līn }

linear algebra The study of vector spaces and linear transformations. { 'lin·ē·ər 'al·jə·brə }

linear algebraic equation An equation in some algebraic system where the unknowns occur linearly, that is, to the first power. { 'lin·ē·ər ˌal·jə¦brā·ik i'kwā·zhən }

linear combination A linear combination of vectors $\mathbf{v}_1, \ldots, \mathbf{v}_n$ in a vector space is any expression of the form $a_1\mathbf{v}_1 + a_2\mathbf{v}_2 + \cdots + a_n\mathbf{v}_n$, where the a_i are scalars. { 'lin·ē·ər ˌkäm·bə'nā·shən }

linear congruence The relation between two quantities that have the same remainder on division by a given integer, where the quantities are polynomials of, at most, the first degree in the variables involved. { 'lin·ē·ər kəŋ'grü·əns }

linear dependence The property of a set of vectors $\mathbf{v}_{,} \ldots, \mathbf{v}_n$ in a vector space for which there exists a linear combination such that $a_1\mathbf{v}_1 + \cdots + a_n\mathbf{v}_n = 0$, and at least one of the scalars a_i is not zero. { 'lin·ē·ər di'pen·dəns }

linear differential equation A differential equation in which all derivatives occur linearly, and all coefficients are functions of the independent variable. { 'lin·ē·ər ˌdif·ə¦ren·chəl i'kwā·zhən }

linear element On a surface determined by equations $x = f(u,v)$, $y = g(u,v)$, and $z = h(u,v)$, the element of length ds given by $ds^2 = E\,du^2 + 2F\,du\,dv + G\,dv^2$, where E, F, and G are functions of u and v. { 'lin·ē·ər 'el·ə·mənt }

linear equation A linear equation in the variables $x_1, \ldots, x_n$, and y is any equation of the form $a_1x_1 + a_2x_2 + \cdots + a_nx_n = y$. { 'lin·ē·ər i'kwā·zhən }

linear form A homogeneous polynomial of the first degree. { 'lin·ē·ər 'fȯrm }

linear fractional transformations *See* Möbius transformations. { 'lin·ē·ər ¦frak·shən·əl ˌtranz·fər'mā·shənz }

linear function *See* linear transformation. { 'lin·ē·ər 'fəŋk·shən }

linear functional A linear transformation from a vector space to its scalar field. { 'lin·ē·ər 'fəŋk·shən·əl }

linear independence The property of a set of vectors $\mathbf{v}_1,\ldots,\mathbf{v}_n$ in a vector space where if $a_1\mathbf{v}_1 + a_2\mathbf{v}_2 + \cdots + a_n\mathbf{v}_n = 0$, then all the scalars $a_i = 0$. { 'lin·ē·ər ˌin·də'pen·dəns }

linear inequalities A collection of relations among variables x_i, where at least one relation has the form $\Sigma_i a_i x_i \geq 0$. { 'lin·ē·ər ˌin·i'kwäl·əd·ēz }

linear interpolation A process to find a value of a function between two known values under the assumption that the three plotted points lie on a straight line. { 'lin·ē·ər inˌtər·pə'lā·shən }

linearity The property whereby a mathematical system is well behaved (in the context of the given system) with regard to addition and scalar multiplication. { ˌlin·ē'ar·əd·ē }

linearly dependent quantities Quantities that satisfy a homogeneous linear equation in which at least one of the coefficients is not zero. { 'lin·ē·ər·lē di¦pen·dənt 'kwän·təˌtēz }

linearly disjoint extensions Two extension fields E and F of a field k contained in a common field L, such that any finite set of elements in E that is linearly independent when E is regarded as a vector space over k remains linearly independent when E is regarded as a vector space over F. { ¦lin·ē·ər·lē ¦dis¦jȯint ik'sten·chənz }

linearly independent quantities Quantities which do not jointly satisfy a homogeneous linear equation unless all coefficients are zero. { 'lin·ē·ər·lē ˌin·də¦pen·dənt 'kwän·əd·ēz }

linearly ordered set A set with an ordering $\leq$ such that for any two elements a and b either $a \leq b$ or $b \leq a$. Also known as chain; serially ordered set; simply ordered set. { 'lin·ē·ər·lē ¦ȯr·dərd 'set }

linear manifold A subset of a vector space which is itself a vector space with the induced operations of addition and scalar multiplication. { 'lin·ē·ər 'man·əˌfōld }

linear operator *See* linear transformation. { 'lin·ē·ər 'äp·ə,rād·ər }

linear order Any order $<$ on a set S with the property that for any two elements a and b in S either $a < b$ or $b < a$. Also known as complete order; simple order; total order. { 'lin·ē·ər 'ȯr·dər }

linear programming The study of maximizing or minimizing a linear function $f(x_1, \ldots, x_n)$ subject to given constraints which are linear inequalities involving the variables x_i. { 'lin·ē·ər 'prō,gram·iŋ }

linear space *See* vector space. { 'lin·ē·ər 'spās }

linear system A system where all the interrelationships among the quantities involved are expressed by linear equations which may be algebraic, differential, or integral. { 'lin·ē·ər 'sis·təm }

linear topological space *See* topological vector space. { ¦lin·ē·ər ,täp·ə¦läj·ə·kəl 'spās }

linear transformation A function T defined in a vector space E and having its values in another vector space over the same field, such that if f and g are vectors in E, and c is a scalar, then $T(f + g) = Tf + Tg$ and $T(cf) = c(Tf)$. Also known as homogeneous transformation; linear function; linear operator. { 'lin·ē·ər ,tranz·fər'mā·shən }

line at infinity *See* ideal line. { 'līn at in'fin·əd·ē }

line graph A graph in which successive points representing the value of a variable at selected values of the independent variable are connected by straight lines. { 'līn ,graf }

line integral **1.** For a curve in a vector space defined by $\mathbf{x} = \mathbf{x}(t)$, and a vector function $\mathbf{V}$ defined on this curve, the line integral of $\mathbf{V}$ along the curve is the integral over t of the scalar product of $\mathbf{V}[\mathbf{x}(t)]$ and $d\mathbf{x}/dt$; this is written $\int \mathbf{V} \cdot d\mathbf{x}$. **2.** For a curve which is defined by $x = x(t)$, $y = y(t)$, and a scalar function f depending on x and y, the line integral of f along the curve is the integral over t of

$$f[x(t),y(t)] \cdot \sqrt{(dx/dt)^2 + (dy/dt)^2};$$

this is written $\int f ds$, where

$$ds = \sqrt{(dx)^2 + (dy)^2}$$

is an infinitesimal element of length along the curve. **3.** For a curve in the complex plane defined by $z = z(t)$, and a function f depending on z, the line integral of f along the curve is the integral over t of $f[z(t)]\ (dz/dt)$; this is written $\int f dz$. { 'līn ¦int·ə·grəl }

line of curvature A curve on a surface whose tangent lies along a principal direction at each point. { 'līn əv 'kər·və·chər }

line of striction The locus of the central points of the rulings of a given ruled surface. { 'līn əv 'strik·shən }

line of support Relative to a convex region in a plane, a line that contains at least one point of the region but is such that a half-plane on one side of the line contains no points of the region. { 'līn əv sə'pȯrt }

line segment A connected piece of a line. { 'līn ,seg·mənt }

Liouville function A function $\lambda(n)$ on the positive integers such that $\lambda(1) = 1$, and for $n \geqq 2$, $\lambda(n)$ is -1 raised to the number of prime factors of n, with repeated factors counted the number of times they appear. { 'lyü‚vēl ‚fəŋk·shən }

Liouville-Neumann series An infinite series of functions constructed from the given functions in the Fredholm equation which under certain conditions provides a solution. Also known as Neumann series. { 'lyü‚vēl 'nȯi‚män ‚sir·ēz }

Liouville number An irrational number x such that for any integer n there exist integers p and q for which the absolute value of $x - (p/q)$ is less than q^n. { 'lyü‚vēl ‚nəm·bər }

Liouville's theorem Every function of a complex variable which is bounded and analytic in the entire complex plane must be constant. { 'lyü‚vēlz ‚thir·əm }

Lipschitz condition A function f satisfies such a condition at a point b if $|f(x) - f(b)| \leq K\,|x - b|$, with K a constant, for all x in some neighborhood of b. { 'lip‚shits kən‚dish·ən }

literal constant A letter denoting a constant. { 'lid·ə·rəl 'kän·stənt }

literal expression An expression or equation in which the constants are represented by letters. { 'lid·ə·rəl ik'spresh·ən }

literal notation The use of letters to denote numbers, known or unknown. { 'lid·ə·rəl nō'tā·shən }

Littlewood conjecture The statement that there exists a number C such that, whenever $n_1, n_2, \ldots, n_N$ are N distinct integers, the integral over x from $-\pi$ to π of the absolute value of the sum from $k = 1$ to $k = N$ of the exponential functions of in_kx is greater than $2\pi\, C \log N$. { 'lid·əl‚wu̇d kən‚jek·chər }

lituus The trumpet-shaped plane curve whose points in polar coordinates (r, θ) satisfy the equation $r^2 = a/\theta$. { 'lich·ə·wəs }

Lobachevski geometry A system of planar geometry in which the euclidean parallel postulate fails; any point p not on a line L has at least two lines through it parallel to L. Also known as Bolyai geometry; hyperbolic geometry. { lō·bə'chef·skē jē'äm·ə·trē }

local algebra An algebra A over a field F which is the sum of the radical of A and the subalgebra consisting of products of elements of F with the multiplicative identity of A. { ¦lō·kəl 'al·jə·brə }

local base For a point x in a topological space, a family of neighborhoods of x such that every neighborhood of x contains a member of the family. Also known as base for the neighborhood system. { ¦lō·kəl 'bās }

local coefficient By using fiber bundles where the fiber is a group, one may generalize cohomology theory for spaces; one uses such bundles as the algebraic base for such a theory and calls the bundle a system of local coefficients. { 'lō·kəl ‚kō·i'fish·ənt }

local coordinate system The coordinate system about a point which is induced when the global space is locally euclidean. { 'lō·kəl kō'ȯrd·ən·ət ‚sis·təm }

local distortion The absolute value of the derivative of an analytic function at a given point. { 'lō·kəl di'stȯr·shən }

locally arcwise connected topological space A topological space in which every point has an arcwise connected neighborhood, that is, an open set any two points of which can be joined by an arc. { 'lō·kə·lē 'ärk‚wīz kə‚nek·təd ¦täp·ə¦läj·ə·kəl ¦spās }

locally compact topological space A topological space in which every point lies in a compact neighborhood. { 'lō·kə·lē kəm'pakt ¦täp·ə¦läj·ə·kəl ¦spās }

locally connected topological space A topological space in which every point has a connected neighborhood. { 'lō·kə·lē kə'nek·təd ¦täp·ə¦läj·ə·kəl ¦spās }

locally convex space A Hausdorff topological vector space E such that every neighborhood of any point x belonging to E contains a convex neighborhood of x. { 'lō·kə·lē 'kän‚veks ‚spās }

locally euclidean topological space A topological space in which every point has a neighborhood which is homeomorphic to a euclidean space. { 'lō·kə·lē yü'klid·ē·ən ¦täp·ə¦läj·ə·kəl ¦spās }

locally finite family of sets A family of subsets of a topological space such that each point of the topological space has a neighborhood that intersects only a finite number of these subsets. { ¦lō·kə·lē ¦fī‚nīt 'fam·lē əv 'sets }

locally integrable function A function is said to be locally integrable on an open set S in n-dimensional euclidean space if it is defined almost everywhere in S and has a finite integral on compact subsets S. { ¦lō·kə·lē ¦int·ə·grə·bəl 'fəŋk·shən }

locally one to one A function is locally one to one if it is one to one in some neighborhood of each point. { 'lō·kə·lē 'wən tə 'wən }

locally trivial bundle A bundle for which each point in the base has a neighborhood U whose inverse image under the projection map is isomorphic to a cartesian product of U with a space isomorphic to the fibers of the bundle. { ¦lō·kə·lē ¦triv·ē·əl 'bən·dəl }

local maximum A local maximum of a function f is a value $f(c)$ of f where $f(x) \leq f(c)$ for all x in some neighborhood of c; if $f(c)$ is a local maximum, f is said to have a local maximum at c. { 'lō·kəl 'mak·sə·məm }

local minimum A local minimum of a function f is a value $f(c)$ of f where $f(x) \geq f(c)$ for all x in some neighborhood of c; if $f(c)$ is a local minimum, f is said to have a local minimum at c. { 'lō·kəl 'min·ə·məm }

local quasi-F martingale A stochastic process $\{X_t\}$ such that the process obtained from $\{X_t\}$ by stopping it when it reaches n or $-n$ is a quasi-F martingale for each integer n. { 'lō·kəl 'kwä·zē¦ef 'märt·ən‚gāl }

local ring A ring with only one maximal ideal. { 'lō·kəl 'riŋ }

local solution A function which solves a system of equations only in a neighborhood of some point. { 'lō·kəl sə'lü·shən }

located vector An ordered pair of points in n-dimensional euclidean space. { ¦lō‚kād·əd 'vek·tər }

location principle A principle useful in locating the roots of an equation stating that if a continuous function has opposite signs for two values of the independent variable, then it is zero for some value of the variable between these two values. { lō'kā·shən ‚prin·sə·pəl }

locus A collection of points in a euclidean space whose coordinates satisfy one or more algebraic conditions. { 'lō·kəs }

logarithm 1. The real-valued function log u defined by log $u = v$ if $e^v = u$, e^v denoting the exponential function. Also known as hyperbolic logarithm; Naperian logarithm; natural logarithm. **2.** An analog in complex variables relative to the function e^z. { 'läg·ə‚rith·əm }

logarithmically convex function A function whose logarithm is a convex function. { ‚läg·ə¦rith·mik·lē ¦kän¦veks ‚fəŋk·shən }

logarithmic coordinate paper Paper ruled with two sets of mutually perpendicular, parallel lines spaced according to the logarithms of consecutive numbers, rather than the numbers themselves. { 'läg·ə‚rith·mik kō'ȯrd·ən·ət ‚pā·pər }

logarithmic coordinates In the plane, logarithmic coordinates are defined by two coordinate axes, each marked with a scale where the distance between two points is the difference of the logarithms of the two numbers. { 'läg·ə‚rith·mik kō'ȯrd·ən·əts }

logarithmic curve A curve whose equation in cartesian coordinates is $y = \log ax$, where a is greater than 1. { 'läg·ə‚rith·mik 'kərv }

logarithmic derivative The logarithmic derivative of a function $f(z)$ of a real (complex) variable is the ratio $f'(z)/f(z)$, that is, the derivative of log $f(z)$. { 'läg·ə‚rith·mik də'riv·əd·iv }

logarithmic differentiation A technique often helpful in computing the derivatives of a differentiable function $f(x)$; set $g(x) = \log f(x)$ where $f(x) \neq 0$, then $g'(x) = f'(x)/f(x)$, and if there is some other way to find $g'(x)$, then one also finds $f'(x)$. { 'läg·ə‚rith·mik ‚dif·ə‚ren·chē'ā·shən }

logarithmic equation An equation which involves a logarithmic function of some variable. { 'läg·ə‚rith·mik i'kwā·zhən }

logarithmic scale A scale in which the distances that numbers are at from a reference point are proportional to their logarithms. { 'läg·ə‚rith·mik 'skāl }

logarithmic spiral The spiral plane curve whose points in polar coordinates (r,θ) satisfy the equation $\log r = a\theta$. Also known as equiangular spiral. { 'läg·ə‚rith·mik 'spī·rəl }

logarithmic trigonometric function The logarithm of the corresponding trigonometric function. { 'läg·ə‚rith·mik 'trig·ə·nə‚me·trik ‚fəŋk·shən }

logic The subject that investigates, formulates, and establishes principles of valid reasoning. { 'läj·ik }

logical addition The additive binary operation of a Boolean algebra. { 'läj·ə·kəl ə'dish·ən }

logical connectives Symbols which link mathematical statements; these symbols represent the terms "and," "or," "implication," and "negation." { 'läj·ə·kəl kə'nek·tivz }

logical function *See* predicate. { 'läj·ə·kəl 'fəŋk·shən }

logical multiplication The multiplicative binary operation of a Boolean algebra. { 'läj·ə·kəl məl·tə·plə'kā·shən }

long radius The distance from the center of a regular polygon to a vertex. { 'lȯŋ ˌrād·ē·əs }

lookahead tree *See* game tree. { 'lu̇k·əˌhed ˌtrē }

loop A line which begins and ends at the same point of the graph. { lüp }

Lorentz group The group of all Lorentz transformations of euclidean four-space with composition as the operation. { 'lȯrˌens ˌgrüp }

Lorentz transformation Any linear transformation of euclidean four space which preserves the quadratic form $q(x,y,z,t) = t^2 - x^2 - y^2 - z^2$. { 'lȯrˌens ˌtranz·fərˌmā·shən }

loss function In decision theory, the function, dependent upon the decision and the true underlying distributions, which expresses the loss produced in taking the decision. { 'lȯs ˌfəŋk·shən }

lower bound A lower bound of a subset A of a set S is a point of S which is smaller than every element of A. { 'lō·ər ¦bau̇nd }

lower semicontinuous function A real-valued function $f(x)$ is lower semicontinuous at a point x_0 if, for any small positive number ϵ, $f(x)$ is always greater than $f(x_0) - \epsilon$ for all x in some neighborhood of x_0. { ¦lō·ər ˌsem·ē·kən'tin·yə·wəs ˌfənk·shən }

loxodromic spiral A curve on a surface of revolution which cuts the meridians at a constant angle other than 90°. { ¦lak·sə¦dräm·ik 'spī·rəl }

lub *See* least upper bound.

Lucas number A number in the Fibonacci sequence whose first two terms are $L_1 = 1$ and $L_2 = 3$. { 'lü·kəs ˌnəm·bər }

lune A section of a plane bounded by two circular arcs, or of a sphere bounded by two great circles. { lün }

Lusin's theorem Given a measurable function f which is finite almost everywhere in a euclidean space, then for every number $\epsilon > 0$ there is a continuous function g which agrees with f, except on a set of measure less than ϵ. { ˌlü·zēnz ˌthir·əm }

Lyapunov function A function of a vector and of time which is positive-definite and has a negative-definite derivative with respect to time for nonzero vectors, is identically zero for the zero vector, and approaches infinity as the norm of the vector approaches infinity; used in determining the stability of control systems. Also spelled Liapunov function. { lē'ap·əˌnȯf ˌfəŋk·shən }

M

m *See* milli-.

MacDonald functions *See* modified Hankel functions. { mək'dän·əld ˌfəŋk·shənz }

Maclaurin-Cauchy test *See* Cauchy's test for convergence. { mə'klȯr·ən kō'shē ˌtest }

Maclaurin expansion The power series representation of a function arising from Maclaurin's theorem. { mə'klȯr·ən ikˌspan·chən }

Maclaurin series The power series in the Maclaurin expansion. { mə'klȯr·ən ˌsir·ēz }

Maclaurin's theorem The theorem giving conditions when a function, which is infinitely differentiable, may be represented in a neighborhood of the origin as an infinite series with *n*th term $(1/n!)\cdot f^{(n)(0)}\cdot x^n$, where $f^{(n)}$ denotes the *n*th derivative. { mə'klȯr·ənz ˌthir·əm }

magic square **1.** A square array of integers where the sum of the entries of each row, each column, and each diagonal is the same. **2.** A square array of integers where the sum of the entries in each row and each column (but not necessarily each diagonal) is the same. { 'maj·ik 'skwer }

magnitude *See* absolute value. { 'mag·nəˌtüd }

main diagonal *See* principal diagonal. { ¦mān dī'ag·ən·əl }

major arc The longer of the two arcs produced by a secant of a circle. { 'mā·jər 'ärk }

major axis The longer of the two axes with respect to which an ellipse is symmetric. { 'mā·jər 'ak·səs }

majority A logic operator having the property that if P, Q, R are statements, then the function (P, Q, R, ...) is true if more than half the statements are true, or false if half or less are true. { mə'jär·əd·ē }

manifold A topological space which is locally euclidean; there are four types: topological, piecewise linear, differentiable, and complex, depending on whether the local coordinate systems are obtained from continuous, piecewise linear, differentiable, or complex analytic functions of those in euclidean space; intuitively, a surface. { 'man·əˌfōld }

mantissa The positive decimal part of a common logarithm. { man'tis·ə }

map *See* mapping. { map }

mapping 1. Any function or multiple-valued relation. Also known as map. **2.** In topology, a continuous function. { 'map·iŋ }

Markov chain A Markov process whose state space is finite or countably infinite. { 'mar‚kóf ‚chān }

Markov process A stochastic process which assumes that in a series of random events the probability of an occurrence of each event depends only on the immediately preceding outcome. { 'mär‚kóf prä·səs }

marriage theorem The proposition that a family of n subsets of a set S with n elements is a system of distinct representatives for S if any k of the subsets, $k = 1, 2, \ldots, n$, together contain at least k distinct elements. Also known as Hall's theorem. { 'mar·ij ‚thir·əm }

Mascheroni's constant *See* Euler's constant. { ‚mäsk·ə'rō‚nēz 'kän·stənt }

match *See* biconditional operation. { mach }

material implication *See* implication. { mə¦tir·ē·əl ‚im·plə'kā·shən }

mathematical analysis *See* analysis. { ¦math·ə¦mad·ə·kəl ə'nal·ə·səs }

mathematical induction A general method of proving statements concerning a positive integral variable: if a statement is proven true for $x = 1$, and if it is proven that, if the statement is true for $x = 1, \ldots, n$, then it is true for $x = n + 1$, it follows that the statement is true for any integer. Also known as complete induction. { ¦math·ə¦mad·ə·kəl in'dək·shən }

mathematical logic The study of mathematical theories from the viewpoint of model theory, recursive function theory, proof theory, and set theory. { ¦math·ə¦mad·ə·kəl 'läj·ik }

mathematical model 1. A mathematical representation of a process, device, or concept by means of a number of variables which are defined to represent the inputs, outputs, and internal states of the device or process, and a set of equations and inequalities describing the interaction of these variables. **2.** A mathematical theory or system together with its axioms. { ¦math·ə¦mad·ə·kəl 'mäd·əl }

mathematical probability The ratio of the number of mutually exclusive, equally likely outcomes of interest to the total number of such outcomes when the total is exhaustive. Also known as a priori probability. { ¦math·ə¦mad·ə·kəl ‚präb·ə'bil·əd·ē }

mathematical programming *See* optimization theory. { ¦math·ə¦mad·ə·kəl 'prō ‚gram·iŋ }

mathematical table A listing of the values of a function of one or several variables at a series of values of the arguments, usually equally spaced. { ¦math·ə¦mad·ə·kəl 'tā·bəl }

Mathieu equation A differential equation of the form $y'' + (a + b \cos 2x)y = 0$, whose solution depends on periodic functions. { ma'tyü i‚kwā·zhən }

Mathieu functions Any solution of the Mathieu equation which is periodic and an even or odd function. { ma'tyü ‚fəŋk·shənz }

matrix A rectangular array of numbers or scalars from a vector space. { 'mā·triks }

matrix algebra An algebra whose elements are matrices and whose operations are addition and multiplication of matrices. { 'mā·triks 'al·jə·brə }

matrix calculus The treatment of matrices whose entries are functions as functions in their own right with a corresponding theory of differentiation; this has application to the study of multidimensional derivatives of functions of several variables. { 'mā·triks 'kal·kyə·ləs }

matrix element One of the set of numbers which form a matrix. { 'mā·triks ˌel·ə·mənt }

matrix game A game involving two persons, which gives rise to a matrix representing the amount received by the two players. Also known as rectangular game. { 'mā·triks ˌgām }

matrix of a linear transformation A unique matrix *A*, such that for a specified linear transformation *L* from one vector space to another, and for specified finite bases in each space, *L* applied to a vector is equal to *A* times that vector. { 'mā·triks əv ə 'lin·ē·ər ˌtranz·fər'mā·shən }

matrix theory The algebraic study of matrices and their use in evaluating linear processes. { 'mā·triks ˌthē·ə·rē }

max *See* maximum. { maks }

maximal element *See* maximal member. { 'mak·sə·məl 'el·ə·mənt }

maximal ideal An ideal *I* in a ring *R* which is not equal to *R*, and such that there is no ideal containing *I* and not equal to *I* or *R*. { ¦mak·sə·məl ī'dēl }

maximal member In a partially ordered set a maximal member is one for which no other element follows it in the ordering. Also known as maximal element. { 'mak·sə·məl 'mem·bər }

maximal planar graph A planar graph to which no new arcs can be added without forcing crossings and hence violating planarity. { 'mak·sə·məl ¦plān·ər ˌgraf }

maximax criterion In decision theory, one of several possible prescriptions for making a decision under conditions of uncertainty; it prescribes the strategy which will maximize the maximum possible profit. { 'mak·səˌmaks krīˌtir·ē·ən }

maxim criterion One of several prescriptions for making a decision under conditions of uncertainty; it prescribes the strategy which will maximize the minimum profit. Also known as maximin criterion. { 'mak·səm krīˌtir·ē·ən }

maximin **1.** The maximum of a set of minima. **2.** In the theory of games, the largest of a set of minimum possible gains, each representing the least advantageous outcome of a particular strategy. { 'mak·səˌmin }

maximin criterion *See* maxim criterion. { 'mak·səˌmin krīˌtir·ē·ən }

maximizing a function Finding the largest value assumed by a function. { 'mak·səˌmīz·iŋ ə 'fəŋk·shən }

maximum The maximum of a real-valued function is the greatest value it assumes. Abbreviated max. { 'mak·sə·məm }

maximum-minimum principle *See* min-max theorem. { 'mak·sə·məm 'min·ə·məm ˌprin·sə·pəl }

maximum-modulus theorem For a complex analytic function in a closed bounded simply connected region its modulus assumes its maximum value on the boundary of the region. { 'mak·sə·məm 'mäj·ə·ləs ˌthir·əm }

mean A single number that typifies a set of numbers, such as the arithmetic mean, the geometric mean, or the expected value. Also known as mean value. { mēn }

mean curvature Half the sum of the principal curvatures at a point on a surface. { ¦mēn 'kər·və·chər }

mean deviation *See* average deviation. { 'mēn ˌdē·vē'ā·shən }

mean evolute The envelope of the planes that are orthogonal to the normals of a given surface and cut the normals halfway between the centers of principal curvature of the surface. { ¦mēn 'ev·əˌlüt }

mean proportional For two numbers a and b, a number x, such that $x/a = b/x$. { 'mēn prə'pȯr·shən·əl }

mean terms The second and third terms of a proportion. { 'mēn 'tərmz }

mean value **1.** For a function $f(x)$ defined on an interval (a,b), the integral from a to b of $f(x)\,dx$ divided by $b - a$. **2.** *See* mean. { 'mēn 'val·yü }

mean value theorem The proposition that, if a function $f(x)$ is continuous on the closed interval $[a,b]$ and differentiable on the open interval (a,b), then there exists x_0, $a < x_0 < b$, such that $f(b) - f(a) = (b - a)f'(x_0)$. Also known as first law of the mean; Lagrange's formula; law of the mean. { 'mēn 'val·yü ˌthir·əm }

measurable function **1.** A real valued function f defined on a measurable space X, where for every real number a all those points x in X for which $f(x) \geq a$ form a measurable set. **2.** A function on a measurable space to a measurable space such that the inverse image of a measurable set is a measurable set. { 'mezh·rə·bəl 'fəŋk·shən }

measurable set A member of the sigma-algebra of subsets of a measurable space. { 'mezh·rə·bəl 'set }

measurable space A set together with a sigma-algebra of subsets of this set. { 'mezh·rə·bəl 'spās }

measure A nonnegative real valued function m defined on a sigma-algebra of subsets of a set S whose value is zero on the empty set, and whose value on a countable union of disjoint sets is the sum of its values on each set. { 'mezh·ər }

measure-preserving transformation A transformation T of a measure space S into itself such that if E is a measurable subset of S then so is $T^{-1}E$ (the set of points mapped into E by T) and the measure of $T^{-1}E$ is then equal to that of E. { ¦mezh·ər pri¦zərv·iŋ ˌtranz·fər'mā·shən }

measure space A set together with a sigma-algebra of subsets of the set and a measure defined on this sigma-algebra. { 'mezh·ər ˌspās }

measure theory The study of measures and their applications, particularly the integration of mathematical functions. { 'mezh·ər ˌthē·ə·rē }

measure zero **1.** A set has measure zero if it is measurable and the measure of it is zero. **2.** A subset of euclidean n-dimensional space which has the property that for any positive number ϵ there is a covering of the set by n-dimensional rectangles such that the sum of the volumes of the rectangles is less than ϵ. { 'mezh·ər ˌzir·ō }

mechanic's rule A rule for estimating the square root of a number x whereby an estimate e is made of $\sqrt{x}$, a new estimate is made by taking the quantity $e' = (1/2)[e + (x/e)]$, and this procedure is repeated as many times as required to achieve the desired accuracy. { mi'kan·iks ¦rül }

median **1.** Any line in a triangle which joins a vertex to the midpoint of the opposite side. **2.** The line that joins the midpoints of the nonparallel sides of a trapezoid. Also known as midline. { 'mē·dē·ən }

median point The point at which all three medians of a triangle intersect. { 'med·ē·ən ˌpȯint }

Meijer transform The Meijer transform of a function $f(x)$ is the function $F(y)$ defined as the integral from 0 to ∞ of $\sqrt{xy}K_n(xy)f(x)dx$, where K_n is a modified Bessel function. { 'mā·ər ˌtranzˌfȯrm }

Mellin transform The transform $F(s)$ of a function $f(t)$ defined as the integral over t from 0 to ∞ of $f(t)t^{s-1}$. { me'lēn ˌtranzˌfȯrm }

member **1.** An element of a set. **2.** For an equation, the expression on either side of the equality sign. { 'mem·bər }

membership function The characteristic function of a fuzzy set, which assigns to each element in a universal set a value between 0 and 1. { 'mem·bərˌship ˌfəŋk·shən }

Menelaus' theorem If ABC is a triangle and PQR is a straight line that cuts AB, CA, and the extension of BC at P, Q, and R respectively, then $(AP/PB)(CQ/QA)(BR/CR) = 1$. { ¦men·ə¦lā·əs ˌthir·əm }

Menger's theorem A theorem in graph theory which states that if G is a connected graph and A and B are disjoint sets of points of G, then the minimum number of points whose deletion separates A and B is equal to the maximum number of disjoint paths between A and B. { 'meŋ·ərz ˌthir·əm }

mensuration The measurement of geometric quantities; for example, length, area, and volume. { ˌmen·sə'rā·shən }

meridian section The intersection of a surface of revolution with a plane that contains the axis of revolution. { mə'rid·ē·ən ˌsek·shən }

meromorphic function A function of complex variables which is analytic in its domain of definition save at a finite number of points which are poles. { ¦mer·ə¦mȯr·fik 'fəŋk·shən }

Mersenne number A number of the form 2;$2^p - 1$, where p is a prime number. { mər'sen ¦nəm·bər }

Mersenne prime A Mersenne number that is also a prime number. { mər'sen ¦prīm }

metamathematics The study of the principles of deductive logic as they are used in mathematical logic. { ¦med·əˌmath·ə'mad·iks }

metric A real valued "distance" function on a topological space X satisfying four rules:

for x, y, and z in X, the distance from x to itself is zero; the distance from x to y is positive if x and y are different; the distance from x to y is the same as the distance from y to x; and the distance from x to y is less than or equal to the distance from x to z plus the distance from z to y (triangle inequality). { 'me·trik }

metric space Any topological space which has a metric defined on it. { 'me·trik 'spās }

metric tensor A second rank tensor of a Riemannian space whose components are functions which help define magnitude and direction of vectors about a point. Also known as fundamental tensor. { 'me·trik 'ten·sər }

metrizable space A topological space on which can be defined a metric whose topological structure is equivalent to the original one. { mə'trīz·ə·bəl 'spās }

Meusnier's theorem A theorem stating that the curvature of a surface curve equals the curvature of the normal section through the tangent to the curve divided by the cosine of the angle between the plane of this normal section and the osculating plane of the curve. { mən'yāz ˌthir·əm }

micro- A prefix representing 10^{-6}, or one-millionth. { 'mī·krō }

micromicro- *See* pico- . { ¦mī·krō¦mī·krō }

midline *See* median. { 'midˌlīn }

midpoint The midpoint of a line segment is the point which separates the segment into two equal parts. { 'midˌpȯint }

mil A unit of angular measure which, due to nonuniformity of usage, may have any one of three values: 0.001 radian or approximately 0.0572958°; 1/6400 of a full revolution or 0.5625°; 1/1000 of a right angle or 0.09°. { mil }

milli- A prefix representing 10^{-3}, or one-thousandth. Abbreviated m. { 'mil·ē }

milli-micro- *See* nano- . { ¦mil·ə¦mī·krō }

million The number 10^6, or 1,000,000. { 'mil·yən }

Milne method A technique which provides numerical solutions to ordinary differential equations. { 'miln ˌmeth·əd }

minimal equation **1.** An algebraic equation whose zeros define a minimal surface. **2.** *See* reduced characteristic equation. { 'min·ə·məl i'kwā·zhən }

minimal polynomial The polynomial of least degree which both divides the characteristic polynomial of a matrix and has the same roots. { 'min·ə·məl ˌpäl·ə'nō·mē·əl }

minimal surface A surface that has assumed a geometric configuration of least area among those into which it can readily deform. { 'min·ə·məl 'sər·fəs }

minimal transformation group A transformation group such that every orbit is dense in the phase space. { ¦min·ə·məl ˌtranz·fər'mā·shən ˌgrüp }

minimax **1.** The minimum of a set of maxima. **2.** In the theory of games, the smallest of a set of maximum possible losses, each representing the most unfavorable outcome of a particular strategy. { 'min·əˌmaks }

minimax technique *See* min-max technique. { 'min·ēˌmaks tekˌnēk }

minimax theorem A theorem of games that the lowest maximum expected loss in a two-person zero-sum game equals the highest minimum expected gain. { 'min·ə,maks ,thir·əm }

minimization The determination of the simplest expression of a Boolean function equivalent to a given one. { ,min·ə·mə'zā·shən }

minimum The least value that a real valued function assumes. { 'min·ə·məm }

minimum-modulus theorem The theorem that a nonvanishing, complex analytic function in a closed, bounded, simply connected region assumes its minimum absolute value on the boundary of the region. { ¦min·ə·məm 'mäj·ə·ləs ,thir·əm }

Minkowski distance function Relative to a convex body with the origin O in its interior, the function whose value at a point P is the distance ratio OP/OQ, where Q is the point of the convex body on the ray OP that is furthest from O. { miŋ'kəf·skē *or* min'kaů·skē 'dis·təns ,fəŋk·shən }

Minkowski's inequality **1.** An inequality involving powers of sums of sequences of real or complex numbers, a_k and b_k:

$$\left[\sum_{k=1} |a_k + b_k|^s\right]^{1/s} \leq \left[\sum_{k=1} |a_k|^s\right]^{1/s} + \left[\sum_{k=1}^{\infty} |b_k|^s\right]^{1/s}$$

provided $s \geq 1$. **2.** An inequality involving powers of integrals of real or complex functions, f and g, over an interval or region R:

$$\left[\int_R |f(x) + g(x)|^s \, dx\right]^{1/s} \leq \left[\int_R |f(x)|^s \, dx\right]^{1/s} + \left[\int_R |g(x)|^s \, dx\right]^{1/s}$$

provided $s \geq 1$ and the integrals involved exist. { miŋ'kȯf·skēz ,in·i'kwäl·əd·ē }

min-max technique A method of approximation of a function f by a function g from some class where the maximum of the modulus of $f - g$ is minimized over this class. Also known as Chebyshev approximation; minimax technique. { 'min 'maks tek,nēk }

min-max theorem The theorem that provides information concerning the nth eigenvalue of a symmetric operator on an inner product space without necessitating knowledge of the other eigenvalues. Also known as maximum-minimum principle. { 'min 'maks ,thir·əm }

minor The minor of an entry of a matrix is the determinant of the matrix obtained by removing the row and column containing the entry. Also known as cofactor; complementary minor. { 'mīn·ər }

minor arc The smaller of the two arcs on a circle produced by a secant. { 'mīn·ər 'ärk }

minor axis The smaller of the two axes of an ellipse. { 'mīn·ər 'ak·səs }

minuend The quantity from which another quantity is to be subtracted. { 'min·yə,wend }

minus A minus B means that the quantity B is to be subtracted from the quantity A. { 'mī·nəs }

minus sign *See* subtraction sign. { 'mī·nəs ,sīn }

minute A unit of measurement of angle that is equal to $^1/_{60}$ of a degree. Symbolized $'$. Also known as arcmin. { 'min·ət }

mirror plane of symmetry *See* plane of mirror symmetry. { 'mir·ər 'plān əv 'sim·ə·trē }

Mittag-Leffler's theorem A theorem that enables one to explicitly write down a formula for a meromorphic complex function with given poles; for a function $f(z)$ with poles at $z = z_i$, having order m_i and principal parts

$$\sum_{j=1}^{m_i} a_{ij}(z - z_i)^{-j},$$

the formula is

$$f(z) = \sum_k \left[\sum_{j=1}^{m_i} a_{ij}(z - z_i)^{-j} + p_i(z) \right] + g(z)$$

where the $p_i(z)$ are polynomials, $g(z)$ is an entire function, and the series converges uniformly in every bounded region where $f(z)$ is analytic. { 'mi,täk 'lef·lərz ,thir·əm }

mixed-base notation A computer number system in which a single base, such as 10 in the decimal system, is replaced by two number bases used alternately, such as 2 and 5. { 'mikst ¦bās nō'tā·shən }

mixed-base number A number in mixed-base notation. Also known as mixed-radix number. { 'mikst ¦bās 'nəm·bər }

mixed decimal Any decimal plus an integer. { 'mikst 'des·məl }

mixed number The sum of an integer and a fraction. { 'mikst 'nəm·bər }

mixed radix Pertaining to a numeration system using more than one radix, such as the biquinary system. { 'mikst 'rā·diks }

mixed-radix number *See* mixed-base number. { 'mikst ¦rā·diks 'nəm·bər }

mixed strategy A method of playing a matrix game in which the player attaches a probability weight to each of the possible options, the probability weights being nonnegative numbers whose sum is unity, and then operates a chance device that chooses among the options with probabilities equal to the corresponding weights. { ¦mikst 'strad·ə·jē }

mixed surd A surd containing a rational factor or term, as well as irrational numbers. { ¦mikst 'sərd }

mixed tensor A tensor with both contravariant and covariant indices. { ¦mikst 'ten·sər }

mixing transformation A function of a measure space which moves the measurable sets in such a manner that, asymptotically as regards measure, any measurable set is distributed uniformly throughout the space. { 'mik·siŋ ,tranz·fər'mā·shən }

Möbius band The nonorientable surface obtained from a rectangular strip by twisting it once and then gluing the two ends. Also known as Möbius strip. { 'mər·bē·əs ,band }

Möbius function The function μ of the positive integers where $\mu(1) = 1$, $\mu(n) = (-1)^r$ if n factors into r distinct primes, and $\mu(n) = 0$ otherwise; also, $\mu(n)$ is the sum of the primitive nth roots of unity. { 'mər·bē·əs ,fəŋk·shən }

Möbius strip *See* Möbius band. { 'mər·bē·əs ˌstrip }

Möbius transformations These are the most commonly used conformal mappings of the complex plane; their form is $f(z) = (az + b)/(cz + d)$ where the real numbers a, b, c, and d satisfy $ad - bc \neq 0$. Also known as bilinear transformations; homographic transformations; linear fractional transformations. { 'mər·bē·əs ˌtranz·fər'mā·shənz }

model theory The general qualitative study of the structure of a mathematical theory. { 'mäd·əl ˌthē·ə·rē }

modern algebra The study of algebraic systems such as groups, rings, modules, and fields. { 'mäd·ərn 'al·jə·brə }

modified Bessel equation The differential equation $z^2f''(z) + zf'(z) - (z^2 + n^2)f(z) = 0$, where z is a variable that can have real or complex values and n is a real or complex number. { ¦mōd·əˌfīd 'bes·əl iˌkwā·zhən }

modified Bessel functions The functions defined by $I_v(x) = \exp(-iv\pi/2)\, J_v(ix)$, where J_v is the Bessel function of order v, and x is real and positive. { 'mäd·əˌfīd 'bes·əl ˌfəŋk·shənz }

modified Hankel functions The functions defined by $K_v(x) = (i\pi/2)\exp(-iv\pi/2)\, H_v^{(1)}(ix)$, where $H_v^{(1)}$ is the first Hankel function of order v, and x is real and positive. Also known as MacDonald functions. { 'mäd·əˌfīd 'häŋk·əl ˌfəŋk·shənz }

module A vector space in which the scalars are a ring rather than a field. { 'mäj·ül }

modulo **1.** A group G modulo a subgroup H is the quotient group G/H of cosets of H in G. **2.** A technique of identifying elements in an algebraic structure in such a manner that the resulting collection of identified objects is the same type of structure. { 'mäj·əˌlō }

modulo *N* Two integers are said to be congruent modulo N (where N is some integer) if they have the same remainder when divided by N. { 'mäj·əˌlō 'en }

modulo *N* arithmetic Calculations in which all integers are replaced by their remainders after division by N (where N is some fixed integer.) { 'mäj·əˌlō ¦en ə'rith·mə·tik }

modulus **1.** The modulus of a logarithm with a given base is the factor by which a logarithm with a second base must be multiplied to give the first logarithm. **2.** *See* absolute value. { 'mäj·ə·ləs }

modulus of a congruence A number a, such that two specified numbers b and c give the same remainder when divided by a; b and c are then said to be congruent, modulus a (or congruent, modulo a). { 'mäj·ə·ləs əv ə kən'grü·əns }

modulus of continuity For a real valued continuous function f, this is the function whose value at a real number r is the maximum of the modulus of $f(x) - f(y)$ where the modulus of $x - y$ is less than r; this function is useful in approximation theory. { 'mäj·ə·ləs əv ˌkänt·ən'ü·əd·ē }

molding surface A surface generated by a plane curve as its plane rolls without slipping over a cylinder. { 'mōld·iŋ ˌsər·fəs }

Monge form An equation of a surface of the form $z = f(x,y)$, where x, y, and z are cartesian coordinates. { 'mònzh ˌfòrm }

Monge's theorem For three coplanar circles, and for radii of these circles which are parallel to each other, the three outer centers of similitude of the circles taken in pairs lie on a single straight line, and any two inner centers of similitude lie on a straight line with one of the outer centers. { 'mōnzh·əz ˌthir·əm }

monic equation A polynomial equation with integer coefficients, where the coefficient of the term of highest degree is +1. { ¦mō·nik i'kwā·zhən }

monodromy theorem If a complex function is analytic at a point of a bounded simply connected domain and can be continued analytically along every curve from the point, then it represents a single-valued analytic function in the domain. { 'män·əˌdrō·mē ˌthir·əm }

monogenic analytic function An analytic function whose domain of definition has been extended directly or indirectly by analytic continuation as far as theoretically possible. { ¦män·əˌjen·ik ˌan·ə¦lid·ik 'fəŋk·shən }

monoid A semigroup which has an identity element. { 'mäˌnȯid }

monomial A polynomial of degree one. { mə'nō·mē·əl }

monomial factor A single factor that can be divided out of every term in a given expression. { mə'nō·mē·əl ˌfak·tər }

monotone convergence theorem The integral of the limit of a monotone increasing sequence of nonnegative measurable functions is equal to the limit of the integrals of the functions in the sequence. { 'män·əˌtōn kən'vər·jəns ˌthir·əm }

monotone function A function which is either monotone nondecreasing or monotone nonincreasing. Also known as monotonic function. { ˌmän·əˌtōn ˌfəŋk·shən }

monotone nondecreasing function A function which never decreases, that is, if $x \leq y$ then $f(x) \leq f(y)$. Also known as increasing function; monotonically nondecreasing function. { ˌmän·əˌtōn ¦nän·di'krēs·iŋ ˌfəŋk·shən }

monotone nonincreasing function A function which never increases, that is, if $x \leq y$ then $f(x) \geq f(y)$. Also known as decreasing function; monotonically nonincreasing function. { ˌmän·əˌtōn ¦nän·in'krēs·iŋ ˌfəŋk·shən }

monotonically nondecreasing function *See* monotone nondecreasing function. { ¦män·ə¦tän·ik·lē ¦nän·di'krēs·iŋ ˌfəŋk·shən }

monotonically nonincreasing function *See* monotone nonincreasing function. { ¦män·ə¦tän·ik·lē ¦nän·in'krēs·iŋ ˌfəŋk·shən }

monotonic function *See* monotone function. { ¦män·ə¦tän·ik 'fəŋk·shən }

monotonic system of sets *See* nested sets. { ¦män·ə¦tän·ik ¦sis·təm əv 'sets }

Moore-Smith convergence Convergence of a net. { 'mu̇r 'smith kən'vər·jəns }

Morera's theorem If a function of a complex variable is continuous in a simply connected domain D, and if the integral of the function about every simply connected curve in D vanishes, then the function is analytic in D. { mȯ'rer·əz ˌthir·əm }

morphism The class of elements which together with objects form a category; in most cases, morphisms are functions which preserve some structure on a set. { 'mȯrˌfiz·əm }

Morse theory The study of differentiable mappings of differentiable manifolds, which by examining critical points shows how manifolds can be constructed from one another. { 'mȯrs ˌthē·ə·rē }

moving trihedral For a space curve, a configuration consisting of the tangent, principal normal, and binormal of the curve at a variable point on the curve. { 'müv·iŋ trī'hē·drəl }

Muller method A method for finding zeros of a function $f(x)$, in which one repeatedly evaluates $f(x)$ at three points, x_1, x_2, and x_3, fits a quadratic polynomial to $f(x_1)$, $f(x_2)$, and $f(x_3)$, and uses x_2, x_3, and the root of this quadratic polynomial nearest to x_3 as three new points to repeat the process. { 'məl·ər ˌmeth·əd }

multidimensional derivative The generalized derivative of a function of several variables which is usually represented as a matrix involving the various partial derivatives of the function. { ¦məl·tə·di'men·shən·əl də'riv·əd·iv }

multifoil A plane figure consisting of congruent arcs of a circle arranged around a regular polygon, with the end points of each arc located at the midpoints of adjacent sides of the polygon, and the tangents to the arcs at these points perpendicular to the sides. { 'məl·tēˌfȯil }

multilinear algebra The study of functions of several variables which are linear relative to each variable. { 'məl·təˌlin·ē·ər 'al·jə·brə }

multilinear form A multilinear form of degree n is a polynomial expression which is linear in each of n variables. { 'məl·təˌlin·ē·ər 'fȯrm }

multinomial An algebraic expression which involves the sum of at least two terms. { ¦məl·tə¦nō·mē·əl }

multinomial distribution The joint distribution of the set of random variables which are the number of occurrences of the possible outcomes in a sequence of multinomial trials. { ¦məl·tə¦nō·mē·əl ˌdi·strə'byü·shən }

multinomial theorem The rule for expanding $(x_1 + x_2 + \cdots + x_m)^n$, where m and n are positive integers; a generalization of the binomial theorem. { ˌməl·tə¦nō·mē·əl 'thir·əm }

multiple integral An integral over a subset of n-dimensional space. { 'məl·tə·pəl 'int·ə·grəl }

multiple point A point of a curve through which passes more than one arc of the curve. { 'məl·tə·pəl 'pȯint }

multiple root A polynomial $f(x)$ has c as a multiple root if $(x - c)^n$ is a factor for some $n > 1$. Also known as repeated root. { 'məl·tə·pəl 'rüt }

multiple-valued A relation between sets is multiple valued if it associates to an element of one more than one element from the other; sometimes functions are allowed to be multiple-valued. { 'məl·tə·pəl 'valˌyüd }

multiple-valued logic A form of logic in which statements can have values other than the two values "true" and "false." { 'məl·tə·pəl ¦valˌyüd 'läj·ik }

multiplicand If a number x is to be multiplied by a number y, then x is called the multiplicand. { ˌməl·tə·pli'kand }

multiplication Any algebraic operation analogous to multiplication of real numbers. { ˌməl·tə·pli'kā·shən }

multiplication on the left *See* premultiplication. { məl·tə·pli'kā·shən ȯn thə 'left }

multiplication on the right *See* postmultiplication. { ˌməl·tə·pli'kā·shən ȯn thə 'rīt }

multiplication sign The symbol × or ·, used to indicate multiplication. Also known as times sign. { ˌməl·tə·pli'kā·shən ˌsīn }

multiplicative identity In a mathematical system with an operation of multiplication, denoted ×, an element 1 such that $1 \times e = e \times 1 = e$ for any element e in the system. { ˌməl·tə'plik·əd·iv ī'den·əd·ē }

multiplicative inverse In a mathematical system with an operation of multiplication, denoted ×, the multiplicative inverse of an element e is an element $\bar{e}$ such that $e \times \bar{e} = \bar{e} \times e = 1$, where 1 is the multiplicative identity. { ˌməl·tə'plik·əd·iv 'inˌvərs }

multiplicative subset A subset S of a commutative ring such that if x and y are in S then so is xy. { ˌməl·tə'plik·əd·iv 'səbˌset }

multiplicity 1. A root of a polynomial $f(x)$ has multiplicity n if $(x - a)^n$ is a factor of $f(x)$ and n is the largest possible integer for which this is true. **2.** The geometric multiplicity of an eigenvalue λ of a linear transformation T is the dimension of the null space of the transformation $T - \lambda I$, where I denotes the identity transformation. **3.** The algebraic multiplicity of an eigenvalue λ of a linear transformation T on a finite-dimensional vector space is the multiplicity of λ as a root of the characteristic polynomial of T. { ˌməl·tə'plis·əd·ē }

multiplier If a number x is to be multiplied by a number y, then y is called the multiplier. { 'məl·təˌplī·ər }

multiply connected region An open set in the plane which has holes in it. { 'məl·tə·plē kə¦nek·təd 'rē·jən }

multiply perfect number An integer such that the sum of all its factors is a multiple of the integer itself. { ¦məl·tə·plē ¦pər·fikt 'nəm·bər }

N

nabla *See* del operator. { ˈnab·lə }

Nakayama's lemma The proposition that, if R is a commutative ring, I is an ideal contained in all maximal ideals of R, and M is a finitely generated module over R, and if $IM = M$, where IM denotes the set of all elements of the form am with a in I and m in M, then $M = 0$. { ˌnä·käˌyä·mäz ˈlem·ə }

NAND A logic operator having the characteristic that if P, Q, R, ... are statements, then the NAND of P, Q, R, ... is true if at least one statement is false, false if all statements are true. Derived from NOT-AND. Also known as sheffer stroke. { nand }

nano- A prefix representing 10^{-9}, which is 0.000000001 or one-billionth of the unit adjoined. Also known as milli-micro- (deprecated usage). { ˈnan·ō }

Naperian logarithm *See* logarithm. { nāˈpir·ē·ən ˈläg·əˌrith·əm }

Napier's analogies Formulas which enable one to study the relationships between the sides and the angles of a spherical triangle. { ˈnā·pē·ərz əˈnal·ə·jēz }

Napier's rules Two rules which give the formulas necessary in the solution of right spherical triangles. { ˈnā·pē·ərz ˌrülz }

nappe One of the two parts of a conical surface defined by the vertex. { nap }

***n*-ary composition** A function that associates an element of a set with every sequence of n elements of the set. { ˈen·ə·rē ˌkäm·pəˈzish·ən }

***n*-ary tree** A rooted tree in which each vertex has at most n successors. { ˈen·ə·rē ˌtrē }

natural boundary Those points of the boundary of a region where an analytic function is defined through which the function cannot be continued analytically. { ˈnach·rəl ˈbaůn·drē }

natural function A trigonometric function, as opposed to its logarithm. { ˈnach·rəl ˈfəŋk·shən }

natural logarithm *See* logarithm. { ˈnach·rəl ˈläg·əˌrith·əm }

natural number One of the integers 1, 2, 3, ... { ˈnach·rəl ˈnəm·bər }

navel point *See* umbilical point. { ˈnā·vəl ˌpȯint }

***n*-cell** A set that is homeomorphic either with the set of points in n-dimensional euclidean space ($n = 1, 2, \ldots$) whose distance from the origin is less than unity, or with the set of points whose distance from the origin is less than or equal to unity. { 'en ˌsel }

n-connected graph A connected graph for which the removal of n points is required to disconnect the graph. { 'en kəˌnek·təd 'graf }

n-dimensional space A vector space whose basis has n vectors. { 'en di'men·shən·əl 'spās }

near ring An algebraic system with two binary operations called multiplication and addition; the system is a group (not necessarily commutative) relative to addition, and multiplication is associative, and is left-distributive with respect to addition, that is, $x(y + z) = xy + xz$ for any x, y, and z in the near ring. { ¦nir ¦riŋ }

negation The negation of a proposition P is a proposition which is true if and only if P is false; this is often written P. { nə'gā·shən }

negative angle The angle subtended by moving a ray in the clockwise direction. { 'neg·əd·iv 'aŋ·gəl }

negative integer The additive inverse of a positive integer relative to the additive group structure of the integers. { 'neg·əd·iv 'int·ə·jər }

negative part For a real-valued function f, this is the function, denoted f, for which $f(x) = f(x)$ if $f(x) \leq 0$ and $f(x) = 0$ if $f(x)0$. { 'neg·əd·iv 'pärt }

negative pedal **1.** The negative pedal of a curve with respect to a point O is the envelope of the line drawn through a point P of the curve perpendicular to OP. Also known as first negative pedal. **2.** Any curve that can be derived from a given curve by repeated application of the procedure specified in the first definition. { 'neg·əd·iv 'ped·əl }

negative series A series whose terms are all negative real numbers. { 'neg·əd·iv 'sirˌēz }

negative skewness Skewness in which the mean is smaller than the mode. { 'neg·əd·iv 'skü·nəs }

negative with respect to a measure A set A is negative with respect to a signed measure m if, for every measurable set B, the intersection of A and B, $A \cap B$, is measurable and $m(A \cap B) \leq 0$. { ¦neg·əd·iv with ri¦spekt tü ə 'mezh·ər }

neighborhood of a point A set in a topological space which contains an open set which contains the point; in euclidean space, an example of a neighborhood of a point is an open (without boundary) ball centered at that point. { 'nā·bərˌhu̇d əv ə 'pȯint }

Neil's parabola The graph of the equation $y = ax^{3/2}$, where a is a constant. { 'nēlz pə'rab·ə·lə }

nephroid An epicycloid for which the diameter of the fixed circle is two times the diameter of the rolling circle. { 'neˌfrȯid }

nephroid of Freeth *See* Freeth's nephroid. { 'neˌfrȯid əv 'frēth }

nested intervals A sequence of intervals, each of which is contained in the preceding interval. { 'nes·təd 'in·tər·vəlz }

nested sets A family of sets where, given any two of its sets, one is contained in the other. Also known as monotonic system of sets. { 'nes·təd 'sets }

net 1. A set whose members are indexed by elements from a directed set; this is a generalization of a sequence. **2.** A nondegenerate partial plane satisfying the parallel axiom. { net }

network The name given to a graph in applications in management and the engineering sciences; to each segment linking points in the graph, there is usually associated a direction and a capacity on the flow of some quantity. { 'net,wərk }

Neumann boundary condition The boundary condition imposed on the Neumann problem in potential theory. { 'nȯi,män 'bau̇n·drē kən,dish·ən }

Neumann function 1. One of a class of Bessel functions arising in the study of the solutions to Bessel's differential equation. **2.** A harmonic potential function in potential theory occurring in the study of Neumann's problem. { 'nȯi,män ,fəŋk·shən }

Neumann line The generalization of the concept of a line occurring in Neumann's study of continuous geometry. { 'nȯi,män ,līn }

Neumann problem The determination of a harmonic function within a finite region of three-dimensional space enclosed by a closed surface when the normal derivatives of the function on the surface are specified. { 'nȯi,män ,präb·ləm }

Neumann series *See* Liouville-Neumann series. { 'nȯi,män ,sir·ēz }

Newton-Cotes formulas Approximation formulas for the integral of a function along a small interval in terms of the values of the function and its derivatives. { 'nüt·ən 'kōts ,fȯr·myə·ləz }

Newton-Raphson formula If c is an approximate value of a root of the equation $f(x) = 0$, then a better approximation is the number $c - [f(c)/f'(c)]$. { 'nüt·ən 'raf·sən ,fȯr·myə·lə }

Newton's identities Relations between the coefficients of a polynomial and quantities s_k defined by $s_k = r_1^k + r_2^k + \cdots + r_n^k$, where $r_1, ..., r_n$ are the roots of the polynomial; for the polynomial $a_0x^n + a_1x^{n-1} + \cdots + a_n$, the relations are $a_0s_k + a_1s_{k-1} + \cdots + a_{k-1}s_1 + ka_k = 0$, for $k < n$, and $a_0s_k + a_1s_{k-1} + \cdots + a_ns_{k-n} = 0$, for $k > n$. { 'nüt·ənz ī'den·əd,ēz }

Newton's inequality For any set of n numbers ($n = 0, 1, 2, \ldots$), the inequity $p_{r-1}\,p_{r+1} \leqq p_r^2$, for $1 \leqq r < n$, where p_r is the average value of the terms constituting the rth elementary symmetric function of the numbers. { ¦nüt·ənz ,in·i'kwäl·əd·ē }

Newton's method A technique to approximate the roots of an equation by the methods of the calculus. { 'nüt·ənz ,meth·əd }

Newton's square-root method A technique for the estimation of the roots of an equation exhibiting faster convergence than Newton's method; this involves calculus methods and the square-root function. { 'nüt·ənz 'skwer ,rüt ,meth·əd }

nilmanifold The factor space of a connected nilpotent Lie group by a closed subgroup. { ¦nil'man·ə,fōld }

nilpotent An element of some algebraic system which vanishes when raised to a certain power. { ¦nil'pōt·ənt }

n-net A finite net in which n lines pass through each point. { 'en ˌnet }

node *See* crunode. { nōd }

Noetherian module A module in which every nested family of submodules possesses only a finite number of members. { ˌnō·əˈthir·ē·ən 'mäj·əl }

Noetherian ring A ring in which every nested family of ideals possesses only a finite number of members. { ˌnō·ə'thir·ē·ən'riŋ }

nomogram *See* nomograph. { 'näm·əˌgram }

nomograph A chart which represents an equation containing three variables by means of three scales so that a straight line cuts the three scales in values of the three variables satisfying the equation. Also known as abac; alignment chart; nomogram. { 'näm·əˌgraf }

nonagon A nine-sided polygon. { 'nän·əˌgän }

nonassociative algebra A generalization of the concept of an algebra; it is a nonassociative ring R which is a vector space over a field F satisfying $a(xy) = (ax)y = x(ay)$ for all a in F and x and y in R. { ˌnän·əˈsō·shəd·əv 'al·jə·brə }

nonassociative ring A generalization of the concept of a ring; it is an algebraic system with two binary operations called addition and multiplication such that the system is a commutative group relative to addition, and multiplication is distributive with respect to addition, but multiplication is not assumed to be associative. { ˌnän·əˈsō·shəd·əv 'riŋ }

nonatomic Boolean algebra A Boolean algebra in which there is no element x with the property that if $y \cdot x = y$ for some y, then $y = 0$. { ˈnän·ə'täm·ik ˈbü·lē·ən 'al·jə·brə }

nonatomic measure space A measure space in which no point has positive measure. { ˈnän·ə'täm·ik ˈmezh·ər ˌspās }

noncentral quadric A quadric surface that does not have a point about which the surface is symmetrical; namely, an elliptic or hyperbolic paraboloid, or a quadric cylinder. { 'nänˌsen·trəl 'kwäˌdrik }

nondegenerate plane In projective geometry, a plane in which to every line L there are at least two distinct points that do not lie on L, and to every point p there are at least two distinct lines which do not pass through p. { ˈnän·di'jen·ə·rət 'plān }

nondenumerable set A set that cannot be put into one-to-one correspondence with the positive integers or any subset of the positive integers. { ˌnän·diˈnüm·rə·bəl 'set }

nondifferentiable programming The branch of nonlinear programming which does not require the objective and constraint functions to be differentiable. { ˈnänˌdif·ə'ren·chə·bəl 'prōˌgram·iŋ }

nondimensional parameter *See* dimensionless number. { ˈnän·di'men·chən·əl pə'ram·əd·ər }

noneuclidean geometry A geometry in which one or more of the axioms of euclidean geometry are modified or discarded. { ˈnän·yü'klid·ē·ən jē'äm·ə·trē }

nonholonomic constraint One of a nonintegrable set of differential equations which describe the restrictions on the motion of a system. { ¦nän‚häl·ə'näm·ik kən 'stränt }

nonlinear equation An equation in variables $x_1, ..., x_n, y$ which cannot be put into the form $a_1x_1 + \cdots + a_nx_n = y$. { 'nän‚lin·ē·ər i'kwā·zhən }

nonlinear programming A branch of applied mathematics concerned with finding the maximum or minimum of a function of several variables, when the variables are constrained to yield values of other functions lying in a certain range, and either the function to be maximized or minimized, or at least one of the functions whose value is constrained, is nonlinear. { 'nän‚lin·ē·ər 'prō‚gram·iŋ }

nonlinear system A system in which the interrelationships among the quantities involved are expressed by equations, some of which are not linear. { 'nän‚lin·ē·ər 'sis·təm }

nonorientable surface *See* one-sided surface. { ‚nän‚ȯr·ē¦ent·ə·bəl 'sər·fəs }

nonremovable discontinuity A point at which a function is not continuous or is undefined, and cannot be made continuous by being given a new value at the point. { ¦nän·ri'müv·ə·bəl dis‚känt·ən'ü·əd·ē }

nonsingular matrix A matrix which has an inverse; equivalently, its determinant is not zero. { 'nän‚siŋ·gyə·lər 'mā·triks }

nonsingular transformation A linear transformation which has an inverse; equivalently, it has null space kernel consisting only of the zero vector. { 'nän‚siŋ·gyə·lər ‚tranz·fər'mā·shən }

nonterminal vertex A vertex in a rooted tree that has at least one successor. { ¦nän'tər·mən·əl 'vər‚teks }

nonterminating continued fraction A continued fraction that has an infinite number of terms. { ‚nän¦tər·mə‚nād·iŋ kən¦tin·yüd 'frak·shən }

nonterminating decimal A decimal for which there is no digit to the right of the decimal point such that all digits farther to the right are zero. { ‚nän¦tər·mə‚nād·iŋ 'des·məl }

nontrivial solution A solution of a set of homogeneous linear equations in which at least one of the variables has a value different from zero. { ¦nän'triv·ē·əl sə'lü·shən }

NOR A logic operator having the property that if P, Q, R, ... are statements, then the NOR of P, Q, R, ... is true if all statements are false, false if at least one statement is true. Derived from NOT-OR. Also known as Peirce stroke relationship. { nȯr }

norm **1.** A scalar valued function on a vector space with properties analogous to those of the modulus of a complex number; namely: the norm of the zero vector is zero, all other vectors have positive norm, the norm of a scalar times a vector equals the absolute value of the scalar times the norm of the vector, and the norm of a sum is less than or equal to the sum of the norms. **2.** For a matrix, the square root of the sum of the squares of the moduli of the matrix entries. **3.** *See* absolute value. { nȯrm }

normal bundle If A is a manifold and B is a submanifold of A, then the normal bundle of B in A is the set of pairs (x,y), where x is in B, y is a tangent vector to A, and y is orthogonal to B. { ¦nȯr·məl ¦bən·dəl }

normal curvature The normal curvature at a point on a surface is the curvature of the normal section to the point. { 'nȯr·məl 'kər·və·chər }

normal derivative The directional derivative of a function at a point on a given curve or surface in the direction of the normal to the curve or surface. { 'nȯr·məl di 'riv·əd·iv }

normal divisor *See* normal subgroup. { 'nȯr·məl di'vīz·ər }

normal extension An algebraic extension of K of a field k, contained in the algebraic closure $\bar{k}$ of k, such that every injective homomorphism of K into $\bar{k}$, inducing the identity on k, is an automorphism of K. { 'nȯr·məl ik'sten·chən }

normal family A family of complex functions analytic in a common domain where every sequence of these functions has a subsequence converging uniformly on compact subsets of the domain to an analytic function on the domain or to $+\infty$. { 'nȯr·məl 'fam·lē }

normal function *See* normalized function. { 'nȯr·məl 'fəŋk·shən }

normalize To multiply a quantity by a suitable constant or scalar so that it then has norm one; that is, its norm is then equal to one. { 'nȯr·mə͵līz }

normalized function A function with norm one; the norm is usually given by an integral $(\int |f|^p d\mu)^{1/p}$, $1 \leq p < \infty$. Also known as normal function. { 'nȯr·mə͵līzd 'fəŋk·shən }

normalized support function The function that results from restricting the domain of the independent variable of the support function to the unit sphere. { ¦nȯr·mə͵līzd sə'pȯrt ͵fəŋk·shən }

normalizer The normalizer of a subset S of a group G is the subgroup of G consisting of all elements x such that xsx^{-1} is in S whenever s is in S. { 'nȯr·mə͵līz·ər }

normal map A planar map in which no more than three regions meet any one point and no region completely encloses another. Also known as regular map. { 'nȯr·məl 'map }

normal matrix A matrix is normal if multiplying it on the right by its adjoint is the same as multiplying it on the left. { 'nȯr·məl 'mā͵triks }

normal number A number whose expansion with respect to a given base (not necessarily 10) is such that all the digits occur with equal frequency, and all blocks of digits of the same length occur equally often. { 'nȯr·məl 'nəm·bər }

normal operator A linear operator where composing it with its adjoint operator in either order gives the same result. Also known as normal transformation. { 'nȯr·məl 'äp·ə͵rād·ər }

normal pedal curve The normal pedal curve of a given curve C with respect to a fixed point P is the locus of the foot of the perpendicular from P to the normal to C. { 'nȯr·məl 'ped·əl ͵kərv }

normal plane For a point P on a curve in space, the plane passing through P which is perpendicular to the tangent to the curve at P. { 'nȯr·məl 'plān }

normal section Relative to a surface, this is a planar section produced by a plane containing the normal to a point. { 'nȯr·məl 'sek·shən }

normal series A normal series of a group G is a normal tower of subgroups of G, G_0, $G_1, \ldots, G_n$ in which $G_0 = G$ and G_n is the trivial group containing only the identity element. { 'nȯr·məl 'sir‚ēz }

normal space A topological space in which any two disjoint closed sets may be covered respectively by two disjoint open sets. { 'nȯr·məl 'spās }

normal subgroup A subgroup N of a group G where every expression $g^{-1}ng$ is in N for every g in G and every n in N. Also known as invariant subgroup; normal divisor. { 'nȯr·məl 'səb‚grüp }

normal to a curve The normal to a curve at a point is the line perpendicular to the tangent line at the point. { 'nȯr·məl tü ə 'kərv }

normal to a surface The normal to a surface at a point is the line perpendicular to the tangent plane at that point. { 'nȯr·məl tü ə 'sər·fəs }

normal tower A tower of subgroups, $G_0, G_1, \ldots, G_n$, such that each G_{i+1} is normal in G_i, $i = 1, 2, \ldots, n-1$. { 'nȯr·məl 'tau̇·ər }

normal transformation *See* normal operator. { 'nȯr·məl ‚tranz·fər'mā·shən }

normed linear space A vector space which has a norm defined on it. Also known as normed vector space. { 'nȯrmd 'lin·ē·ər 'spās }

normed vector space *See* normed linear space. { 'nȯrmd 'vek·tər 'spās }

notation **1.** The use of symbols to denote quantities or operations. **2.** *See* positional notation. { nō'tā·shən }

NOT function A logical operator having the property that if P is a statement, then the NOT of P is true if P is false, and false if P is true. { 'nät ‚fəŋk·shən }

nowhere dense set A set in a topological space whose closure has empty interior. { 'nō‚wer 'dens 'set }

***n* space** A vector space over the real numbers whose basis has n vectors. { 'en ‚spās }

***n*-sphere** The set of all points in $(n+1)$-dimensional euclidean space whose distance from the origin is unity, where n is a positive integer. { 'en ‚sfir }

null Indicating that an object is nonexistent or a quantity is zero. { nəl }

nullary composition The selection of a particular element of a set. { 'nəl·ə·rē ‚käm·pə'zish·ən }

null geodesic In a Riemannian space, a minimal geodesic curve. { 'nəl ‚jē·ə'des·ik }

nullity The dimension of the null space of a linear transformation. { 'nəl·əd·ē }

null matrix The matrix all of whose entries are zero. { 'nəl 'mā·triks }

null sequence A sequence of numbers or functions which converges to the number zero or the zero function. { 'nəl 'sē·kwəns }

null set The empty set; the set which contains no elements. { 'nəl 'set }

null space For a linear transformation, the vector subspace of all vectors which the transformation sends to the zero vector. Also known as kernel. { 'nəl 'spās }

null vector A vector whose invariant length, that is, the sum over the coordinates of the vector space of the product of its covariant component and contravariant component, is equal to zero. { 'nəl 'vek·tər }

number 1. Any real or complex number. 2. The number of elements in a set is the cardinality of the set. { 'nəm·bər }

number class modulo *N* The class of all numbers which differ from a given number by a multiple of *N*. { 'nəm·bər ¦klas ¦mäj·ə·lō 'en }

number field Any set of real or complex numbers that includes the sum, difference, product, and quotient (except division by zero) of any two members of the set. { 'nəm·bər ˌfēld }

number line *See* real line. { 'nəm·bər ˌlīn }

number scale Representation of points on a line with numbers arranged in some order. { 'nəm·bər ˌskāl }

number theory The study of integers and relations between them. { 'nəm·bər 'thē·ə·rē }

numeral A symbol used to denote a number. { 'nüm·rəl }

numeral system *See* numeration system. { 'nüm·rəl ˌsis·təm }

numeration The listing of numbers in their natural order. { ˌnü·mə'rā·shən }

numeration system An orderly method of representing numbers by numerals in which each numeral is associated with a unique number. Also known as numeral system. { ˌnü·mə'rā·shən ˌsis·təm }

numerator In a fraction a/b, the numerator is the quantity a. { 'nü·məˌrād·ər }

numerical Pertaining to numbers. { nü'mer·i·kəl }

numerical analysis The study of approximation techniques using arithmetic for solutions of mathematical problems. { nü'mer·i·kəl ə'nal·ə·səs }

numerical equation An equation all of whose constants and coefficients are numbers. { nü'mer·i·kəl i'kwā·zhən }

numerical integration The process of using a set of approximate values of a function to calculate its integral to comparable accuracy. { nü'mer·i·kəl ˌint·ə'grā·shən }

numerical range For a linear operator T of a Hilbert space into itself, the set of values assumed by the inner product of Tx with x as x ranges over the set of vectors with norm equal to 1. { nü'mer·i·kəl 'rānj }

numerical tensor A tensor whose components are the same in all coordinate systems. { nü'mer·i·kəl 'ten·sər }

numerical value *See* absolute value. { nü'mer·i·kəl 'val·yü }

numeric character *See* digit. { nü'mer·ik 'kar·ik·tər }

O

obelisk A frustrum of a regular, rectangular pyramid. { 'äb·ə,lisk }

objective function In nonlinear programming, the function, expressing given conditions for a system, which one seeks to minimize subject to given constraints. { äb'jek·tiv 'fəŋk·shən }

oblate ellipsoid *See* oblate spheroid. { 'ä,blāt i'lip,sȯid }

oblate spheroid The surface or ellipsoid generated by rotating an ellipse about one of its axes so that the diameter of its equatorial circle exceeds the length of the axis of revolution. Also known as oblate ellipsoid. { 'ä,blāt 'sfir,ȯid }

oblate spheroidal coordinate system A three-dimensional coordinate system whose coordinate surfaces are the surfaces generated by rotating a plane containing a system of confocal ellipses and hyperbolas about the minor axis of the ellipses, together with the planes passing through the axis of rotation. { ¦ō,blāt sfir¦ȯid·əl kō'ȯrd·ən·ət ,sis·təm }

oblique angle An angle that is neither a right angle nor a multiple of a right angle. { ə'blēk 'aŋ·gəl }

oblique circular cone A circular cone whose axis is not perpendicular to its base. { ə¦blēk ¦sər·kyə·lər 'kōn }

oblique coordinates Magnitudes defining a point relative to two intersecting nonperpendicular lines, called axes; the magnitudes indicate the distance from each axis, measured along a parallel to the other axis; oblique coordinates are a form of cartesian coordinates. { ə'blēk kō'ȯrd·ən·əts }

oblique lines Lines that are neither perpendicular nor parallel. { ə'blēk 'līnz }

oblique parallelepiped A parallelepiped whose lateral edges are not perpendicular to its bases. { ə¦blēk ,par·ə,lel·ə'pī,ped }

oblique spherical triangle A spherical triangle that has no right angle. { ə¦blēk ¦sfer·ə·kəl 'trī,aŋ·gəl }

oblique strophoid A plane curve derived from a straight line L and two points called the pole and the fixed point, where the fixed point lies on L but is not the foot of the perpendicular from the pole to the line; it consists of the locus of points on a rotating line L' passing through the pole whose distance from the intersection of L and L' is equal to the distance of this intersection from the fixed point. { ə'blēk 'strō,fȯid }

oblique triangle A triangle that does not contain a right angle. { ə'blēk 'trī‚aŋ·gəl }

obtuse angle An angle of more than 90° and less than 180°. { äb'tüs 'aŋ·gəl }

obtuse triangle A triangle having one obtuse angle. { äb'tüs 'trī‚aŋ·gəl }

octagon A polygon with eight sides. { 'äk·tə‚gän }

octahedron A polyhedron having eight faces, each of which is an equilateral triangle. { ‚ak·tə'hē·drən }

octal Pertaining to the octal number system. { 'äkt·əl }

octal digit The symbol 0, 1, 2, 3, 4, 5, 6, or 7 used as a digit in the octal number system. { 'äkt·əl 'dij·ət }

octal number system A number system in which a number r is written as $n_k n_{k-1} \ldots n_1$ where $r = n_1 8^0 + n_2 8^1 + \cdots + n_k 8^{k-1}$. { 'äkt·əl 'nəm·bər ‚sis·təm }

octant **1.** One of the eight regions into which three-dimensional euclidean space is divided by the coordinate planes of a cartesian coordinate system. **2.** A unit of plane angle equal to 45° or $\pi/8$ radians. { 'äk·tənt }

octillion **1.** The number 10^{27}. **2.** In British and German usage, the number 10^{48}. { äk'til·yən }

octonions *See* Cayley numbers. { äk'tän·yənz }

odd function A function $f(x)$ is odd if, for every x, $f(-x) = -f(x)$. { 'äd ‚fəŋk·shən }

odd number A natural number not divisible by 2. { 'äd 'nəm·bər }

odd permutation A permutation that may be represented as the result of an odd number of transpositions. { 'äd ‚pər·myə'tā·shən }

one-dimensional strain A transformation that elongates or compresses a configuration in a given direction, given by $x' = kx$, $y' = y$, $z' = z$, where k is a constant, when the direction is that of the x axis. { 'wən di‚men·chən·əl 'strān }

one-parameter semigroup A semigroup with which there is associated a bijective mapping from the positive real numbers onto the semigroup. { ¦wən pə¦ram·əd·ər 'sem·i‚grüp }

one-point compactification The one-point compactification $\bar{X}$ of a topological space X is the union of X with a set consisting of a single element, with the topology of $\bar{X}$ consisting of the open subsets of X and all subsets of $\bar{X}$ whose complements in $\bar{X}$ are closed compact subsets in X. Also known as Alexandroff compactification. { 'wən ‚pȯint kəm‚pak·tə·fə'kā·shən }

one-sided limit Either a limit on the left or a limit on the right. { 'wən ‚sīd·əd 'lim·ət }

one-sided surface A surface such that an object resting on one side can be moved continuously over the surface to reach the other side without going around an edge; the Möbius band and the Klein bottle are examples. Also known as nonorientable surface. { 'wən ‚sīd·əd 'sər·fəs }

one-to-one correspondence A pairing between two classes of elements whereby each element of either class is made to correspond to one and only one element of the other class. { ¦wən tə ¦wən ˌkär·ə'spän·dəns }

one-to-one mapping *See* injection. { ¦wən tə ¦wən 'map·iŋ }

open ball In a metric space, an open set about a point x which consists of all points within a fixed distance from x. { 'ō·pən 'bȯl }

open covering For a set S in a topological space, a collection of open sets whose union contains S. { 'ō·pən 'kəv·ər·iŋ }

open interval An open interval of real numbers, denoted by (a,b), consists of all numbers strictly greater than a and strictly less than b. { 'ō·pən 'in·tər·vəl }

open map A function between two topological spaces which sends each open set of one to an open set of the other. { 'ō·pən 'map }

open mapping theorem A continuous linear function between Banach spaces which has closed range must be an open map. { 'ō·pən 'map·iŋ ˌthir·əm }

open *n*-cell A set that is homeomorphic with the set of points in n-dimensional euclidean space ($n = 1, 2, \ldots$) whose distance from the origin is less than unity. { ¦ō·pən 'en ˌsel }

open set A set included in a topology; equivalently, a set which is a neighborhood of each of its points; a topology on a space is determined by a collection of subsets which are called open. { 'ō·pən ˌset }

open simplex A modification of a simplex based on points $p_0, p_1, \ldots, p_n$, in which the points $a_0p_0 + \cdots + a_np_n$ are excluded if one or more of the a_i are zero. { 'ō·pən 'simˌpleks }

operation An operation of a group G on a set S is a mapping which associates to each ordered pair (g,s), where g is in G and s is in S, another element in S, denoted gs, such that, for any g,h in G and s in S, $(gh)s = g(hs)$, and $es = s$, where e is the identity element of G. { ˌäp·ə'rā·shən }

operational analysis *See* operational calculus. { ˌäp·ə'rā·shən·əl ə'nal·ə·səs }

operational calculus A technique by which problems in analysis, in particular differential equations, are transformed into algebraic problems, usually the problem of solving a polynomial equation. Also known as operational analysis. { ˌäp·ə'rā·shən·əl 'kal·kyə·ləs }

operations research The mathematical study of systems with input and output from the viewpoint of optimization subject to given constraints. { ˌäp·ə'rā·shənz riˌsərch }

operator A function between vector spaces. { 'äp·əˌrād·ər }

operator algebra An algebra whose elements are functions and in which the multiplication of two elements f and g is defined by composition; that is, $(fg)(x) = (f \circ g)(x) = f[g(x)]$. { 'äp·əˌrād·ər ˌal·jə·brə }

operator theory The general qualitative study of operators in terms of such concepts as eigenvalues, range, domain, and continuity. { 'äp·əˌrād·ər ˌthē·ə·rē }

opposite side 1. One of two sides of a polygon with an even number of sides that have the same number of sides between them along either path around the polygon from one of the sides to the other. **2.** For a given vertex of a polygon with an odd number of sides, a side of the polygon that has the same number of sides between it and the vertex along either path around the polygon. { 'äp·ə·zət 'sīd }

opposite vertices Two vertices of a polygon with an even number of sides that have the same number of sides between them along either path around the polygon from one vertex to the other. { 'äp·ə·zət 'vərd·əˌsēz }

optimal policy In optimization problems of systems, a sequence of decisions changing the states of a system in such a manner that a given criterion function is minimized. { 'äp·tə·məl 'päl·ə·sē }

optimal strategy One of the pair of mixed strategies carried out by the two players of a matrix game when each player adjusts strategy so as to minimize the maximum loss that an opponent can inflict. { 'äp·tə·məl 'strad·ə·jē }

optimal system A system where the variables representing the various states are so determined that a given criterion function is minimized subject to given constraints. { 'äp·tə·məl 'sis·təm }

optimization The maximizing or minimizing of a given function possibly subject to some type of constraints. { ˌäp·tə·mə'zā·shən }

optimization theory The specific methodology, techniques, and procedures used to decide on the one specific solution in a defined set of possible alternatives that will best satisfy a selected criterion; includes linear and nonlinear programming, stochastic programming, and control theory. Also known as mathematical programming. { ˌäp·tə·mə'zā·shən ˌthē·ə·rē }

or A logical operation whose result is false (or zero) only if every one of its operands is false, and true (or one) otherwise. Also known as inclusive or. { ȯr }

orbit Let G be a group which operates on a set S; the orbit of an element s of S under G is the subset of S consisting of all elements gs where g is in G. { 'ȯr·bət }

orbit space The orbit space of a G space X is the topological space whose points are equivalence classes obtained by identifying points in X which have the same G orbit and whose topology is the largest topology that makes the function which sends x to its orbit continuous. { 'ȯr·bət ˌspās }

order 1. A differential equation has order n if the derivatives of a function appear up to the nth derivative. **2.** The number of elements contained within a given group. **3.** A square matrix with n rows and n columns has order n. **4.** The number of poles a given elliptic function has in a parallelogram region where it repeats its values. **5.** A characteristic of infinitesimals used in their comparison. **6.** For a polynomial, the largest exponent appearing in the polynomial. { 'ȯrd·ər }

ordered field A field with an ordering as a set analogous to the properties of less than or equal for real numbers relative to addition and multiplication. { 'ȯrd·ərd 'fēld }

ordered pair A pair of elements x and y from a set, written (x,y), where x is distinguished as first and y as second. { 'ȯrd·ərd 'per }

ordered rings Rings which have an ordering on them as sets in a manner analogous to the behavior of the usual ordering of the real numbers relative to addition and multiplication. { 'ȯrd·ərd 'riŋz }

ordering A binary relation, denoted $\leq$, among the elements of a set such that $a \leq b$ and $b \leq c$ implies $a \leq c$, and $a \leq b$, $b \leq a$ implies $a = b$; it need not be the case that either $a \leq b$ or $b \leq a$. Also known as order relation; partial ordering. { 'ȯrd·ə·riŋ }

order of degeneracy *See* degree of degeneracy. { 'ȯrd·ər əv di'jen·ə·rə·sē }

order relation *See* ordering. { 'ȯrd·ər ri'lā·shən }

ordinal number A generalized number which expresses the size of a set, in the sense of "how many" elements. { 'ȯrd·nəl 'nəm·bər }

ordinary differential equation An equation involving functions of one variable and their derivatives. { 'ȯrd·ən‚er·ē ‚dif·ə'ren·chəl i'kwā·zhən }

ordinary point A point of a curve where a curve does not cross itself and where there is a smoothly turning tangent. { 'ȯrd·ən‚er·ē 'pȯint }

ordinary singular point A singular point at which the tangents to all branches at the point are distinct. { ¦ȯrd·ən‚er·ē ¦siŋ·gyə·lər 'pȯint }

ordinate The perpendicular distance of a point (x,y) of the plane from the x axis. { 'ȯrd·ən·ət }

orientable surface A surface for which an object resting on one side of it cannot be moved continuously over it to get to the other side without going around an edge. { ‚ȯr·ē‚en·tə·bəl 'sər·fəs }

orientation **1.** A choice of sense or direction in a topological space. **2.** An ordering $p_0, p_1, \ldots, p_n$ of the vertices of a simplex, two such orderings being regarded as equivalent if they differ by an even permutation. { ‚ȯr·ē·ən'tā·shən }

oriented graph A directed graph in which there is no pair of points a and b such that there is both an arc directed from a to b and an arc directed from b to a. { 'ȯr·ē·ent·əd 'graf }

origin The point of a coordinate system at which all coordinate axes meet. { 'är·ə·jən }

orthocenter The point at which the altitudes of a triangle intersect. { ¦ȯr·thō'sen·tər }

orthogonal Perpendicular, or some concept analogous to it. { ȯr'thäg·ən·əl }

orthogonal basis A basis for an inner product space consisting of mutually orthogonal vectors. { ȯr'thäg·ən·əl 'bā·səs }

orthogonal complement In an inner product space, the orthogonal complement of a vector $\mathbf{v}$ consists of all vectors orthogonal to $\mathbf{v}$; the orthogonal complement of a subset S consists of all vectors orthogonal to each vector in S. { ȯr'thäg·ən·əl 'käm·plə·mənt }

orthogonal family *See* orthogonal system. { ȯr'thäg·ən·əl'fam·lē }

orthogonal functions Two real-valued functions are orthogonal if their inner product vanishes. { ȯr'thäg·ən·əl 'fəŋk·shənz }

orthogonal group The group of matrices arising from the orthogonal transformations of a euclidean space. { ȯr'thäg·ən·əl 'grüp }

orthogonality Two geometric objects have this property if they are perpendicular. { ȯr‚thäg·ə'nal·əd·ē }

orthogonalization A procedure in which, given a set of linearly independent vectors in an inner product space, a set of orthogonal vectors is recursively obtained so that each set spans the same subspace. { ȯr‚thäg·ə·nə·lə'zā·shən }

orthogonal Latin square One of two $n \times n$ Latin squares which, when superposed, have the property that the n^2 cells contain each of the n^2 possible pairs of symbols exactly once. { ȯr¦thäg·ən·əl ¦lat·ən‚skwer }

orthogonal lines Lines which are perpendicular. { ȯr'thäg·ən·əl 'līnz }

orthogonal matrix A matrix whose inverse and transpose are identical. { ȯr'thäg·ən·əl 'mā·triks }

orthogonal polynomial Orthogonal polynomials are various families of polynomials, which arise as solutions to differential equations related to the hypergeometric equation, and which are mutually orthogonal as functions. { ȯr'thäg·ən·əl päl·ə'nō·mē·əl }

orthogonal projection Also known as orthographic projection. **1.** A continuous linear map P of a Hilbert space H onto a subspace M such that if h is any vector in H, $h = Ph + w$, where w is in the orthogonal complement of M. **2.** A mapping of a configuration into a line or plane that associates to any point of the configuration the intersection with the line or plane of the line passing through the point and perpendicular to the line or plane. { ȯr'thäg·ən·əl prə'jek·shən }

orthogonal series An infinite series each term of which is the product of a member of an orthogonal family of functions and a coefficient; the coefficients are usually chosen so that the series converges to a desired function. { ȯr¦thäg·ən·əl 'sir‚ēz }

orthogonal spaces Two subspaces F and F' of a vector space E with a scalar product g such that $g(x,x') = 0$ for any x in F and x' in F'. { ȯr¦thäg·ən·əl 'spās·əz }

orthogonal sum **1.** A vector space E with a scalar product is said to be the orthogonal sum of subspaces F and F' if E is the direct sum of F and F' and if F and F' are orthogonal spaces. **2.** A scalar product g on a vector space E is said to be the orthogonal sum of scalar products f and f' on subspaces F and F' if E is the orthogonal sum of F and F' (in the sense of the first definition) and if $g(x + x', y + y') = f(x,y) + f'(x',y')$ for all x,y in F and x',y' in F'. { ȯr¦thäg·ən·əl 'səm }

orthogonal system **1.** A system made up of n families of curves on an n-dimensional manifold in an $n + 1$ dimensional euclidean space, such that exactly one curve from each family passes through every point in the manifold, and, at each point, the tangents to the n curves that pass through that point are mutually perpendicular. **2.** A set of real-valued functions, the inner products of any two of which vanish. Also known as orthogonal family. { ȯr'thäg·ən·əl 'sis·təm }

orthogonal trajectory A curve that intersects all the curves of a given family at right angles. { ȯr'thäg·ən·əl trə'jek·tə·rē }

orthogonal transformation A linear transformation between real inner product spaces which preserves the length of vectors. { ȯr'thäg·ən·əl ‚tranz‚fər'mā·shən }

orthogonal vectors In an inner product space, two vectors are orthogonal if their inner product vanishes. { ȯr'thäg·ən·əl 'vek·tərz }

orthographic projection *See* orthogonal projection. { ¦ȯr·thə¦graf·ik prə'jek·shən }

orthonormal coordinates In an inner product space, the coordinates for a vector expressed relative to an orthonormal basis. { ¦ȯr·thə¦nȯr·məl kō'ȯrd·ən·əts }

orthonormal functions Orthogonal functions f_1, f_2, ... with the additional property that the inner product of $f_n(x)$ with itself is 1. { ¦ȯr·thə¦nȯr·məl 'fəŋk·shənz }

orthonormal vectors A collection of mutually orthogonal vectors, each having length 1. { ¦ȯr·thə¦nȯr·məl 'vek·tərz }

orthoptic The locus of the intersection of tangents to a given curve that meet at a right angle. { ȯr'thäp·tik }

orthotomic The orthotomic of a curve with respect to a point is the envelope of the circles which pass through the point and whose centers lie on the curve. { ˌȯr·thə'täm·ik }

oscillating series A series that is divergent but not properly divergent; that is, the partial sums do not approach a limit, or become arbitrarily large or arbitrarily small. { 'äs·əˌlād·iŋ ˌsirˌēz }

oscillation **1.** The oscillation of a real-valued function on an interval is the difference between its least upper bound and greatest lower bound there. **2.** The oscillation of a real-valued function at a point x is the limit of the oscillation of the function on the interval $[x - e, x + e]$ as e approaches 0. Also known as saltus. { ˌäs·ə'lā·shən }

osculating circle For a plane curve C at a point p, the limiting circle obtained by taking the circle that is tangent to C at p and passes through a variable point q on C, and then letting q approach p. { ¦äs·kyəˌlād·iŋ 'sər·kəl }

osculating plane For a curve C at some point p, this is the limiting plane obtained from taking planes through the tangent to C at p and containing some variable point p' and then letting p' approach p along C. { 'äs·kyəˌlād·iŋ 'plān }

osculating sphere For a curve C at a point p, the limiting sphere obtained by taking the sphere that passes through p and three other points on C and then letting these three points approach p independently along C. { ¦äs·kyəˌlād·iŋ 'sfir }

outer automorphism Any element of the quotient group formed from the group of automorphisms of a group and the subgroup of inner automorphisms. { 'au̇d·ər ¦ȯd·ō'mȯrˌfiz·əm }

outer measure A function with the same properties as a measure except that it is only countably subadditive rather than countably additive; usually defined on the collection of all subsets of a given set. { 'au̇d·ər 'mezh·ər }

outer product For any two tensors R and S, a tensor T each of whose indices corresponds to an index of R or an index of S, and each of whose components is the product of the component of R and the component of S with identical values of the corresponding indices. { ¦au̇d·ər 'präd·əkt }

oval A curve shaped like a section of an egg. { 'ō·vəl }

oval of Cassini An ovallike curve similar to a lemniscate obtained as the locus corresponding to a general type of quadratic equation in two variables x and y; it is expressed as $[(x + a)^2 + y^2]\,[(x - a)^2 + y^2] = k^4$, where a and k are constants. Also known as Cassinian oval. { 'ō·vəl əv kə'sē·nē }

over a map A map f from a set A to a set L is said to be over a map g from a set B to L if B is a subset of A and the restriction of f to B equals g. { ¦ō·vər ə 'map }

over a set A map f from a set A to a set L is said to be over a set B if B is a subset of both A and L and if the restriction of F to B is the identity map on B. { ¦ō·vər ə 'set }

P

p *See* pico-.

Pade table A table associated to a power series having in its pth row and qth column the ratio of a polynomial of degree q by one of degree p so that this fraction expanded into a power series agrees with the original up to the $p + q$ term. { 'päd·ə ˌtā·bəl }

Pappian plane Any projective plane in which points and lines satisfy Pappus' theorem (third definition). { ¦pap·ē·ən 'plān }

Pappus' theorem **1.** The proposition that the area of a surface of revolution generated by rotating a plane curve about an axis in its own plane which does not intersect it is equal to the length of the curve multiplied by the length of the path of its centroid. **2.** The proposition that the volume of a solid of revolution generated by rotating a plane area about an axis in its own plane which does not intersect it is equal to the area multiplied by the length of the path of its centroid. **3.** A theorem of projective geometry which states that if A, B, and C are collinear points and A', B' and C' are also collinear points, then the intersection of AB' with $A'B$, the intersection of AC' with $A'C$, and the intersection of BC' with $B'C$ are collinear. **4.** A theorem of projective geometry which states that if A, B, C, and D are fixed points on a conic and P is a variable point on the same conic, then the product of the perpendiculars from P to AB and CD divided by the product of the perpendiculars from P to AD and BC is constant. { 'pap·əs ˌthir·əm }

parabola The plane curve given by an equation of the form $y = ax^2 + bx + c$. { pə'rab·ə·lə }

parabolic coordinate system **1.** A two-dimensional coordinate system determined by a system of confocal parabolas. **2.** A three-dimensional coordinate system whose coordinate surfaces are the surfaces generated by rotating a plane containing a system of confocal parabolas about the axis of symmetry of the parabolas, together with the planes passing through the axis of rotation. { ¦par·ə¦bäl·ik kō'ȯrd·ən·ət ˌsis·təm }

parabolic cylinder A cylinder whose directrix is a parabola. { ¦par·ə¦bäl·ik 'sil·ən·dər }

parabolic cylinder functions Solutions to the Weber differential equation, which results from separation of variables of the Laplace equation in parabolic cylindrical coordinates. { ¦par·ə¦bäl·ik 'sil·ən·dər ˌfəŋk·shənz }

parabolic cylindrical coordinate system A three-dimensional coordinate system in which two of the coordinates depend on the x and y coordinates in the same manner as parabolic coordinates and are independent of the z coordinate, while the third coordinate is directly proportional to the z coordinate. { ¦par·ə¦bäl·ik si¦lin·drə·kəl kō'ȯrd·ən·ət ˌsis·təm }

parabolic differential equation A general type of second-order partial differential equation which includes the heat equation and has the form

$$\sum_{ij=1}^{n} A_{ij}(\partial^2 u/\partial x_i \partial x_j) + \sum_{i=1}^{n} B_i(\partial u/\partial x_i) + Cu + F = 0$$

where the A_{ij}, B_i, C, and F are suitably differentiable real functions of $x_1, x_2, ..., x_n$, and there exists at each point $(x_1, ..., x_n)$ a real linear transformation on the x_i which reduces the quadratic form

$$\sum_{ij=1}^{n} A_{ij} x_i x_j$$

to a sum of $n - 1$ squares all of which have the same sign, while the same transformation does not reduce the B_i to 0. Also known as parabolic partial differential equation. { ¦par·ə¦bäl·ik ˌdif·ə'ren·chəl i'kwā·zhən }

parabolic partial differential equation *See* parabolic differential equation. { ¦par·ə¦bäl·ik ¦pär·shəl ˌdif·ə'ren·chəl iˌkwā·zhən }

parabolic point A point on a surface where the total curvature vanishes. { ¦par·ə¦bäl·ik 'pȯint }

parabolic segment The line segment given by a chord perpendicular to the axis of a parabola. { ¦par·ə¦bäl·ik 'seg·mənt }

parabolic spiral The curve whose equation in polar coordinates is $r^2 = a\theta$. { ¦par·ə¦bäl·ik 'spī·rəl }

parabolic type A type of simply connected Riemann surface that can be mapped conformally on the complex plane, excluding the origin and the point at infinity. { ¦par·ə¦bäl·ik ˌtīp }

paraboloid A surface where sections through one of its axes are ellipses or hyperbolas, and sections through the other are parabolas. { pə'rab·əˌlȯid }

paraboloidal coordinate system A three-dimensional coordinate system in which the coordinate surfaces form families of confocal elliptic and hyperbolic paraboloids. { pə¦rab·ə¦lȯid·əl kō'ȯrd·ən·ət ˌsis·təm }

paraboloid of revolution The surface obtained by rotating a parabola about its axis. { pə'rab·əˌlȯid əv ˌrev·ə'lü·shən }

paracompact space A Hausdorff space with the property that for every open covering F there is a locally finite open covering G, such that every element of G is a subset of an element of F. { ¦par·ə¦kämˌpakt ˌspās }

parallel **1.** Lines are parallel in a euclidean space if they lie in a common plane and do not intersect. **2.** Planes are parallel in a euclidean three-dimensional space if they do not intersect. **3.** A circle parallel to the primary great circle of a sphere or spheroid. **4.** A curve is parallel to a given curve C if it consists of points that are a fixed distance from C along lines perpendicular to C. { 'par·əˌlel }

parallel axiom The axiom of an affine plane which states that if p and L are a point and line in the plane such that p is not on L, then there exists exactly one line that passes through p and does not intersect L. { ¦par·əˌlel 'ak·sē·əm }

parallel curves Two curves such that one curve is the locus of points on the normals to the other curve at a fixed distance along the normals. { 'par·ə‚lel ‚kərvz }

parallel displacement A vector A at a point P of an affine space is said to be obtained from a vector B at a point Q of the space by a parallel displacement with respect to a curve connecting A and B if a vector $V(X)$ can be associated with each point X on the curve in such a manner that $A = V(P)$, $B = V(Q)$, and the values of V at neighboring points of the curve are parallel as specified by the affine connection. { 'par·ə‚lel di'splās·mənt }

parallelepiped A polyhedron all of whose faces are parallelograms. { ‚par·ə‚lel·ə'pī·pəd }

parallelogram A four-sided polygon with each pair of opposite sides parallel. { ‚par·ə'lel·ə‚gram }

parallelogram of vectors A parallelogram whose sides form two vectors to be added and whose diagonal is the sum of the two vectors. { ‚par·ə'lel·ə‚gram əv 'vek·tərz }

parallelotope A parallelepiped with sides in proportion of 1, 1/2, and 1/4. { ‚par·ə'lel·ə‚tōp }

parameter An arbitrary constant or variable so appearing in a mathematical expression that changing it gives various cases of the phenomenon represented. { pə'ram·əd·ər }

parameter of distribution For a fixed line on a ruled surface, a quantity whose magnitude is the limit, as a variable line on the surface approaches the fixed line, of the ratio of the minimum distance between the two lines to the angle between them; and whose sign is positive or negative according to whether the motion of the tangent plane to the surface is left- or right-handed as the point of tangency moves along the fixed line in a positive direction. { pə¦ram·əd·ər əv ‚dis·trə'byü·shən }

parametric curves On a surface determined by equations $x = f(u,v)$, $y = g(u,v)$, and $z = h(u,v)$, these are families of curves obtained by setting the parameters u and v equal to various constants. { ¦par·ə¦me·trik 'kərvz }

parametric equation An equation where coordinates of points appear dependent on parameters such as the parametric equation of a curve or a surface. { ¦par·ə¦me·trik i'kwā·zhən }

parity Two integers have the same parity if they are both even or both odd. { 'par·əd·ē }

Parseval's equation The equation which states that the square of the length of a vector in an inner product space is equal to the sum of the squares of the inner products of the vector with each member of a complete orthonormal base for the space. Also known as Parseval's identity; Parseval's relation. { 'pär·sə·vəlz i‚kwā·zhən }

Parseval's identity *See* Parseval's equation. { 'pär·sə·vəlz ī'den·əd·ē }

Parseval's relation *See* Parseval's equation. { 'pär·sə·vəlz re'lā·shən }

Parseval's theorem A theorem that gives the integral of a product of two functions, $f(x)$ and $F(x)$, in terms of their respective Fourier coefficients; if the coefficients are defined by

$$a_n = (1/\pi) \int_0^{2\pi} f(x) \cos nx dx$$

$$b_n = (1/\pi) \int_0^{2\pi} f(x) \sin nx dx$$

and similarly for $F(x)$, the relationship is

$$\int_0^{2\pi} f(x)F(x)dx = \pi \left[\tfrac{1}{2} a_0 A_0 + \sum_{n=1}^{\infty} (a_n A_n + b_n B_n) \right]$$

{ 'pär·sə·vəlz 'thir·əm }

partial derivative A derivative of a function of several variables taken with respect to one variable while holding the others fixed. { 'pär·shəl də'riv·əd·iv }

partial differential equation An equation that involves more than one independent variable and partial derivatives with respect to those variables. { 'pär·shəl ˌdif·ə'ren·chəl iˌkwā·zhən }

partial fractions A collection of fractions which when added are a given fraction whose numerator and denominator are usually polynomials; the partial fractions are usually constants or linear polynomials divided by factors of the denominator of the given fraction. { 'pär·shəl 'frak·shənz }

partially ordered set A set on which a partial order is defined. { 'pär·shə·lē ¦ȯr·dərd 'set }

partial ordering *See* ordering. { 'pär·shəl 'ȯr·də·riŋ }

partial plane In projective geometry, a plane in which at most one line passes through any two points. { 'pär·shəl 'plān }

partial product The product of a multiplicand and one digit of a multiplier that contains more than one digit. { 'pär·shəl 'präd·əkt }

partial sum A partial sum of an infinite series is the sum of its first n terms for some n. { 'pär·shəl 'səm }

particular integral *See* particular solution. { pər¦tik·yə·lər 'int·ə·grəl }

particular solution A solution to an ordinary differential equation obtained by assigning numerical values to the parameters in the general solution. Also known as particular integral. { pər¦tik·yə·lər sə'lü·shən }

partition **1.** For an integer n, any collection of positive integers whose sum equals n. **2.** For a set A, a finite collection of disjoint sets whose union is A. { pär'tish·ən }

partition of unity On a topological space X, this is a covering by open sets $U\alpha$ with continuous functions $f\alpha$ from X to [0,1], where each $f\alpha$ is zero on all but a finite number of the $U\alpha$, and the sum of all these $f\alpha$ at any point equals 1. { pär'tish·ən əv 'yü·nəd·ē }

Pascal's limaçon *See* limaçon. { pa'skalz ˌlē·mə'son }

Pascal's theorem The theorem that when one inscribes a simple hexagon in a conic, the three pairs of opposite sides meet in collinear points. { pa'skalz ˌthir·əm }

Pascal's triangle A triangular array of the binomial coefficients, bordered by ones, where the sum of two adjacent entries from a row equals the entry in the next row directly below. Also known as binomial array. { pa'skalz 'trī,aŋ·gəl }

path 1. In a topological space, a path is a continuous curve joining two points. 2. In graph theory, a walk whose vertices are all distinct. Also known as simple path. 3. *See* walk. { path }

path curve A curve whose equation is given in parametric form. { 'path ,kərv }

payoff matrix A matrix arising from certain two-person games which gives the amount gained by a player. { 'pā,ȯf ,mā·triks }

Peano continuum A compact, connected, and locally connected metric space. { pā'än·ō kən,tin·yü·əm }

Peano curve 1. A continuous curve that passes through each point of the unit square. 2. *See* Peano space. { pā'än·ō ,kərv }

Peano space Any Hausdorff topological space that is the image of the closed unit interval under a continuous mapping. Also known as Peano curve. { pā'än·ō ,spās }

Peano's postulates The five axioms by which the natural numbers may be formally defined; they state that (1) there is a natural number 1; (2) every natural number n has a successor n^+; (3) no natural number has 1 as its successor; (4) every set of natural numbers which contains 1 and the successor of every member of the set contains all the natural numbers; (5) if $n^+ = m^+$, then $n = m$. { pā'än·ōz 'päs·chə·ləts }

pedal coordinates The coordinates r and p describing a point P on a plane curve C, where r is the distance from a fixed point O to P, and p is the perpendicular distance from O to the tangent to C at P. { 'ped·əl kō'ȯrd·ən·əts }

pedal curve 1. The pedal curve of a given curve C with respect to a fixed point P is the locus of the foot of the perpendicular from P to a variable tangent to C. Also known as first pedal curve; first positive pedal curve; positive pedal curve. 2. Any curve that can be derived from a given curve C by repeated application of the procedure specified in the first definition. { 'ped·əl ,kərv }

pedal equation An equation that characterizes a plane curve in terms of its pedal coordinates. { 'ped·əl i,kwā·zhən }

pedal point 1. The fixed point with respect to which a pedal curve is defined. 2. The fixed point with respect to which the pedal coordinates of a curve are defined. { 'ped·əl ,pȯint }

pedal triangle 1. The triangle whose vertices are located at the feet of the perpendiculars from some given point to the sides of a specified triangle. 2. In particular, the triangle whose vertices are located at the feet of the altitudes of a given triangle. { 'ped·əl 'trī,aŋ·gəl }

Peirce stroke relationship *See* NOR. { 'pirs ¦strōk ri'lā·shən,ship }

Pell equation The diophantine equation $x^2 - Dy^2 = 1$, with D a positive integer that is not a perfect square. { 'pel i,kwā·zhən }

penalty function A function used in treating maxima and minima problems subject to constraints. { 'pen·əl·tē ,fəŋk·shən }

pencil A family of geometric objects which share a common property. { 'pen·səl }

pentadecagon A polygon with 15 sides. { ˌpen·tə'dek·əˌgän }

pentagon A polygon with five sides. { 'pen·təˌgän }

pentagonal prism A prism with two pentagonal sides, parallel and congruent. { pen'tag·ən·əl 'priz·əm }

pentagonal pyramid A pyramid whose base is a pentagon. { pen'tag·ən·əl 'pir·əˌmid }

percent A quantitative term whereby n percent of a number is n one-hundredths of the number. Symbolized %. { pər'sent }

percentage The result obtained by taking a given percent of a given quantity. { pər'sen·tij }

percolation network A lattice constructed of a random mixture of conducting and nonconducting links. { ˌpər·kə'lā·shən 'netˌwərk }

percolation problem The problem of determining the critical threshold concentration of conducting links in a pecolation network at which an infinite cluster of conducting links is formed and the lattice transforms from an insulator to a conductor. { ˌpər·kə'lā·shən ˌpräb·ləm }

perfect cube A number or polynomial which is the exact cube of another number or polynomial. { 'pər·fikt 'kyüb }

perfect group A group that is equal to its commutator subgroup. { 'pər·fikt 'grüp }

perfect number An integer which equals the sum of all its factors other than itself. { 'pər·fikt 'nəm·bər }

perfect set A set in a topological space which equals its set of accumulation points. { 'pər·fikt 'set }

perfect square A number or polynomial which is the exact square of another number or polynomial. { 'pər·fikt 'skwer }

perfect trinomial square A trinomial that is the exact square of a binomial. { ¦pər·fikt trī¦nō·mē·əl 'skwer }

perigon An angle that contains 360° or 2π radians. { 'per·əˌgän }

perimeter The total length of a closed curve; for example, the perimeter of a polygon is the total length of its sides. { pə'rim·əd·ər }

period **1.** A number T such that $f(x + T) = f(x)$ for all x, where $f(x)$ is a specified function of a real or complex variable. **2.** The period of an element a of a group G is the smallest positive integer n such that a^n is the identity element; if there is no such integer, a is said to be of infinite period. { 'pir·ē·əd }

periodic continued fraction *See* recurring continued fraction. { ¦pir·ē¦äd·ik kən¦tin·yüd 'frak·shən }

periodic function A function $f(x)$ of a real or complex variable is periodic with period T if $f(x + T) = f(x)$ for every value of x. { ¦pir·ē¦äd·ik ¦fəŋk·shən }

periodicity The property of periodic functions. { ˌpir·ē·ə'dis·əd·ē }

periodic perturbation A perturbation which is periodic as a function. { ¦pir·ē¦äd·ik pər·tər'bā·shən }

period parallelogram For a doubly periodic function $f(z)$ of a complex variable, a parallelogram with vertices at z_0, $z_0 + a$, $z_0 + a + b$, and $z_0 + b$, where z_0 is any complex number, and a and b are periods of $f(z)$ but are not necessarily primitive periods. { ¦pir·ē·əd ˌpar·ə'lel·əˌgram }

periphery The bounding curve of a surface or the surface of a solid. { pə'rif·ə·rē }

permanently convergent series A series that is convergent for all values of the variable or variables involved in its terms. { ¦pər·mə·nənt·lē kən¦vər·jənt 'sirˌēz }

permissible value A value of a variable for which a given function is defined. { pər¦mis·ə·bəl 'val·yü }

permutation A function which rearranges a finite number of symbols; more precisely, a one-to-one function of a finite set onto itself. { ˌpər·myə'tā·shən }

permutation group The group whose elements are permutations of some set of symbols where the product of two permutations is the permutation arising from successive application of the two. { ˌpər·myə'tā·shən ˌgrüp }

permutation matrix A square matrix whose elements in any row, or any column, are all zero, except for one element that is equal to unity. { ˌpər·myə'tā·shən ˌmāˌtriks }

permutation tensor *See* determinant tensor. { ˌpər·myə'tā·shən ˌten·sər }

perpendicular Geometric objects are perpendicular if they intersect in an angle of 90°. { ¦pər·pən¦dik·yə·lər }

Perron-Frobenius theorem If M is a matrix with positive entries, then its largest eigenvalue λ is positive and simple; moreover, there exist vectors v and w with positive components such that $vM = \lambda v$ and $Mw = \lambda w$, if the inner product of v with w is 1, then the limit of $\lambda - n$ times the i,jth entry of M^n as n goes to infinity is the product of the ith component of w and the jth component of v. { pe'rōn frō'bā·nē·u̇s ˌthir·əm }

Perron-Frobenius theory The study of positive matrices and their eigenvalues; in particular, application of the Perron-Frobenius theorem. { pe'rōn frō'bā·nē·u̇s ˌthē·ə·rē }

perturbation A function which produces a small change in the values of some given function. { ˌpər·tər'bā·shən }

perturbation theory The study of the solutions of differential and partial differential equations from the viewpoint of perturbation of solutions. { ˌpər·tər'bā·shən ˌthē·ə·rē }

Pfaffian differential equation The first-order linear total differential equation $P(x,y,z)dx + Q(x,y,z)dy + R(x,y,z)dz = 0$, where the functions P, Q, and R are continuously differentiable. { 'faf·ē·ən ˌdif·ə¦ren·chəl i'kwā·zhən }

***p*-form** A totally antisymmetric covariant tensor of rank p. { 'pē ˌfȯrm }

phase An additive constant in the argument of a trigonometric function. { fāz }

phase space In a dynamical system or transformation group, the topological space whose points are being moved about by the given transformations. { 'fāz ˌspās }

pi The irrational number which is the ratio of the circumference of any circle to its diameter; an approximation is 3.14159. Symbolized π. { pī }

Picard method A method of successive substitution for solving differential equations. { pi'kär ˌmeth·əd }

Picard's big theorem The image of every neighborhood of an essential singularity of a complex function is dense in the complex plane. Also known as Picard's second theorem. { pi'kärz 'big ¦thir·əm }

Picard's first theorem *See* Picard's little theorem. { pi'kärz 'fərst ¦thir·əm }

Picard's little theorem A nonconstant entire function of the complex plane assumes every value save at most one. Also known as Picard's first theorem. { pi'kärz 'lid·əl ¦thir·əm }

Picard's second theorem *See* Picard's big theorem. { pi'kärz 'sek·ənd 'thir·əm }

pico- A prefix meaning 10^{-12}; used with metric units. Abbreviated p. Also known as micromicro- . { 'pē·kō }

piecewise-continuous function A function defined on a given region, which can be divided into a finite number of pieces such that the function is continuous on the interior of each piece and its value approaches a finite limit as the argument of the function in the interior approaches a boundary point of the piece. { ¦pēsˌwīz kən¦tin·yə·wəs 'fəŋk·shən }

piecewise-linear A continuous curve or function obtained by joining a finite number of linear pieces. { 'pēsˌwīz ˌlin·ē·ər }

piecewise linear topology *See* combinatorial topology. { 'pēsˌwīz ¦lin·ē·ər tə'päl·ə·jē }

pie chart A circle divided by several radii into sectors whose relative areas represent the relative magnitudes of quantities or the relative frequencies of items in a frequency distribution. Also known as circle graph; sectorgram. { 'pī ˌchärt }

piercing point *See* trace. { 'pirs·iŋ ˌpȯint }

pigeonhole principle The principle, that if a very large set of elements is partitioned into a small number of blocks, then at least one block contains a rather large number of elements. Also known as Dirichlet drawer principle. { 'pij·ənˌhōl ˌprin·sə·pəl }

piriform A plane curve whose equation in cartesian coordinates x and y is $y^2 = ax^3 - bx^4$, where a and b are constants. { 'pir·əˌfȯrm }

pivotal condensation A method of evaluating a determinant that is convenient for determinants of large order, especially when digital computers are used, involving a repeated process in which a determinant of order n is reduced to the product of one of its elements raised to a power and a determinant of order $n - 1$. { ¦piv·əd·əl ˌkän·dən'sā·shən }

pivoting In the solution of a system of linear equations by elimination, a method of choosing a suitable equation to eliminate at each step so that certain difficulties are avoided. { 'piv·əd·iŋ }

place A position corresponding to a given power of the base in positional notation. Also known as column. { plās }

place value The value given to a digit by virtue of its location in a numeral. { 'plās ˌval·yü }

planar Lying in or pertaining to a euclidean plane. { 'plā·nər }

planar graph A graph that can be drawn in a plane without any lines crossing. { 'plā·nər ˌgraf }

planar map A plane or sphere divided into connected regions by a topological graph. { 'plān·ər ˌmap }

planar point A point on a surface at which the curvatures of all the normal sections vanish. { 'plā·nər *or* 'plā·när ˌpȯint }

plane **1.** A surface such that a straight line that joins any two of its points lies entirely in that surface. **2.** In projective geometry, a triple of sets (P,L,I) where P denotes the set of points, L the set of lines, and I the incidence relation on points and lines, such that (1) P and L are disjoint sets, (2) the union of P and L is nonnull, and (3) I is a subset of $P \times L$, the cartesian product of P and L. { plān }

plane angle An angle between lines in the euclidean plane. { 'plān ˌaŋ·gəl }

plane curve Any curve lying entirely within a plane. { 'plān ˌkərv }

plane cyclic curve *See* cyclic curve. { 'plān 'sī·klik ˌkərv }

plane field *See* field of planes on a manifold. { 'plān ˌfēld }

plane geometry The geometric study of the figures in the euclidean plane such as lines, triangles, and polygons. { 'plān jē'äm·ə·trē }

plane group One of 17 two-dimensional patterns which can be produced by one asymmetric motif that is repeated by symmetry operations to produce a pattern unit which then is repeated by translation to build up an ordered pattern that fills any two-dimensional area. Also known as plane symmetry group. { 'plān ˌgrüp }

plane of mirror symmetry An imaginary plane which divides an object into two halves, each of which is the mirror image of the other in this plane. Also known as mirror plane of symmetry; plane of symmetry; reflection plane; symmetry plane. { 'plān əv 'mir·ər ˌsim·ə·trē }

plane of reflection *See* plane of mirror symmetry. { 'plān əv ri'flek·shən }

plane of support Relative to a convex body in a three-dimensional space, a plane that contains at least one point of the body but is such that the half-space on one side of the plane contains no points of the body. { 'plān əv sə'pȯrt }

plane of symmetry *See* plane of mirror symmetry. { 'plān əv 'sim·ə·trē }

plane polygon A polygon lying in the euclidean plane. { 'plān 'päl·əˌgän }

plane quadrilateral A four-sided polygon lying in the euclidean plane. { 'plān ˌkwä·drə'lad·ə·rəl }

plane section The intersection of a plane with a surface or a solid. { 'plān ˌsek·shən }

plane symmetry group *See* plane group. { 'plān 'sim·ə·trē ˌgrüp }

plane triangulation The process of adding arcs between pairs of verticles of a planar graph to producer another planar graph, each of whose regions is bounded by three sides. { 'plān trīˌaŋ·gyə'lā·shən }

plane trigonometry The study of triangles in the euclidean plane with the use of functions defined by the ratios of sides of right triangles. { 'plān trig·ə'näm·ə·trē }

plateau problem The problem of finding a minimal surface having as boundary a given curve. { pla'tō ¦präb·ləm }

Plemelj formulas Formulas for the limits of the Cauchy integrals of an arc with respect to a point z as z approaches the arc from either side. { 'plā·mə·lē ˌfȯr·myə·ləz }

plus A mathematical symbol; A plus B, where A and B are mathematical quantities, denotes the quantity obtained by taking their sum in an appropriate context. { pləs }

plus sign *See* addition sign. { 'pləs ˌsīn }

Pockels equation A partial differential equation which states that the Laplacian of an unknown function, plus the product of the value of the function with a constant, is equal to 0; it arises in finding solutions of the wave equation that are products of time-independent and space-independent functions. { 'päk·əlz iˌkwā·zhən }

Poincaré-Birkhoff fixed-point theorem The theorem that a bijective, continuous, area-preserving mapping of the ring between two concentric circles onto itself that moves one circle in the positive sense and the other in the negative sense has at least two fixed points. { ˌpwän·kə¦rā 'bərkˌhȯf ¦fikst ¦pȯint 'thir·əm }

Poincaré conjecture The question as to whether a compact, simply connected three-dimensional manifold without boundary must be homeomorphic to the three-dimensional sphere. { ˌpwänˌkä'rā kənˌjek·chər }

Poincaré recurrence theorem **1.** A volume preserving homeomorphism T of a finite dimensional euclidean space will have, for almost all points x, infinitely many points of the form $T^i(x)$, $i = 1, 2, \ldots$ within any open set containing x. **2.** A measure preserving transformation on a space with finite measure is recurrent. { ˌpwänˌkä'rā ri'kə·rəns ˌthir·əm }

Poinsot's spiral Either of two plane curves whose equations in polar coordinates (r,θ) are $r \cosh n\theta = a$ and $r \sinh n\theta = a$, where a is a constant and n is an integer. { pwän'sōz ¦spī·rəl }

point **1.** An element in a topological space. **2.** One of the basic undefined elements of geometry possessing position but no nonzero dimension. **3.** In positional notation, the character or the location of an implied symbol that separates the integral part of a numerical expression from its fractional part; for example, it is called the binary point in binary notation and the decimal point in decimal notation. { pȯint }

point at infinity *See* ideal point. { 'pȯint at in'fin·əd·ē }

point distal flow A transformation group on a compact metric space for which there exists a distal point with a dense orbit. { ¦pȯint ¦dis·təl 'flō }

point function A function whose values are points. { 'pȯint ˌfəŋk·shən }

point of division The point that divides the line segment joining two given points in a given ratio. { ˌpȯint əv di′vizh·ən }

point of inflection A point where a plane curve changes from the concave to the convex relative to some fixed line; equivalently, if the function determining the curve has a second derivative, this derivative changes sign at this point. Also known as inflection point. { ′pȯint əv in′flek·shən }

point of osculation *See* double cusp. { ′pȯint əv ˌäs·kyə′lā·shən }

point set A collection of points in a geometrical or topological space. { ′pȯint ˌset }

point-set topology *See* general topology. { ′pȯint ˌset tə′päl·ə·jē }

point-slope form The equation of a straight line in the form $y - y_1 = m(x - x_1)$, where m is the slope of the line and (x_1,y_1) are the coordinates of a given point on the line in a cartesian coordinate system. { ′pȯint ˌslōp ˌfȯrm }

point spectrum Those eigenvalues in the spectrum of a linear operator between Banach spaces whose corresponding eigenvectors are nonzero and of finite norm. { ′pȯint ˌspek·trəm }

pointwise convergence A sequence of functions $f_1, f_2, \ldots$ defined on a set S converges pointwise to a function f if the sequence $f_1(x), f_2(x), \ldots$ converges to $f(x)$ for each x in S. { ′pȯintˌwīz kən′vər·jəns }

Poisson formula If the infinite series of functions $f(2\pi k + t)$, k ranging from $-\infty$ to ∞, converges uniformly to a function of bounded variation, then the infinite series with term $f(2\pi k)$, k ranging from $-\infty$ to ∞, is identical to the series with term the integral of $f(x)e^{-iks}dx$, k ranging from $-\infty$ to ∞. { pwä′sōn ˌfȯr·myə·lə }

Poisson integral formula This formula gives a solution function for the Dirichlet problem in terms of integrals; an integral representation for the Bessel functions. { pwä′sōn ′int·ə·grəl ˌfȯr·myə·lə }

Poisson's equation The partial differential equation which states that the Laplacian of an unknown function is equal to a given function. { pwä′sōnz iˌkwā·zhən }

Poisson transform An integral transform which transforms the function $f(t)$ to the function

$$F(x) = (2/\pi) \int_0^\infty [t/(x^2 + t^2)]f(t)dt$$

Also known as potential transform. { pwä′sōn ′tranzˌfȯrm }

polar **1.** For a conic section, the polar of a point is the line that passes through the points of contact of the two tangents drawn to the conic from the point. **2.** For a quadric surface, the polar of a point is the plane that passes through the curve which is the locus of the points of contact of the tangents drawn to the surface from the point. **3.** For a quadric surface, the polar of a line is the line of intersection of the planes which are tangent to the surface at its points of intersection with the original line. { ′pō·lər }

polar angle The angular coordinate θ in a polar coordinate system whose value at a given point p is the angle that a line from the origin to p makes with the polar axis. { ′pō·lər ′aŋ·gəl }

polar axis The directed straight line relative to which the angle is measured for a representation of a point in the plane by polar coordinates. { ′pō·lər ′ak·səs }

polar coordinates A point in the plane may be represented by coordinates (r,θ), where θ is the angle between the positive x-axis and the ray from the origin to the point, and r the length of that ray. { 'pō·lər kō'ȯrd·ən·əts }

polar developable The envelope of the normal planes of a space curve. { ¦pō·lər di¦vel·əp·ə·bəl }

polar equation An equation expressed in polar coordinates. { ¦pō·lər i¦kwā·zhən }

polar form A complex number $x + iy$ has as polar form $re^{i\theta}$, where (r,θ) are the polar coordinates corresponding to the point of the plane with rectangular coordinates (x,y), that is, $r \sqrt{x^2 y^2}$ and θ arc tan y/x. { 'pō·lər 'fȯrm }

polarity Property of a line segment whose two ends are distinguishable. { pə'lar·əd·ē }

polar line For a point on a space curve, the line that is normal to the osculating plane of the curve and passes through the center of curvature at that point. { 'pō·lər ˌlīn }

polar normal For a given point on a plane curve, the segment of the normal between the given point and the intersection of the normal with the radial line of a polar coordinate system that is perpendicular to the radial line to the given point. { 'pō·lər 'nȯr·məl }

polar reciprocal convex bodies Any two convex bodies, each containing the origin in its interior, such that the Minkowski distance function of each is the support function of the other. { ¦pō·lər ri¦sip·rə·kəl ¦känˌveks 'bädˌēz }

polar subnormal For a given point on a plane curve, the projection of the polar normal on the radial line of the polar coordinate system that is perpendicular to the radial line to the given point. { 'pō·lər səb'nȯr·məl }

polar subtangent For a given point on a plane curve, the projection of the polar tangent on the radial line of the polar coordinate system that is perpendicular to the radial line to the given point. { 'pō·lər səb'tan·jənt }

polar tangent For a given point on a plane curve, the segment of the tangent between the given point and the intersection of the tangent with the radial line of a polar coordinate system that is perpendicular to the radial line to the given point. { 'pō·lər 'tan·jənt }

polar triangle A triangle associated to a given spherical triangle obtained from three directed lines perpendicular to the planes associated with the sides of the original triangle. { 'pō·lər 'trīˌaŋ·gəl }

pole **1.** An isolated singular point z_0 of a complex function whose Laurent series expansion about z_0 will include finitely many terms of form $a_n(z - z_0)^{-n}$. **2.** For a great circle on a sphere, the pole of the circle is a point of intersection of the sphere and a line that passes through the center of the sphere and is perpendicular to the plane of the circle. **3.** For a conic section, the pole of a line is the intersection of the tangents to the conic at the points of intersection of the conic with the line. **4.** For a quadric surface, the pole of a plane is the vertex of the cone which is tangent to the surface along the curve where the plane intersects the surface. **5.** The origin of a system of polar coordinates on a plane. **6.** The origin of a system of geodesic polar coordinates on a surface. { pōl }

Polish space A separable metric space which is homeomorphic to a complete metric space. { 'pōl·ish 'spās }

Polya counting formula A formula which counts the number of functions from a finite set D to another finite set, with two functions f and g assumed to be the same if some element of a fixed group of complete permutations of D takes f into g. { ¦pōl·yə 'kau̇n·tiŋ ˌfȯr·myə·lə }

polyalgorithm A set of algorithms together with a strategy for choosing and switching among them. { ¦päl·ē'al·gəˌrith·əm }

polygon A figure in the plane given by points p_1, p_2, ..., p_n and line segments p_1p_2, p_2p_3, ..., $p_{n-1}p_n$, p_np_1. { 'päl·iˌgän }

polygon of vectors A polygon all but one of whose sides represent vectors to be added, directed in the same sense along the perimeter, and whose remaining side represents the sum of these vectors, directed in the opposite sense. { ¦päl·iˌgän əv 'vek·tərz }

polyhedral angle The shape formed by the lateral faces of a polyhedron which have a common vertex. { ¦päl·i¦hē·drəl 'aŋ·gəl }

polyhedron A solid bounded by planar polygons. { ¦päl·i¦hē·drən }

polynomial A polynomial in the quantities x_1, x_2, ..., x_n is an expression involving a finite sum of terms of form $bx_1^{p_1}x_2^{p_2} \ldots x^{p_n}n$, where b is some number, and p_1, ..., p_n are integers. { ¦päl·ə¦nō·mē·əl }

polynomial equation An equation in which a polynomial in one or more variables is set equal to zero. { ¦päl·ə¦nō·mē·əl i'kwā·zhən }

polytope A finite region in n-dimensional space ($n = 2, 3, 4, \ldots$), enclosed by a finite number of hyperplanes; it is the n-dimensional analog of a polygon ($n = 2$) and a polyhedron ($n = 3$). { 'päl·iˌtōp }

Pontryagin's maximum principle A theorem giving a necessary condition for the solution of optimal control problems: let $\theta(\tau)$, $\tau_0 \le \tau \le T$ be a piecewise continuous vector function satisfying certain constraints; in order that the scalar function $S = \Sigma c_i x_i(T)$ be minimum for a process described by the equation $\partial x_i/\partial \tau = (\partial H/\partial z_i)[z(\tau), x(\tau), \theta(\tau)]$ with given initial conditions $x(\tau_0) = x^0$ it is necessary that there exist a nonzero continuous vector function $z(\tau)$ satisfying $dz_i/d\tau = -(\partial H/\partial x_i)[z(\tau), x(\tau), \theta(\tau)]$, $z_i(T) = -c_i$, and that the vector $\theta(\tau)$ be so chosen that $H[z(\tau), x(\tau), \theta(\tau)]$ is maximum for all τ, $\tau_0 \le \tau \le T$. { ˌpän·trē'ä·gənz 'mak·sə·məm ˌprin·sə·pəl }

positional notation Any of several numeration systems in which a number is represented by a sequence of digits in such a way that the significance of each digit depends on its position in the sequence as well as its numeric value. Also known as notation. { pə'zish·ən·əl nō'tā·shən }

position vector The position vector of a point in euclidean space is a vector whose length is the distance from the origin to the point and whose direction is the direction from the origin to the point. Also known as radius vector. { pə'zish·ən ˌvek·tər }

positive Having value greater than zero. { 'päz·əd·iv }

positive angle The angle swept out by a ray moving in a counterclockwise direction. { 'päz·əd·iv 'aŋ·gəl }

positive axis The segment of an axis arising from a cartesian coordinate system which is realized by positive values of the coordinate variables. { 'päz·əd·iv 'ak·səs }

positive definite 1. A square matrix A of order n is positive definite if

$$\sum_{ij=1}^{n} A_{ij}x_i\bar{x}_j > 0$$

for every choice of complex numbers $x_1, x_2, ..., x_n$, not all equal to 0, where $\bar{x}_j$ is the complex conjugate of x_j. **2.** A linear operator T on an inner product space is positive definite if $\langle Tu,u \rangle$ is greater than 0 for all nonzero vectors u in the space. { 'päz·əd·iv 'def·ə·nət }

positive linear functional A linear functional on some vector space of real-valued functions which takes every nonnegative function into a nonnegative number. { ¦päz·əd·iv ¦lin·ē·ər 'fəŋk·shən·əl }

positive part For a real-valued function f, this is the function, denoted f^+, for which $f^+(x) = f(x)$ if $f(x) \geq 0$ and $f^+(x) = 0$ if $f(x) < 0$. { 'päz·əd·iv ˌpärt }

positive pedal curve *See* pedal curve. { 'päz·əd·iv 'ped·əl ˌkərv }

positive real function An analytic function whose value is real when the independent variable is real, and whose real part is positive or zero when the real part of the independent variable is positive or zero. { 'päz·əd·iv 'rēl 'fəŋk·shən }

positive semidefinite Also known as nonnegative semidefinite. **1.** A square matrix A is positive semidefinite if

$$\sum_{ij=1}^{n} A_{ij}x_i\bar{x}_j \geq 0$$

for every choice of complex numbers $x_1, x_2, ..., x_n$, where $\bar{x}_j$ is the complex conjugate of x_j. **2.** A linear operator T on an inner product space is positive semidefinite if $\langle Tu,u \rangle$ is equal to or greater than 0 for all vectors u in the space. { 'päz·əd·iv ¦sem·i'def·ə·nət }

positive series A series whose terms are all positive real numbers. { 'päz·əd·iv 'sirˌēz }

positive with respect to a signed measure A set A is positive with respect to a signed measure m if, for every measurable set B, the intersection of A and B, $A \cap B$, is measurable and $m(A \cap B) \geq 0$. { ¦päz·əd·iv with riˌspekt tü ə ¦sīnd 'mezh·ər }

postmultiplication In multiplying a matrix or operator **B** by another matrix or operator **A**, the operation that results in the matrix or operator **BA**. Also known as multiplication on the right. { ˌpōst·məl·tə·plə'kā·shən }

postulate *See* axiom. { 'päs·chə·lət }

potential theory The study of the functions arising from Laplace's equation, especially harmonic functions. { pə'ten·chəl ˌthē·ə·rē }

potential transform *See* Poisson transform. { pə'ten·chəl 'tranzˌfórm }

power 1. The value that is assigned to a mathematical expression and its exponent. **2.** For a point, with reference to a circle, the power of a set is its cardinality. **3.** The quantity $(x - a)^2 + (y - b)^2 - r^2$, where x and y are the coordinates of the point, a and b are the coordinates of the center of the circle, and r is the radius of the circle. **4.** For a point, with reference to a sphere, the quantity $(x - a)^2 + (y - b)^2 + (z - c)^2 - r^2$, where x, y, and z are the coordinates of the point; a, b, and c are the coordinates of the center of the sphere; and r is the radius of the sphere. { 'paů·ər }

power function A function whose value is the product of a constant and a power of the independent variable. { 'paů·ər ˌfəŋk·shən }

power of the continuum The cardinality of the set of real numbers. { 'paů·ər əv thə kən'tin·yə·wəm }

power series An infinite series composed of functions having nth term of the form $a_n(x - x_0)^n$, where x_0 is some point and a_n some constant. { 'paů·ər ˌsir·ēz }

power set The set consisting of all subsets of a given set. { 'paů·ər ˌset }

precision The number of digits in a decimal fraction to the right of the decimal point. { prə'sizh·ən }

precompact set A set in a metric space which can always be covered by open balls of any diameter about some finite number of its points. Also known as totally bounded set. { prē'kämˌpakt ˌset }

predecessor For a vertex a in a directed graph, any vertex b for which there is an arc between a and b directed from b to a. { 'pred·əˌses·ər }

predicate **1.** To affirm or deny, in mathematical logic, one or more subjects. **2.** A function of one or more variables which takes the values "true" or "false." Also known as logical function; propositional function. { 'pred·əˌkāt }

predicate calculus The mathematical study of logical statements relating to arbitrary sets of objects and involving predicates and quantifiers as well as propositional connectives. { 'pred·ə·kət ˌkal·kyə·ləs }

predictor-corrector methods Methods of calculating numerical solutions of differential equations that employ two formulas, the first of which predicts the value of the solution function at a point x in terms of the values and derivatives of the function at previous points where these have already been calculated, enabling approximations to the derivatives at x to be obtained, while the second corrects the value of the function at x by using the newly calculated values. { 'pri¦dik·tər kə'rek·tər ˌmeth·ədz }

pre-Hilbert space A linear space which has an inner product defined on it. { prē'hil·bərt ˌspās }

premultiplication In multiplying a matrix or operator **B** by another matrix or operator **A**, the operation that results in the matrix or operator **AB**. Also known as multiplication on the left. { ˌprē·məl·tə·plə'kā·shən }

primary decomposition A primary decomposition of a submodule N of a module M is an expression of N as a finite intersection of primary submodules of M. { 'prīˌmer·ē dēˌkäm·pə'zish·ən }

primary submodule A submodule N of a module M over a commutative ring R such that $M \neq N$ and, for any a in R, the principal homomorphism of the factor module M/N associated with a, $a_{M/N}$, is either injective or nilpotent. { 'prīˌmer·ē səb'mä·jəl }

prime *See* prime element. { prīm }

prime element An irreducible element of a unique factorization domain. Also known as prime. { 'prīm 'el·ə·mənt }

prime factor A prime number or prime polynomial that exactly divides a given number or polynomial. { 'prīm 'fak·tər }

prime field For a field K with multiplicative unit element e, the field consisting of elements of the form $(ne)(me)^{-1}$, where $m \neq 0$ and n are integers. { 'prīm 'fēld }

prime ideal A principal ideal of a ring given by a single element that has properties analogous to those of the prime numbers. { 'prīm ī'dēl }

prime number A positive integer having no divisors except itself and the integer 1. { 'prīm 'nəm·bər }

prime number theorem The theorem that the limit of the quantity $[\pi(x)]$ $(\ln x)/x$ as x approaches infinity is 1, where $\pi(x)$ is the number of prime numbers not greater than x and $\ln x$ is the natural logarithm of x. { ¦prīm ¦nəm·bər ˌthir·əm }

prime polynomial A polynomial whose only factors are itself and constants. { 'prīm ˌpäl·i'nō·mē·əl }

prime ring For a field K with multiplicative unit element e, the ring consisting of elements of the form ne, where n is an integer. { ¦prīm 'riŋ }

primitive abundant number An integer that is an abundant number and has no proper divisors that are also abundant numbers. { ¦prim·əd·iv ə¦bən·dənt 'nəm·bər }

primitive circle The stereographic projection of the great circle whose plane is perpendicular to the diameter of the projected sphere that passes through the point of projection. { 'prim·əd·iv 'sər·kəl }

primitive element A member of a finite number field from which all the other members can be generated by repeated multiplication. { 'prim·əd·iv 'el·ə·mənt }

primitive period **1.** A period a of a simply periodic function $f(x)$ such that any period of $f(x)$ is an integral multiple of a. **2.** Either of two periods a and b of a doubly periodic function $f(x)$ such that any period of $f(x)$ is of the form $ma + nb$, where m and n are integers. { 'prim·əd·iv 'pir·ē·əd }

primitive period parallelogram For a doubly periodic function $f(z)$ of a complex variable, a parallelogram with vertices at z_0, $z_0 + a$, $z_0 + a + b$, and $z_0 + b$, where z_0 is any complex number and a and b are primitive periods of $f(z)$. { ¦prim·əd·iv ¦pir·ē·əd ˌpar·ə'lel·əˌgram }

primitive plane A partial plane in which every line passes through at least two points.

primitive polynomial A polynomial with integer coefficients which have 1 as their greatest common divisor. { 'prim·əd·iv ˌpäl·i'nō·mē·əl }

primitive pseudoperfect number An integer that is a pseudoperfect number and has no proper divisors that are also pseudoperfect numbers. { ¦prim·əd·iv ¦süd·ō¦pərˌfikt 'nəm·bər }

primitive root An nth root of unity that is not an mth root of unity for any m less than n. { 'prim·əd·iv 'rüt }

principal axis 1. One of a set of perpendicular axes such that a quadratic function can be written as a sum of squares of coordinates referred to these axes. **2.** For a conic, a straight line that passes through the midpoints of all the chords perpendicular to it. **3.** For a quadric surface, the intersection of two principal planes. { 'prin·sə·pəl 'ak·səs }

principal branch For complex valued functions such as the logarithm which are multiple-valued, a selection of values so as to obtain a genuine single-valued function. { 'prin·sə·pəl 'branch }

principal curvatures For a point on a surface, the absolute maximum and absolute minimum values attained by the normal curvature. { 'prin·sə·pəl 'kər·və·chərz }

principal diagonal For a square matrix, the diagonal extending from the upper left-hand corner to the lower right-hand corner of the matrix, that is, the diagonal containing the elements a_{ij} for which $i = j$. Also known as main diagonal. { 'prin·sə·pəl dī'ag·ən·əl }

principal directions For a point on a surface, the directions in which the normal curvature attains its absolute maximum and absolute minimum values. { 'prin·sə·pəl di'rek·shənz }

principal homomorphism Let a be an element of a ring R and M be a module over R; the principal homomorphism of M associated with a, denoted a_M, is the mapping which takes each element x in M into ax. { 'prin·sə·pəl ˌhō·mō'mȯrˌfiz·əm }

principal ideal The smallest ideal of a ring which contains a given element of the ring. { 'prin·sə·pəl ī'dēl }

principal ideal ring A commutative ring with a unit element in which every ideal is a principal ideal. { ¦prin·sə·pəl i¦del ˌriŋ }

principal normal The line perpendicular to a space curve at some point which also lies in the osculating plane at that point. { 'prin·sə·pəl 'nȯr·məl }

principal normal section For a point on a surface, a normal section in a direction in which the curvature of this section has a maximum or minimum value. { ¦prin·sə·pəl ¦nȯr·məl 'sek·shən }

principal part The principal part of an analytic function $f(z)$ defined in an annulus about a point z_0 is the sum of terms in its Laurent expansion about z_0 with negative powers of $(z - z_0)$. { 'prin·sə·pəl 'pärt }

principal plane For a quadric surface, a plane that passes through the midpoints of all the chords perpendicular to it. { 'prin·sə·pəl 'plān }

principal radii The radii of curvature of the normal sections with maximum and minimum curvature at a given point on a surface; the reciprocals of the principal curvatures. { 'prin·sə·pəl 'rād·ēˌī }

principal radii of normal curvature The reciprocals of the principal curvatures of a surface at a point. { ¦prin·sə·pəl ¦rād·ēˌī əv ¦nȯr·məl 'kər·və·chər }

principal root The positive real root of a positive number, or the negative real root in the case of odd roots of negative numbers. { 'prin·sə·pəl 'rüt }

principal section A normal section at a given point on a surface whose curvature has a maximum or minimum value. { 'prin·sə·pəl 'sek·shən }

principal value 1. The numerically smallest value of the arc sine, arc cosine, or arc tangent of a number, the positive value being chosen when there are values that are numerically equal but opposite in sign. **2.** *See* Cauchy principal value. { 'prin·sə·pəl 'val·yü }

principle of duality *See* duality principle. { 'prin·sə·pəl əv dü'al·əd·ē }

principle of the maximum The principle that for a nonconstant complex analytic function defined in a domain, the absolute value of the function cannot attain its maximum at any interior point of the domain. { 'prin·sə·pəl əv thə 'mak·sə·məm }

principle of the minimum The principle that for a nonvanishing nonconstant complex analytic function defined in a domain, the absolute value of the function cannot attain its minimum at any interior point of the domain. { 'prin·sə·pəl əv thə 'min·ə·məm }

prism A polyhedron with two parallel, congruent faces and all other faces parallelograms. { 'priz·əm }

prismatic surface A surface generated by moving a straight line which always meets a broken line lying in a given plane and which is always parallel to some given line not in that plane. { priz'mad·ik 'sər·fəs }

prismatoid A polyhedron whose vertices all are in one or the other of two parallel planes. { 'priz·mə,tȯid }

prismoid A prismatoid whose two parallel faces are polygons having the same number of sides while the other faces are trapezoids or parallelograms. { 'priz,mȯid }

prismoidal formula A formula that gives the volume of a prismatoid as $(1/6)h(A_1 + 4A_m + A_2)$, where h is the altitude, A_1 and A_2 are the areas of the bases, and A_m is the area of a plane section halfway between the bases. { priz¦mȯid·əl 'fȯr·myə·lə }

probability measure The measure on a probability space. { ,präb·ə'bil·əd·ē ,mezh·ər }

probability space A measure space such that the measure of the entire space equals 1. { ,präb·ə'bil·əd·ē ,spās }

probability theory The study of the mathematical structures and constructions used to analyze the probability of a given set of events from a family of outcomes. { ,präb·ə'bil·əd·ē ,thē·ə·rē }

problem of type The problem of determining whether a given simply connected Riemann surface is of hyperbolic, parabolic, or elliptic type. { ¦präb·ləm əv 'tīp }

product 1. For two integers, m and n, the number of objects in the set formed by combining m sets, each of which has n objects. **2.** For two rational numbers, a/b and c/d, where a, b, c, and d are integers, the number $(ac)/(bd)$. **3.** For any two real numbers, which are the limits of sequences of rational numbers p_n and q_n respectively, the limit of the sequence $p_n q_n$. **4.** The product of two algebraic quantities is the result of their multiplication relative to an operation analogous to multiplication of real numbers. **5.** The product of a collection of sets $A_1, A_2, \ldots, A_n$ is the set of all elements of the form $(a_1, a_2, \ldots, a_n)$ where each a_i is an element of A_i for each $i = 1, 2, \ldots, n$. **6.** For two transformations, the transformation that results from their successive application. **7.** For two fuzzy sets A and B, with membership functions m_A and m_B, the fuzzy set whose membership function $m_{A \cdot B}$ satisfies the equation $m_{A \cdot B}(x) = m_A(x) \cdot m_B(x)$ for every element x. **8.** The product AB of two matrices A and B, where the number n of columns in A equals the number of rows in B, is the matrix whose element c_{ij} in row i and column j is the sum over $k = 1, 2, \ldots, n$ of the product of the elements a_{ik} in A and b_{kj} in B. { 'präd·əkt }

product bundle A bundle whose total space is the cartesian product of the base space B and a topological space F and whose projection map sends (b,a) to b. { 'präd·əkt ˌbən·dəl }

product measure A measure on a product of measure spaces constructed from the measures on each of the individual spaces by taking the measure of the product of a finite number of measurable sets, one from each of the measure spaces in the product, to be the product of the measures of these sets. { 'prä·dəkt ˌmezh·ər }

product topology A topology on a product of topological spaces whose open sets are constructed from cartesian products of open sets from the individual spaces. { 'präˌdəkt tə'päl·ə·jē }

progression A sequence or series of mathematical objects or quantities, each entry determined from its predecessors by some algorithm. { prə'gresh·ən }

projecting cylinder A cylinder whose elements pass through a given curve and are perpendicular to one of the three coordinate planes. { prə¦jekt·iŋ 'sil·ən·dər }

projecting plane A plane that contains a given straight line in space and is perpendicular to one of the three coordinate planes. { prə¦jekt·iŋ 'plān }

projection **1.** The continuous map for a fiber bundle. **2.** Geometrically, the image of a geometric object or vector superimposed on some other. **3.** A linear map P from a linear space to itself such that $P \circ P$ is equal to P. { prə'jek·shən }

projective geometry The study of those properties of geometric objects which are invariant under projection. { prə'jek·tiv jē'äm·ə·trē }

projective group A group of transformations arising in the general theory of projective geometry. { prə'jek·tiv 'grüp }

projective line The line obtained from the stereographic projection of the circle. { prə'jek·tiv 'līn }

projective plane **1.** The topological space obtained from the two-dimensional sphere by identifying antipodal points; the space of all lines through the origin in euclidean space. **2.** More generally, a plane (in the sense of projective geometry) such that (1) every two points lie on exactly one line, (2) every two lines pass through exactly one point, and (3) there exists a four-point. { prə'jek·tiv 'plān }

projective point The point from which a projection by rays is performed, as in stereographic projection. { prə'jek·tiv 'pȯint }

projective space The topological space obtained from the n-dimensional sphere under identification of antipodal points. { prə'jek·tiv 'spās }

prolate cycloid A trochoid in which the distance from the center of the rolling circle to the point describing the curve is greater than the radius of the circle. { 'prōˌlāt 'sīˌklȯid }

prolate ellipsoid *See* prolate spheroid. { 'prō‚lāt i'lip‚sȯid }

prolate spheroid The ellipsoid or surface obtained by revolving an ellipse about one of its axes so that the equatorial circle has a diameter less than the length of the axis of revolution. Also known as prolate ellipsoid. { 'prō‚lāt 'sfir‚ȯid }

prolate spheroidal coordinate system A three-dimensional coordinate system whose coordinate surfaces are the surfaces generated by rotating a plane containing a system of confocal ellipses and hyperbolas about the major axis of the ellipses, together with the planes passing through the axis of rotation. { ¦prō‚lāt sfir¦ȯid·əl kō'ȯrd·ən·ət ‚sis·təm }

proof A deductive demonstration of a mathematical statement. { prüf }

proper fraction **1.** A fraction a/b where the absolute value of a is less than the absolute value of b. **2.** The quotient of two polynomials in which the degree of the numerator is less than the degree of the denominator. { 'präp·ər 'frak·shən }

properly divergent series A series whose partial sums become either arbitrarily large or arbitrarily small (algebraically). { ¦präp·ər·lē də¦vər·jənt 'sir‚ēz }

proper function *See* eigenfunction. { 'präp·ər 'fəŋk·shən }

proper orthogonal transformation An orthogonal transformation such that the determinant of its matrix is +1. { ¦präp·ər ȯr¦thäg·ən·əl ‚tranz·fər'mā·shən }

proper rational function The quotient of a polynomial P by a polynomial Q whose order is greater than P. { 'präp·ər ¦rash·ən·əl ‚fəŋk·shən }

proper subset A set X is a proper subset of a set Y if there is an element of Y which is not in X while X is a subset of Y. { 'präp·ər 'səb‚set }

proper value *See* eigenvalue. { 'präp·ər 'val·yü }

proportion **1.** The proportion of two quantities is their ratio. **2.** The statement that two ratios are equal. { prə'pȯr·shən }

proportional parts Numbers in the same proportion as a set of given numbers; such numbers are used in an auxiliary interpolation table based on the assumption that the tabulated quantity and entering arguments differ in the same proportion. { prə'pȯr·shən·əl 'pärts }

proposition **1.** Any problem or theorem. **2.** A statement that makes an assertion that is either false or true or has been designated as false or true. { ‚präp·ə'zish·ən }

propositional algebra The study of finite configurations of symbols and the interrelationships between them. { ‚prä·pə¦zish·ən·əl 'al·jə·brə }

propositional calculus The mathematical study of logical connectives between propositions and deductive inference. Also known as sentential calculus. { ‚präp·ə'zish·ən·əl 'kal·kyə·ləs }

propositional connectives The symbols , ~, $\wedge$, $\vee$, $\rightarrow$ or $\supset$, and $\leftrightarrow$ or $\equiv$, denoting logical relations that may be expressed by the phrases "it is not the case that," "and," "or," "if . . . , then," and "if and only if." Also known as sentential connectives. { ‚prä·pə¦zish·ənəl kə'nek·tivz }

propositional function An expression that becomes a proposition when the values of certain symbols in the expression are specified. Also known as predicate. { ˌpräp·əˈzish·ən·əl ˈfəŋk·shən }

Prüfer domain An integral domain in which every nonzero finitely generated ideal is invertible. { ˈpru̇f·ər dəˌmān }

pseudoperfect number An integer that is equal to the sum of some of its proper divisors. { ¦süd·ōˌpər·fikt ˈnəm·bər }

pseudometric *See* semimetric. { ¦süd·ə¦me·trik }

pseudosphere The pseudospherical surface generated by revolving a tractrix about its asymptote. { ˈsüd·əˌsfir }

pseudospherical surface A surface whose total curvature has a constant negative value. { ˌsüd·ō¦sfer·ə·kəl ˈsər·fəs }

psi function The special function of a complex variable which is obtained from differentiating the logarithm of the gamma function. { ˈsī ˌfəŋk·shən }

Ptolemy's theorem The theorem that a necessary and sufficient condition for a convex quadrilateral to be inscribed in a circle is that the sum of the products of the two pairs of opposite sides equal the product of the diagonals. { ˈtäl·ə·mēz ˌthir·əm }

pure geometry Geometry studied from the standpoint of its axioms and postulates rather than its objects. { ˈpyu̇r jēˈäm·ə·trē }

pure imaginary number A complex number $z = x + iy$, where $x = 0$. { ˈpyu̇r i¦maj·əˌner·ē ¦nəm·bər }

purely inseparable An element a is said to be purely inseparable over a field F with characteristic p greater than 0 if it is algebraic over F and if there exists a nonnegative integer n such that ap^n lies in F. { ¦pyu̇r·lē inˈsep·rə·bəl }

purely inseparable extension A purely inseparable extension E of a field F is an algebraic extension of F whose separable degree over F equals 1 or, equivalently, an algebraic extension of F in which every element is purely inseparable over F. { ¦pyu̇r·lē in¦sep·rə·bəl ikˈsten·chən }

pure mathematics The intrinsic study of mathematical structures, with no consideration given as to the utility of the results for practical purposes. { ˈpyu̇r ˌmath·əˈmad·iks }

pure projective geometry The axiomatic study of geometric systems which exhibit invariance relative to a notion of projection. { ˈpyu̇r prə¦jek·tiv jēˈäm·ə·trē }

pure strategy In game theory, a predetermined plan covering all possible situations in a game and not involving the use of random devices. { ˈpyu̇r ˈstrad·ə·jē }

pure surd A surd, all of whose terms are irrational numbers. { ˈpyu̇r ˈsərd }

pyramid A polyhedron with one face a polygon and all other faces triangles with a common vertex. { ˈpir·əˌmid }

pyramidal numbers The numbers 1, 4, 10, 20, 35, . . . , which are the number of dots in successive pyramidal arrays and are given by $(1/6)n(n + 1)(n + 2)$, where $n = 1, 2, 3, \ldots$. { ¦pir·əˌmid·əl ˈnəm·bərz }

pyramidal surface A surface generated by a line passing through a fixed point and moving along a broken line in a plane not containing that point. { ¦pir·ə¦mid·əl 'sər·fəs }

Pythagorean numbers Positive integers x, y, and z which satisfy the equation $x^2 + y^2 = z^2$. { pə,thag·ə'rē·ən 'nəm·bərz }

Pythagorean theorem In a right triangle the square of the length of the hypotenuse equals the sum of the squares of the lengths of the other two sides. { pə,thag·ə'rē·ən 'thir·əm }

Q

quadrangle A geometric figure bounded by four straight-line segments called sides, each of which intersects each of two adjacent sides in points called vertices, but fails to intersect the opposite sides. Also known as quadrilateral. { 'kwä͵draŋ·gəl }

quadrant 1. A quarter of a circle; either an arc of 90° or the area bounded by such an arc and the two radii. **2.** Any of the four regions into which the plane is divided by a pair of coordinate axes. { 'kwä·drənt }

quadrantal angle An angle equal to 90° or $\pi/2$ radians multiplied by a positive or negative integer or zero. { kwä¦drant·əl 'aŋ·gəl }

quadrantal spherical triangle A spherical triangle that has one and only one right angle. { kwä¦drant·əl ¦sfir·ə·kəl 'trī͵aŋ·gəl }

quadratic Any second-degree expression. { kwä 'drad·ik }

quadratic congruence A statement that two polynomials of second degree have the same remainder on division by a given integer. { kwä¦drad·ik kən'grü·əns }

quadratic equation Any second-degree polynomial equation. { kwä'drad·ik i'kwā·zhən }

quadratic form Any second-degree, homogeneous polynomial. { kwä'drad·ik 'fȯrm }

quadratic formula A formula giving the roots of a quadratic equation in terms of the coefficients; for the equation $ax^2 + bx + c = 0$, the roots are $x = (-b \pm \sqrt{b^2 - 4ac})/2a$. { kwä'drad·ik 'fȯr·myə·lə }

quadratic polynomial A polynomial where the highest degree of any of its terms is 2. { kwä'drad·ik ͵päl·ə'nō·mē·əl }

quadratic programming A body of techniques developed to find extremal points for systems of quadratic inequalities. { kwä'drad·ik 'prō͵gram·iŋ }

quadratic reciprocity law The law that, if p and q are distinct odd primes, then

$$(p \mid q)\,(q \mid p) = (-\,1)^{(1/4)(p-1)(q-1)}$$

(the vertical line inside parentheses is Legendre's symbol). { kwuä¦drad·ik ͵res·ə'präs·əd·ē ͵lȯ }

quadratic residue A residue of order 2. { kwä¦drad·ik 'rez·ə·dü }

quadratic surd A square root of a rational number that is itself an irrational number. { kwä'drad·ik ͵sərd }

quadratrix of Hippias A plane curve whose equation in cartesian coordinates x and y is $y = x \cot [\pi x/(2a)]$, where a is a constant. { 'kwäd·rə‚triks əv 'hip·ē·əs }

quadrature 1. The construction of a square whose area is equal to that of a given surface. **2.** The process of calculating a definite integral. { 'kwä·drə·chər }

quadric cone A conical surface whose directrices are conic curves. { 'kwä·drik 'kōn }

quadric curve An algebraic curve whose equation is of the second degree. { 'kwäd·rik 'kərv }

quadrics Homogeneous, second-degree expressions. { 'kwä·driks }

quadric surface A surface whose equation is a second-degree algebraic equation. { 'kwä·drik 'sər·fəs }

quadrilateral *See* quadrangle. { ¦kwä·drə'lad·ə·rəl }

quadrillion 1. The number 10^{15}. **2.** In British and German usage, the number 10^{24}. { kwə'dril·yən }

quadruple vector product 1. For any four vectors, the dot product of two derived vectors, one of which is the cross product of two of the original vectors, and the other of which is the cross product of the other two. **2.** For any four vectors, the cross product of two derived vectors, one of which is the cross product of two of the original vectors, and the other of which is the cross product of the other two. { kwə'drüp·əl 'vek·tər ‚präd·əkt }

quadrupole A mass distribution that has unequal components of the moment-of-inertia tensor along the three principal directions. { 'kwä·drə‚pōl }

quantic A homogeneous algebraic polynomial with more than one variable. { 'kwän·tik }

quantifier Either of the phrases "for all" and "there exists"; these are symbolized respectively by an inverted A and a backward E. { 'kwän·tə‚fī·ər }

quantity Any expression which is concerned with value rather than relations. { 'kwän·əd·ē }

quartic *See* biquadratic. { 'kwȯrd·ik }

quartic equation Any fourth-degree polynomial equation. Also known as biquadratic equation. { 'kwȯrd·ik i'kwā·zhən }

quartic surd A fourth root of a rational number that is itself an irrational number. { 'kwärd·ik ‚sərd }

quasi-F martingale A stochastic process which is the sum of an F martingale and an F process having bounded variation on every finite time interval. { ¦kwä·zē ¦ef 'märt·ən‚gāl }

quasi-perfect number An integer that is 1 less than the sum of all its factors other than itself. { ¦kwäz·ē ¦pər·fikt 'nəm·bər }

quaternion The division algebra over the real numbers generated by elements i, j, k subject to the relations $i^2 = j^2 = k^2 = -1$ and $ij = -ji = k, jk = -kj = i$, and $ki = -ik = j$. Also known as hypercomplex number. { kwə'ter·nē·ən }

quatrefoil A multifoil consisting of four congruent arcs of a circle arranged around a square. { 'kwä·trə‚fȯil }

queueing theory The area of stochastic processes emphasizing those processes modeled on the situation of individuals lining up for service. { 'kyü·iŋ ‚thē·ə·rē }

quintic A fifth-degree expression. { 'kwin·tik }

quintic equation A fifth-degree polynomial equation. { 'kwin·tik i'kwā·zhən }

quintic surd A fifth root of a rational number that is itself an irrational number. { 'kwin·tik ‚sərd }

quintillion **1.** The number 10^{18}. **2.** In British and German usage, the number 10^{30}. { kwin'til·yən }

quotient The result of dividing one quantity by another. { 'kwō·shənt }

quotient field The smallest field containing a given integral domain; obtained by formally introducing all quotients of elements of the integral domain. { 'kwō·shənt ‚fēld }

quotient group A group G/H whose elements are the cosets gH of a given normal subgroup H of a given group G, and the group operation is defined as $g_1H \cdot g_2H \equiv (g_1 \cdot g_2)H$. Also known as factor group. { 'kwō·shənt ‚grüp }

quotient ring A ring R/I whose elements are the cosets rI of a given ideal I in a given ring R, where the additive and multiplicative operations have the form: $r_1I + r_2I \equiv (r_1 + r_2)I$ and $r_1I \cdot r_2I \equiv (r_1 \cdot r_2)I$. Also known as factor ring; residue class ring. { 'kwō·shənt ‚riŋ }

quotient set The set of all the equivalence classes relative to a given equivalence relation on a given set. { 'kwō·shənt ‚set }

quotient space The topological space Y which is the set of equivalence classes relative to some given equivalence relation on a given topological space X; the topology of Y is canonically constructed from that of X. Also known as factor space. { 'kwō·shənt ‚spās }

quotient topology If X is a topological space, X/R the quotient space by some equivalence relation on X, the quotient topology on X/R is the smallest topology which makes the function which assigns to each element of X its equivalence class in X/R a continuous function. { 'kwō·shənt tə'päl·ə·jē }

R

Raabe's convergence test An infinite series with positive terms a_n where, for each n, $a_{n+1}/a_n = 1/(1 + b_n)$ will converge if, after a certain term, nb_n always exceeds a fixed number greater than 1 and will diverge if nb_n always is less than a fixed number less than or equal to 1. { 'räb·əz kən'vər·jəns ˌtest }

radial For a plane curve C, the locus of end points of lines, drawn from a fixed point, that are equal and parallel to the radius of curvature of C. { 'rād·ē·əl }

radial distribution function A function $F(r)$ equal to the average of a given function of the three coordinates over a sphere of radius r centered at the origin of the coordinate system. { 'rād·ē·əl ˌdis·trə'byü·shən ˌfəŋk·shən }

radian The central angle of a circle determined by two radii and an arc joining them, all of the same length. { 'rād·ē·ən }

radical **1.** In a ring, the intersection of all maximal ideals. Also known as Jacobson radical. **2.** An indicated root of a quantity. Symbolized $\sqrt{\ }$. { 'rad·ə·kəl }

radical axis The line passing through the two points of intersection of a pair of circles. { 'rad·ə·kəl 'ak·səs }

radical center **1.** For three circles, the point at which the three radical axes of pairs of the circles intersect. **2.** For four spheres, the point at which the six radical planes of pairs of the spheres intersect. { 'rad·ə·kəl 'sen·tər }

radical equation *See* irrational equation. { 'rad·ə·kəl i'kwā·zhən }

radical plane The plane containing the circle of intersection of a pair of spheres. { 'rad·ə·kəl 'plān }

radius **1.** A line segment joining the center and a point of a circle or sphere. **2.** The length of such a line segment. { 'rād·ē·əs }

radius of convergence The positive real number corresponding to a power series expansion about some number a with the property that if $x - a$ has absolute value less than this number the power series converges at x, and if $x - a$ has absolute value greater than this number the power series diverges at x. { 'rād·ē·əs əv kən'vər·jəns }

radius of curvature The radius of the circle of curvature at a point of a curve. { 'rād·ē·əs əv 'kər·və·chər }

radius of geodesic curvature For a point on a curve lying on a surface, the reciprocal of the geodesic curvature at the point. { ¦rād·ē·əs əv ˌjē·ə¦des·ik 'kər·və·chər }

radius of geodesic torsion The reciprocal of the geodesic torsion of a surface at a point in a given direction. { ¦rād·ē·əs əv ˌjē·ə¦des·ik 'tȯr·shən }

radius of gyration The square root of the ratio of the moment of inertia of a plane figure about a given axis to its area. { 'rād·ē·əs əv ji'rā·shən }

radius of normal curvature The reciprocal of the normal curvature of a surface at a point and in a given direction. { ¦rād·ē·əs əv ¦nȯr·məl 'kər·və·chər }

radius of torsion The reciprocal of the torsion of a space curve at a point. { ¦rād·ē·əs əv 'tȯr·shən }

radius of total curvature The quantity $\sqrt{-1/C}$, where C is the total curvature of a surface at a point. { ¦rād·ē·əs əv ¦tōd·əl 'kər·və·chər }

radius vector The coordinate r in a polar coordinate system, which gives the distance of a point from the origin. { 'rād·ē·əs ˌvek·tər }

radix *See* base of a number system; root. { 'rād·iks }

radix approximation The approximation of a number by a number that can be expressed by a specified finite number of digits in radix notation. { 'rād·iks əˌpräk·sə'mā·shən }

radix complement A numeral in positional notation that can be derived from another by subtracting the original numeral from the numeral of highest value with the same number of digits, and adding 1 to the difference. Also known as complement; true complement. { 'rād·iks 'käm·plə·mənt }

radix fraction A generalization of a decimal fraction given by an expression of the form $(a/r) + (b/r^2) + (c/r^3) + \cdots$, where r is an integer and $a, b, c, \ldots$ are integers that are less than r. { 'rād·iks ˌfrak·shən }

radix-minus-one complement A numeral in positional notation of base (or radix) B derived from a given numeral by subtracting the latter from the highest numeral with the same number of digits, that is, from B−1; it is 1 less than the radix complement. { 'rād·iks ¦mī·nəs ¦wən 'käm·plə·mənt }

radix notation A positional notation in which the successive digits are interpreted as coefficients of successive integral powers of a number called the radix or base; the represented number is equal to the sum of this power series. Also known as base notation. { 'rād·iks nōˌtā·shən }

radix point A dot written either on or slightly above the line, used to mark the point at which place values change from positive to negative powers of the radix in a number system; a decimal point is a radix point for radix 10. { 'rād·iks ˌpȯint }

Radon measure *See* regular Borel measure. { 'rāˌdän ˌmezh·ər }

Radon transform A mathematical operation that is roughly equivalent to finding the projection of a function along a given line; useful in computerized tomography. { 'rāˌdän 'tranzˌfȯrm }

ramphoid cusp A cusp of a curve which has both branches of the curve on the same side of the common tangent. Also known as single cusp of the second kind. { 'ramˌfȯid ˌkəsp }

random function A function whose domain is an interval of the extended real numbers and has range in the set of random variables on some probability space; more precisely, a mapping of the cartesian product of an interval in the extended reals with a probability space to the extended reals so that each section is a random variable. { 'ran·dəm 'fəŋk·shən }

random noise A form of random stochastic process arising in control theory. { 'ran·dəm 'nȯiz }

random numbers A listing of numbers which is nonrepetitive and satisfies no algorithm. { 'ran·dəm 'nəm·bərz }

random process *See* stochastic process. { 'ran·dəm 'prä·səs }

random variable A measurable function on a probability space; usually real valued, but possibly with values in a general measurable space. Also known as chance variable; stochastic variable; variate. { 'ran·dəm 'ver·ē·ə·bəl }

random walk A succession of movements along line segments where the direction and possibly the length of each move is randomly determined. { 'ran·dəm 'wȯk }

range The range of a function f from a set X to a set Y consists of those elements y in Y for which there is an x in X with $f(x) = y$. { rānj }

rank **1.** The rank of a matrix is its maximum number of linearly independent rows. **2.** The rank of a system of homogeneous linear equations equals the rank of the matrix of its coefficients. **3.** A tensor in an n-dimensional space is of rank r if it has n^r components. **4.** The rank of a group G is the number of elements in the basis of the quotient group of G over the subgroup consisting of all elements of G having finite period. **5.** The rank of a place or valuation is equal to the number of proper prime ideals in its valuation ring. **6.** The rank of a prime ideal P is the largest number n for which there exists a sequence $P_0 = P, P_1, P_2, \ldots, P_n$ of prime ideals such that P_i is a subset of P_{i-1}. { raŋk }

rate of change *See* derivative. { 'rāt əv 'chānj }

ratio A ratio of two quantities or mathematical objects A and B is their quotient or fraction A/B. { 'rā·shō }

rational fraction **1.** A fraction whose numerator and denominator are both rational numbers. **2.** A fraction whose numerator and denominator are both polynomials. { 'rash·ən·əl 'frak·shən }

rational function A function which is a quotient of polynomials. { 'rash·ən·əl 'fəŋk·shən }

rationalize **1.** To carry out operations on an algebraic equation that remove radicals containing the varible. **2.** To multiply the numerator and denominator of a fraction by a quantity that removes the radicals in the denominator. **3.** To make a substitution in an integral that removes the radicals in the integrand. { 'rash·ən·əl‚īz }

rational number A number which is the quotient of two integers. { 'rash·ən·əl 'nəm·bər }

rational root theorem The theorem that, if a rational number p/q, where p and q have no common factors, is a root of a polynomial equation with integral coefficients, then the coefficient of the term of highest order is divisible by q and the coefficient of the term of lowest order is divisible by p. { 'rash·ən·əl ‚rüt ‚thir·əm }

ratio of similitude The ratio of the lengths of corresponding line segments of similar figures. Also known as homothetic ratio. { ¦rā·shō əv sə'mil·ə‚tüd }

ratio test *See* Cauchy ratio test. { 'rā·shō ‚test }

ray A straight-line segment emanating from a point. Also known as half line. { rā }

Rayleigh-Ritz method An approximation method for finding solutions of functional equations in terms of finite systems of equations. { 'rā·lē 'rits ‚meth·əd }

real axis The horizontal axis of the cartesian coordinate system for the euclidean or complex plane. { 'rēl 'ak·səs }

real closed field A real field which has no algebraic extensions other than itself. { ¦rēl ¦klōzd 'fēld }

real closure A real closure of a real field F is a real closed field which is an algebraic extension of F. { ¦rēl 'klō·zhər }

real line A straight line of infinite extent upon which the real numbers are plotted according to their distance in a positive or negative direction from a point arbitrarily chosen as zero. Also known as number line. { 'rēl 'līn }

real linear group The group of all nonsingular linear transformations of a real vector space whose group operation is composition. { ¦rēl ¦lin·ē·ər 'grüp }

real number Any member of the unique (to within isomorphism) complete ordered field. { 'rēl 'nəm·bər }

real orthogonal group The group composed of orthogonal matrices having real number entries. { 'rēl ȯr'thäg·ən·əl ‚grüp }

real part The real part of a complex number $z = x + iy$ is the real number x. { 'rēl ¦pärt }

real plane A plane whose points are assigned ordered pairs of real numbers for coordinates. { 'rēl ‚plān }

real unimodular group The group of all square $n \times n$ matrices with real number entries and of determinant 1. { 'rēl ‚yün·i'mäj·ə·lər ‚grüp }

real-valued function A function whose values are real numbers. { ¦rēl ¦val·yüd 'fəŋk·shən }

real variable A variable that assumes real numbers for its values. { 'rēl 'ver·ē·ə·bəl }

reciprocal The reciprocal of a number A is the number $1/A$. { ri'sip·rə·kəl }

reciprocal differences An interpolation technique using successive quotients of a function with its values so as to obtain a continued fraction expansion approximating the given function by a rational function. { ri'sip·rə·kəl 'dif·rən·səz }

reciprocal equation An algebraic equation in one variable whose roots are unchanged when the unknown is replaced by its reciprocal. { rə'sip·rə·kəl i'kwā·zhən }

reciprocal polar figures Two plane figures consisting of lines and their points of intersection such that the points of each of them are the poles of the lines of the other with respect to a given conic. { ri¦sip·rə·kəl ¦pō·lər 'fig·yərz }

reciprocal series A series whose terms are reciprocals of the corresponding terms of a given series. { rə′sip·rə·kəl ′sir‚ēz }

reciprocal spiral *See* hyperbolic spiral. { ri′sip·rə·kəl ‚spī·rəl }

reciprocal theorem **1.** In plane geometry, a theorem (which may be true or false) that is obtained from a given theorem by exchanging points and lines, angles and sides, and so forth. **2.** *See* dual theorem. { ri′sip·rə·kəl ′thir·əm }

rectangle A plane quadrilateral having four interior right angles and opposite sides of equal length. { ′rek‚taŋ·gəl }

rectangular cartesian coordinate system *See* cartesian coordinate system. { rek′taŋ·gyə·lər kär¦tē·zhən kō′ȯrd·ən·ət ‚sis·təm }

rectangular game *See* matrix game. { rek′taŋ·gyə·lər ′gām }

rectangular hyperbola A hyperbola whose major and minor axes are equal. { rek′taŋ·gyə·lər hī′pər·bə·lə }

rectangular parallelepiped A parallelepiped with bases as rectangles all perpendicular to its lateral faces. Also known as rectangular solid. { rek′taŋ·gyə·lər ‚par·ə‚lel·ə′pī‚ped }

rectangular solid *See* rectangular parallelepiped. { rek′taŋ·gyə·lər ′säl·əd }

rectifiable curve A curve whose length can be computed and is finite. { ′rek·tə‚fī·ə·bəl ′kərv }

rectifying developable The envelope of the rectifying planes of a space curve. { ′rek·tə‚fī·iŋ də′vel·əp·ə·bəl }

rectifying plane The plane that contains the tangent and binormal to a curve at a given point on the curve. { ′rek·tə‚fī·iŋ ‚plān }

rectilinear Consisting of or bounded by lines. { ¦rek·tə′lin·ē·ər }

rectilinear generators Straight lines which generate ruled surfaces. { ¦rek·tə′lin·ē·ər ′jen·ə‚rād·ərz }

recurrence formula methods Methods of calculating numerical solutions of differential equations in which the equation is written in the form of a recurrence relation between values of the solution function at successive points by replacing the derivatives with corresponding finite difference expressions. { ri¦kər·əns ′fȯr·myə·lə ‚meth·ədz }

recurrent transformation **1.** A measurable function from a measure space T to itself such that for every measurable set A in the space and every point x in A there is a positive integer n such that $T^n(x)$ is also in A. **2.** A continuous function from a topological space T to itself such that for every open set A in the space and every point x in A there is a positive integer n such that $T^n(x)$ is also in A. { ri′kər·ənt ‚tranz·fər′mā·shən }

recurring continued fraction A continued fraction in which a finite sequence of terms is repeated indefinitely. Also known as periodic continued fraction. { ri¦kər·iŋ kən¦tin·yüd ′frak·shən }

recurring decimal *See* repeating decimal. { ri′kər·iŋ ′des·məl }

recursion formula An algorithm allowing computation of a succession of quantities. Also known as recursion relation. { ri'kər·zhən ˌfȯr·myə·lə }

recursion relation *See* recursion formula. { ri'kər·zhən riˌlā·shən }

recursive Pertaining to a process that is inherently repetitive, with the results of each repetition usually depending upon those of the previous repetition. { ri'kər·siv }

recursive functions Functions that can be obtained by a finite number of operations, computations, or algorithms. { ri'kər·siv 'fəŋk·shənz }

reduced characteristic equation The polynomial equation of lowest degree that is satisfied by a given matrix. Also known as minimal equation. { ri'düst ˌkär·ik·tə'ris·tik iˌkwā·zhən }

reduced cubic equation A cubic equation in a variable x, where the coefficient of x^2 is zero. { ri¦düst ¦kyü·bik i'kwā·zhən }

reduced equation *See* auxiliary equation. { ri'düst i'kwā·zhən }

reduced form A lambda expression that has no subexpressions of the form (λxMA), where M and A are lambda expressions, is said to be in reduced form. { ri'düst 'fȯrm }

reduced residue system modulo *n* A set of integers that are relatively prime to n (where n is a positive integer), and that includes one and only one member of each number class modulo n whose integers are relatively prime to n. { ri¦düst ¦res·əˌdü ¦sis·təm ¦mäj·əˌlō 'en }

reducible configuration A graph such that the four-colorability of any planar graph containing the configuration can be deduced from the four-colorability of planar graphs with fewer vertices. { ri'dü·sə·bəl kənˌfig·yə'rā·shən }

reducible curve A curve that can be shrunk to a point by a continuous deformation without passing outside a given region. { ri¦düs·ə·bəl 'kərv }

reducible polynomial A polynomial relative to some field which can be written as the product of two polynomials of degree at least 1. { ri'dü·sə·bəl ˌpäl·i'nō·mē·əl }

reducible transformation A linear transformation T on a vector space V that can be completely specified by describing its effect on two subspaces, M and N, that are each transformed into themselves by T and are such that any vector of V can be uniquely represented as the sum of a vector of M and a vector of N. { ri¦düs·ə·bəl ˌtranz·fər'mā·shən }

reductio ad absurdum A method of proof in which it is first supposed that the fact to be proved is false, and then it is shown that this supposition leads to the contradiction of accepted facts. { ri¦dək·tē·ō äd ab'sərd·əm }

reduction formula **1.** An equation that expresses an integral as the sum of certain functions and a simpler integral. **2.** An identity that expresses the values of a trigonometric function of an angle greater than 90° in terms of a function of an angle less than 90°. { ri'dək·shən ˌfȯr·myə·lə }

reduction sequence A sequence of applications of the reduction rule to a lambda expression. { ri'dək·shən ˌsē·kwəns }

redundancy A repetitive statement. { ri'dən·dən·sē }

redundant equation An equation with roots that have been introduced in the process of solving another equation but that are not solutions of the equation to be solved. { ri'dən·dənt i'kwā·zhən }

reentrant angle An interior angle of a polygon that is greater than 180°. { rē'en·trənt ˌaŋ·gəl }

refinement A tower that can be obtained by inserting a finite number of subsets in a given tower. { ri'fīn·mənt }

reflection The reflection of a configuration in a line, in a plane, or in the origin of a coordinate system is the replacement of each point in the configuration by a point that is symmetric to the given point with respect to the line, plane, or origin. { ri'flek·shən }

reflection plane *See* plane of mirror symmetry; plane of reflection. { ri'flek·shən ˌplān }

reflex angle An angle greater than 180° and less than 360°. { 'rēˌfleks ˌaŋ·gəl }

reflexive Banach space A Banach space B such that for every continuous linear functional F on the conjugate space B^*, there corresponds a point x_0 of B such that $F(f) = f(x_0)$ for each element f of B^*. Also known as regular Banach space. { ri'flek·siv 'bäˌnäk ˌspās }

reflexive relation A relation among the elements of a set such that every element stands in that relation to itself. { ri'flek·siv riˌlā·shən }

region *See* domain. { 'rē·jən }

regret criterion *See* Savage principle. { ri'gret krīˌtir·ē·ən }

regula falsi A method of calculating an unknown quantity by first making an estimate and then using this and the properties of the unknown to obtain it. Also known as rule of false position. { 'reg·yə·lə 'fäl·sē }

regular Baire measure A Baire measure such that the measure of any Baire set E is equal to both the greatest lower bound of measures of open Baire sets containing E, and to the least upper bound of closed, compact sets contained in E. { 'reg·yə·lər 'bār ˌmezh·ər }

regular Banach space *See* reflexive Banach space. { 'reg·yə·lər 'bäˌnäk ˌspās }

regular Borel measure A Borel measure such that the measure of any Borel set E is equal to both the greatest lower bound of measures of open Borel sets containing E, and to the least upper bound of measures of compact sets contained in E. Also known as Radon measure. { 'reg·yə·lər bə'rel ˌmezh·ər }

regular curve A curve that has no singular points. { 'reg·yə·lər 'kərv }

regular dodecahedron A regular polyhedron of 12 faces. { 'reg·yə·lər dōˌdek·ə'hē·drən }

regular extension An extension field K of a field F such that F is algebraically closed in K and K is separable over F; equivalently, an extension field K of a field F such that K and $\overline{F}$ are linearly disjoint over F, where $\overline{F}$ is the algebraic closure of F. { 'reg·yə·lər ik'sten·chən }

regular function An analytic function of one or more complex variables. { 'reg·yə·lər 'fəŋk·shən }

regular map *See* normal map. { 'reg·yə·lər 'map }

regular icosahedron A 20-sided regular polyhedron, having five equilateral triangles meeting at each face. { 'reg·yə·lər ī,käs·ə'hē·drən }

regular octahedron A regular polyhedron of eight faces. { 'reg·yə·lər ,äk·tə'hē·drən }

regular permutation group A permutation group of order n on n objects, where n is a positive integer. { 'reg·yə·lər ,pər·myə'tā·shən ,grüp }

regular polygon A polygon with congruent sides and congruent interior angles. { 'reg·yə·lər 'päl·i,gän }

regular polyhedron A polyhedron all of whose faces are regular polygons, and whose polyhedral angles are congruent. { 'reg·yə·lər ,päl·i'hē·drən }

regular prism A right prism whose bases are regular polygons. { ¦reg·yə·lər 'priz·əm }

regular representation A regular representation of a finite group is an isomorphism of it with a group of permutations. { 'reg·yə·lər ,rep·rə·zən'tā·shən }

regular singular point A regular singular point of a differential equation is a singular point of the equation at which none of the solutions has an essential singularity. { 'reg·yə·lər ¦siŋ·gyə·lər 'pȯint }

regular tetrahedron A regular polyhedron of four faces. { 'reg·yə·lər ,te·trə'hē·drən }

regular topological space A topological space where any point and a closed set not containing it can be enclosed in disjoint open sets. { 'reg·yə·lər ¦täp·ə¦läj·ə·kəl 'spās }

related angle The acute angle at which trigonometric functions have the same absolute values as at a given angle outside the first quadrant. { ri'lād·əd ,aŋ·gəl }

relation A set of ordered pairs. { ri'lā·shən }

relative coordinates Coordinates given as offsets from some point whose location can be adjusted. { 'rel·əd·iv kō'ȯrd·ən·əts }

relative error The absolute error in estimating a quantity divided by its true value. { 'rel·əd·iv 'er·ər }

relatively closed set A subset of a topological space is relatively closed if it is a closed set in some relative topology of a subset. { 'rel·ə,tiv·lē ¦klōzd ,set }

relatively compact set *See* conditionally compact set. { 'rel·ə,tiv·lē ¦käm,pakt ,set }

relatively open set A subset of a topological space is relatively open if it is an open set in some relative topology of a subset. { 'rel·ə,tiv·lē ¦ō·pən ,set }

relatively prime Integers m and n are relatively prime if there are integers p and q so that $pm + qn = 1$; equivalently, if they have no common factors other than 1. { 'rel·ə,tiv·lē 'prīm }

relative maximum A value of a function at a point x_0 which is equal to or greater than the values of the function at all points in some neighborhood of x_0. { 'rel·ə·tiv 'mak·sə·məm }

relative minimum A value of a function at a point x_0 which is equal to or less than the values of the function at all points in some neighborhood of x_0. { 'rel·ə·tiv 'min·ə·məm }

relative primes Two positive integers with no common positive divisor other than 1. { 'rel·ə·tiv 'prīmz }

relative topology In a topological space X any subset A has a topology on it relative to the given one by intersecting the open sets of X with A to obtain open sets in A. { 'rel·əd·iv tə'päl·ə·jē }

relaxation *See* relaxation method. { ˌrēˌlak'sā·shən }

relaxation method A successive approximation method for solving systems of equations where the errors from an initial approximation are viewed as constraints to be minimized or relaxed within a toleration limit. Also known as relaxation. { ˌrēˌlak'sā·shən ˌmeth·əd }

remainder **1.** The remaining integer when a division of an integer by another is performed; if $l = m \cdot p + r$, where l, m, p, and r are integers and r is less than p, then r is the remainder when l is divided by p. **2.** The remaining polynomial when division of a polynomial is performed; if $l = m \cdot p + r$, where l, m, p, and r are polynomials, and the degree of r is less than that of p, then r is the remainder when l is divided by p. **3.** The remaining part of a convergent infinite series after a computation, for some n, of the first n terms. { ri'mān·dər }

remainder formula A formula by which the remainder resulting from an approximation of a function by a partial sum of a power series can be computed or analyzed. { ri'mān·dər ˌfȯr·myə·lə }

remainder theorem Dividing a polynomial $p(x)$ by $(x - a)$ gives a remainder equaling the number $p(a)$. { ri'mān·dər ˌthir·əm }

removable discontinuity A point where a function is discontinuous, but it is possible to redefine the function at this point so that it will be continuous there. { ri'müv·ə·bəl ¦disˌkänt·ən'ü·əd·ē }

renaming rule A transformation rule in the lambda calculus that allows conflicts of variables to be eliminated; it states that a bound variable x in a lambda expression M may be uniformly replaced by some other bound variable y, provided y does not occur in M. Also known as alpha rule. { rē'nām·iŋ ˌrül }

renormalization transformation A transformation of a mathematical function involving a change of scale. { rēˌnȯr·mə·lə'zā·shən ˌtranz·fərˌmā·shən }

repeated root *See* multiple root. { ri'pēd·əd 'rüt }

repeating decimal A decimal that is either finite or infinite with a finite block of digits repeating indefinitely. Also known as recurring decimal. { ri'pēd·iŋ 'des·məl }

representation A representation of a group is given by a homomorphism of it onto some group either of matrices or unitary operators of a Hilbert space. { ˌrep·riˌzen'tā·shən }

representation theory **1.** The study of groups by the use of their representations.

2. The determination of representations of specific groups. { ˌrep·riˌzen'tā·shən ˌthē·ə·rē }

residual set In a topological space, the complement of a set which is a countable union of nowhere dense sets. { rə'zij·ə·wəl 'set }

residual spectrum Those members λ of the spectrum of a linear operator A on a Banach space X for which $(A - \lambda I)^{-1}$, I being the identity operator, is unbounded with domain not dense in X. { rə'zij·ə·wəl 'spek·trəm }

residue **1.** The residue of a complex function $f(z)$ at an isolated singularity z_0 is given by $(1/2\pi i) \int f(z)dz$ along a simple closed curve interior to an annulus about z_0; equivalently, the coefficient of the term $(z - z_0)^{-1}$ in the Laurent series expansion of $f(z)$ about z_0. **2.** In general, a coset of an ideal in a ring. **3.** A residue of m of order n, where m and n are integers, is a remainder that results from raising some integer to the nth power and dividing by m. { 'rez·əˌdü }

residue class A set of numbers satisfying a congruency relation. { 'rez·əˌdü ˌklas }

residue class ring *See* quotient ring. { 'rez·əˌdü ¦klas ˌriŋ }

residue theorem The value of the integral of a complex function, taken along a simple closed curve enclosing at most a finite number of isolated singularities, is given by $2\pi i$ times the sum of the residues of the function at each of the singularities. { 'rez·əˌdü ˌthir·əm }

resolution For a vector, the determination of vectors parallel to specified (usually perpendicular) axes such that their sum equals the given vector. { ˌrez·ə'lü·shən }

resolution of the identity A family of linear projection operators on a Banach space used in studying the spectra of linear operators. { ˌrez·ə'lü·shən əv thē ə'den·əd·ē }

resolvable balanced incomplete block design A balanced incomplete block design such that the blocks themselves are partitioned into r families of v/k blocks, such that every element occurs in exactly one block of each of these families. { ri¦zäl·və·bəl ¦bal·ənst ˌiŋ·kəmˌplēt 'bläk diˌzīn }

resolvent For a linear operator T on a Banach space, the function, defined on the complement of the spectrum of T given by $(T - \lambda I)^{-1}$ for each λ in this complement, where I is the identity operator; this enables a study of T relative to its eigenvalues. { ri'zäl·vənt }

resolvent kernel A function appearing as an integrand in an integral representation for a solution of a linear integral equation which often completely determines the solutions. { ri'zäl·vənt 'kər·nəl }

resolvent set Those scalars λ for which the operator $T - \lambda I$ has a bounded inverse, where T is some linear operator on a Banach space, and I is the identity operator. { ri'zäl·vənt 'set }

resultant *See* vector sum. { ri'zəl·tənt }

reticular density The number of points per unit area in a two-dimensional lattice, such as the plane of a crystal lattice. { re'tik·yə·lər 'den·səd·ē }

retract A subset R of a topological space X is a retract of X if there is a continuous map f from X to R, with $f(r) = r$ for all points r of R. { 'rēˌtrakt }

reverse curve An S-shaped curve, that is, one having two arcs with their centers on opposite sides of the curve. Also known as S curve. { ri'vərs 'kərv }

rhodonea *See* rose. { ˌrōd·ən'ē·ə }

rhombohedron A prism with six parallelogram faces. { ¦räm·bō¦hē·drən }

rhomboid A parallelogram whose adjacent sides are not equal. { 'rämˌbȯid }

rhombus A parallelogram with all sides equal. { 'räm·bəs }

ribbon The plane figure generated by a straight line which moves so that it is always perpendicular to the path traced by its middle point. { 'rib·ən }

Riccati-Bessel functions Solutions of a second-order differential equation in a complex variable which have the form $zf(z)$, where $f(z)$ is a function in terms of polynomials and cos (z), sin (z). { ri'käd·ē 'bes·əl ˌfəŋk·shənz }

Riccati equation **1.** A first-order differential equation having the form $y' = A_0(x) + A_1(x)y + A_2(x)y^2$; every second-order linear differential equation can be transformed into an equation of this form. **2.** A matrix equation of the form $dP(t)/dt + P(t)F(t) + F^T(t)P(t) - P(t)G(t)R^{-1}(t)G^T(t)P(t) + Q(t) = 0$, which frequently arises in control and estimation theory. { ri'käd·ē iˌkwā·zhən }

Ricci equations Equations relating the components of the Ricci tensor, the curvature tensor, and an arbitrary tensor of a Riemann space. Also known as Ricci identities. { 'rēˌchē iˌkwā·zhənz }

Ricci identities *See* Ricci equations. { 'rēˌchē īˌden·ə·dēz }

Ricci tensor *See* contracted curvature tensor. { 'rēˌchē ˌten·sər }

Ricci theorem The covariant derivative vanishes for either of the fundamental tensors of a Riemann space. { 'rēˌchē ˌthir·əm }

Riemann-Christoffel tensor The basic tensor used for the study of curvature of a Riemann space; it is a fourth-rank tensor, formed from Christoffel symbols and their derivatives, and its vanishing is a necessary condition for the space to be flat. Also known as curvature tensor. { 'rēˌmän 'kris·tə·fəl ˌten·sər }

Riemann function A type of Green's function used in solving the Cauchy problem for a real hyperbolic partial differential equation. { 'rēˌmän ˌfəŋk·shən }

Riemann hypothesis The conjecture that the only zeros of the Riemann zeta function with positive real part must have their real part equal to $^1/_2$. { 'rēˌmän hīˌpäth·ə·səs }

Riemannian curvature A general notion of space curvature at a point of a Riemann space which is directly obtained from orthonormal tangent vectors there. { rē'män·ē·ən 'kər·və·chər }

Riemannian geometry *See* elliptic geometry. { rē'män·ē·ən jē'äm·ə·trē }

Riemannian manifold A differentiable manifold where the tangent vectors about each point have an inner product so defined as to allow a generalized study of distance and orthogonality. { rē'män·ē·ən 'man·əˌfōld }

Riemann integral The Riemann integral of a real function $f(x)$ on an interval (a,b) is the unique limit (when it exists) of the sum of $f(a_i)(x_i - x_{i-1})$, $i = 1, ..., n$, taken over all partitions of (a,b), $a = x_0 < a_1 < x_1 < \cdots < a_n < x_n = b$, as the maximum distance between x_i and x_{i-1} tends to zero. { 'rē,män ,int·ə·grəl }

Riemann-Lebesgue lemma If the absolute value of a function is integrable over the interval where it has a Fourier expansion, then its Fourier coefficients a_n tend to zero as n goes to infinity. { 'rē,män lə'beg ,lem·ə }

Riemann mapping theorem Any simply connected domain in the plane with boundary containing more than one point can be conformally mapped onto the interior of the unit disk. { 'rē,män 'map·iŋ ,thir·əm }

Riemann method A method of solving the Cauchy problem for hyperbolic partial differential equations. { 'rē,män ,meth·əd }

Riemann P function A scheme for exhibiting the singular points of a second-order ordinary differential equation, and the orders at these points of solutions of the equation. { 'rē·män 'pē ,fəŋk·shən }

Riemann space A Riemannian manifold or subset of a euclidean space where tensors can be defined to allow a general study of distance, angle, and curvature. { 'rē,män ,spās }

Riemann sphere The two-sphere whose points are identified with all complex numbers by a stereographic projection. Also known as complex sphere. { 'rē,män ,sfir }

Riemann surfaces Sheets or surfaces obtained by analyzing multiple-valued complex functions and the various choices of principal branches. { 'rē,män ,sər·fə·səz }

Riemann tensors Various types of tensors used in the study of curvature for a Riemann space. { 'rē,män ,ten·sərz }

Riemann zeta function The complex function $\zeta(z)$ defined by an infinite series with nth term $e^{-z \log n}$. Also known as zeta function. { 'rē,män 'zād·ə ,fəŋk·shən }

Riesz-Fischer theorem The vector space of all real- or complex-valued functions whose absolute value squared has a finite integral constitutes a complete inner product space. { 'rēsh 'fish·ər ,thir·əm }

right angle An angle of 90°. { 'rīt 'aŋ·gəl }

right circular cone A circular cone whose axis is perpendicular to its base. { 'rīt 'sər·kyə·lər 'kōn }

right circular cylinder A solid bounded by two parallel planes and by a cylindrical surface consisting of the straight lines perpendicular to the planes and passing through a circle in one of them. { 'rīt 'sər·kyə·lər 'sil·ən·dər }

right coset A right coset of a subgroup H of a group G is a subset of G consisting of all elements of the form ha, where a is a fixed element of G and h is any element of H. { 'rīt 'kō,set }

right-handed coordinate system 1. A three-dimensional rectangular coordinate system such that when the thumb of the right hand extends in the positive direction of the first (or x) axis the fingers fold in the direction in which the second (or y) axis could be rotated about the first axis to coincide with the third (or z) axis. 2. A Riemann space which has negative scalar density function. { 'rīt ¦han·dəd kō'ȯrd·ən,ət ,sis·təm }

right-handed curve A space curve whose torsion is negative at a given point. Also known as dextrorse curve; dextrorsum. { ¦rīt ¦hand·əd 'kərv }

right-hand limit *See* limit on the right. { 'rīt ¦hand 'lim·ət }

right helicoid The surface that is swept out by a ray that originates at an axis and remains perpendicular to this axis while the ray is rotated about the axis and is translated in the direction of the axis, both at a constant rate. { 'rīt 'hel·ə‚kȯid }

right identity In a set on which a binary operation $\circ$ is defined, an element e with the property that $a \circ e = a$ for every element a in the set. { 'rīt ī'den·əd·ē }

right parallelepiped A parallelepiped whose lateral faces are perpendicular to its bases. { 'rīt ‚par·ə‚lel·ə'pī·pəd }

right prism A prism whose lateral edges are perpendicular to the bases. { 'rīt ‚priz·əm }

right section A plane section by a plane perpendicular to the elements of a given cylinder, or to the lateral faces of a given prism. { 'rīt 'sek·shən }

right spherical triangle A spherical triangle that has at least one right angle. { ¦rīt ¦sfir·ə·kəl 'trī‚aŋ·gəl }

right strophoid A plane curve derived from a straight line L and a point called the pole, consisting of the locus of points on a rotating line L' passing through the pole whose distance from the intersection of L and L' is equal to the distance of this intersection from the foot of the perpendicular from the pole to L. { 'rīt 'strä‚fȯid }

right triangle A triangle one of whose angles is a right angle. { 'rīt 'trī‚aŋ·gəl }

right truncated prism A truncated prism, in which one of the cutting planes is perpendicular to the lateral edges. { ¦rīt ¦trəŋ·kād·əd 'priz·əm }

ring **1.** An algebraic system with two operations called multiplication and addition; the system is a commutative group relative to addition, and multiplication is associative, and is distributive with respect to addition. **2.** A ring of sets is a collection of sets where the union and difference of any two members is also a member. { riŋ }

ring isomorphism An isomorphism between rings. { 'riŋ ‚ī·sō'mȯr‚fiz·əm }

ring of operators *See* von Neumann algebra. { ¦riŋ əv 'äp·ə‚rād·ərz }

ring theory The study of the structure of rings in algebra. { 'riŋ ‚thē·ə·rē }

Ritz method A method of solving boundary value problems based upon reformulating the given problem as a minimization problem. { 'ritz ‚meth·əd }

Rodrigues formula **1.** The equation giving the nth function in a class of special functions in terms of the nth derivatives of some polynomial. **2.** The formula $d\mathbf{n} + k\,d\mathbf{r} = 0$, expressing the difference $d\mathbf{n}$ in the unit normals to a surface at two neighboring points on a line of curvature, in terms of the difference $d\mathbf{r}$ in the position vectors of the two points and the principal curvature k. **3.** The formula for a matrix that is used to transform the cartesian coordinates of a vector in three-space under a rotation through a specified angle about an axis with specified direction cosines. { rə'drē·gəs ‚fȯr·myə·lə }

Rolle's theorem If a function $f(x)$ is continuous on the closed interval $[a,b]$ and differentiable on the open interval (a,b) and if $f(a) = f(b)$, then there exists x_0, ax_0b, such that $f'(x^0) = 0$. { 'rȯlz ˌthir·əm }

root 1. A root of a given real or complex number is a number which when raised to some exponent equals that number. Also known as radix. **2.** A root of a polynomial $p(x)$ is a number a such that $p(a) = 0$. **3.** A root of an equation is a number or quantity that satisfies that equation. { rüt }

rooted ordered tree A rooted tree in which the order of the subtrees formed by deleting the root vertex is significant. { 'rüd·əd 'ȯr·dərd 'trē }

rooted tree A directed tree graph in which one vertex has no predecessor, and each of the remaining vertices has a unique predecessor. { 'rüd·əd 'trē }

root field *See* Galois field. { 'rüt ˌfēld }

root of unity A root of unity in a field F is an element a in F such that $a^n = 1$ for some positive integer n. { ¦rüt əv 'yü·nəd·ē }

root-squaring methods Methods of solving algebraic equations which involve calculating the coefficients in a sequence of equations, each of which has roots which are the squares of the roots in the previous equation. { 'rüt ˌskwer·iŋ ˌmeth·ədz }

root test An infinite series of nonnegative terms a_n converges if, after some term, the ith root of a_i is less than a fixed number smaller than 1. { 'rüt ˌtest }

root vertex The vertex of a rooted tree that has no predecessor. { 'rüt 'vərˌteks }

rose A graph consisting of loops shaped like rose petals arising from the equations in polar coordinates $r = a \sin n\theta$ or $r = a \cos n\theta$. Also known as rhodonea. { rōz }

rotation *See* curl. { rō'tā·shən }

rotation group The group consisting of all orthogonal matrices or linear transformations having determinant 1. { rō'tā·shən ˌgrüp }

Rouche's theorem If analytic functions $f(z)$ and $g(z)$ in a simply connected domain satisfy on the boundary $g(z)$ $f(z)$, then $f(z)$ and $f(z) + g(z)$ have the same number of zeros in the domain. { 'rüsh·əz ˌthir·əm }

roulette The curve traced out by a point attached to a given curve that rolls without slipping along another given curve that remains fixed. { rü'let }

rounding Dropping or neglecting decimals after some significant place. Also known as truncation. { 'rau̇nd·iŋ }

rounding error The computational error due to always rounding numbers in a calculation. Also known as round-off error. { 'rau̇nd·iŋ ˌer·ər }

round off To truncate the least significant digit or digits of a numeral, and adjust the remaining numeral to be as close as possible to the original number. { 'rau̇nˌdȯf }

round-off error *See* rounding error. { 'rau̇nˌdȯf ˌer·ər }

Routh's rule The number of roots with positive real parts of an algebraic equation is equal to the number of changes of algebraic sign of a sequence whose terms are formed from coefficients of the equation in a specified manner. Also known as Routh test. { 'rau̇ths ˌrül }

Routh table An array of numbers each of which is formed from coefficients of an algebraic equation in a specified manner; the first row of this array constitutes the sequence used in Routh's rule. { 'rau̇th ˌtā·bəl }

Routh test *See* Routh's rule. { 'rau̇th ˌtest }

rule An antecedent condition and a consequent proposition that can support deductive processes. { rül }

ruled surface A surface that can be generated by the motion of a straight line. { 'rüld 'sər·fəs }

rule of detachment The rule that if an implication is true and its antecedent is true, then the consequent is true. { 'rül əv diˌtach·mənt }

ruling One of the positions of the straight line that generates a ruled surface. { 'rül·iŋ }

rule of false position *See* regula falsi. { 'rül əv 'fȯls pə'zish·ən }

Runge-Kutta method A numerical approximation technique for solving differential equations. { 'rəŋ·ə 'ku̇d·ə ˌmeth·əd }

Russell's paradox The paradox concerning the concept of all sets which are not members of themselves which forces distinctions in set theory between sets and classes. { 'rəs·əlz 'par·əˌdäks }

S

saddle point A point where all the first partial derivatives of a function vanish but which is not a local maximum or minimum. { 'sad·əl ˌpȯint }

saddle-point method *See* steepest-descent method. { 'sad·əl ¦pȯint ˌmeth·əd }

saddle-point theory The study of differentiable functions and their derivatives from the viewpoint of saddle points, especially applicable to the calculus of variations. { 'sad·əl ¦pȯint ˌthē·ə·rē }

sagitta The distance between the midpoint of an arc and the midpoint of its chord. { sə'jid·ə }

salient angle An interior angle of a polygon that is less than 180°. { ¦sāl·yənt ¦aŋ·gəl }

salient point A point at which two branches of a curve with different tangents meet and terminate. { ¦sāl·yənt ¦pȯint }

saltus *See* oscillation. { 'sal·təs }

sample path If $\{X_t\colon t \text{ in } T\}$ is a stochastic process, a sample path for the process is the function on T to the range of the process which assigns to each t the value $X_t(w)$, where w is a previously given fixed point in the domain of the process. { 'sam·pəl ˌpath }

Savage principle A technique used in decision theory; a criterion is used to construct a regret matrix in which each outcome entry represents a regret defined as the difference between best possible outcome and the given outcome; the matrix is then used as in decision making under risk with expected regret as the decision-determining quality. Also known as regret criterion. { 'sav·ij ˌprin·sə·pəl }

scalar One of the algebraic quantities which form a field, usually the real or complex numbers, by which the vectors of a vector space are multiplied. { 'skā·lər }

scalar field **1.** The field consisting of the scalars of a vector space. **2.** A function on a vector space into the scalars of the vector space. { 'skā·lər 'fēld }

scalar function A function from a vector space to its scalar field. { 'skā·lər 'fəŋk·shən }

scalar gradient The gradient of a function. { 'skā·lər 'grā·dē·ənt }

scalar multiplication The multiplication of a vector from a vector space by a scalar from the associated field; this usually contracts or expands the length of a vector. { 'skā·lər ˌməl·tə·pli'kā·shən }

scalar product 1. A symmetric, alternating, or Hermitian form. 2. *See* inner product. { 'skā·lər 'präd·əkt }

scalar triple product The scalar triple product of vectors $\mathbf{v}_1$, $\mathbf{v}_2$, and $\mathbf{v}_3$ from euclidean three-dimensional space determines the volume of the parallelepiped with these vectors as edges; it is given by the determinant of the 3×3 matrix whose rows are the components of $\mathbf{v}_1$, $\mathbf{v}_2$, and $\mathbf{v}_3$. Also known as triple scalar product. { 'skā·lər 'trip·əl ˌpräd·əkt }

scalene spherical triangle A spherical triangle no two of whose sides are equal. { ¦skāˌlēn ¦sfir·ə·kəl 'trīˌaŋ·gəl }

scalene triangle A triangle where no two angles are equal. { 'skāˌlēn 'trīˌaŋ·gəl }

scaling symmetry The property of an object each part of which is identical to the whole seen at a different magnification; the property that characterizes a fractal. { 'skāl·iŋ ˌsim·ə·trē }

Schauder's fixed-point theorem A continuous mapping from a closed, compact, convex set in a Banach space into itself has at least one fixed point. { ¦shau̇d·ərz ¦fikst ˌpȯint 'thir·əm }

schlicht function *See* simple function. { 'shlikt ˌfəŋ·shən }

Schroeder-Bernstein theorem If a set A has at least as many elements as another set B and B has at least as many elements as A, then A and B have the same number of elements. { 'shrād·ər 'bərnˌstīn ˌthir·əm }

Schur-Cohn test A test to determine whether all the coefficients of a polynomial have magnitude less than one; the polynomial has this property only if each of a series of determinants formed from the coefficients of the polynomial in a specified manner is positive for determinants of even degree and negative for determinants of odd degree. { 'shür 'kōn ˌtest }

Schur's lemma For certain types of modules M, the ring consisting of all homomorphisms of M to itself will be a division ring. { 'shürz ˌlem·ə }

Schwartz's theory of distributions A theory that treats distributions as continuous linear functionals on a vector space of continuous functions which have continuous derivatives of all orders and vanish appropriately at infinity. { ¦shwȯrts ¦thē·ə·rē əv ˌdis·trə'byü·shənz }

Schwarz-Christoffel transformations Those complex transformations which conformally map the interior of a given polygon onto the portion of the complex plane above the real axis. { 'shvärts 'kris·tə·fel ˌtranz·fər'mā·shənz }

Schwarz' inequality *See* Cauchy-Schwarz inequality. { 'shvärts ˌin·iˌkwäl·əd·ē }

Schwarz reflection principle To obtain the analytic continuation of a given function $f(z)$ analytic in a region R, whose boundary contains a segment of the real axis, into a region reflected from R through this segment, one takes the complex conjugate function $f(\bar{z})$. { 'shvärts ri'flek·shən ˌprin·sə·pəl }

Schwarz's lemma If an analytic function of the unit disk to itself sends the origin to the origin then it must be distance-decreasing. { 'shvärt·səz ˌlem·ə }

S curve *See* reverse curve. { 'es ˌkərv }

sec *See* secant; second.

secant 1. The function given by the reciprocal of the cosine function. Abreviated sec. **2.** The secant of an angle A is $1/\cos A$. **3.** A line of unlimited length that intersects a given curve. { 'sē,kant }

sech *See* hyperbolic secant. { sek }

second A unit of plane angle, equal to 1/60 minute, or 1/3,600 degree, or $\pi/648{,}000$ radian. { 'sek·ənd }

secondary diagonal The elements of a square matrix that lie on the straight line extending from the lower left-hand corner to the upper right-hand corner of the matrix. { ¦sek·ən,der·ē dī'ag·ən·əl }

second category A set is of second category if it cannot be expressed as a countable union of nowhere dense sets. { 'sek·ənd 'kad·ə,gȯr·ē }

second countable topological space A topological space that has a countable base. { ¦sek·ənd ¦kau̇n·tə·bəl ,täp·ə¦läj·ə·kəl 'spās }

second mean-value theorem The theorem that for two functions $f(x)$ and $g(x)$ that are continuous on a closed interval $[a,b]$ and differentiable on the open interval (a,b), such that $g(b) \neq g(a)$, there exists a number x_1 in (a,b) such that either $[f(b) - f(a)]/[g(b) - g(a)] = f'(x_1)/g'(x_1)$ or $f'(x_1) = g'(x_1) = 0$. Also known as Cauchy's mean-value theorem; double law of the mean; extended mean-value theorem; generalized mean-value theorem. { ¦sek·ənd ¦mēn ¦val·yü ,thir·əm }

second moment of area *See* geometric moment of inertia. { ¦sek·ənd ¦mō·mənt əv 'er·ē·ə }

second-order difference One of the first-order differences of the sequence formed by taking the first-order differences of a given sequence. { 'sek·ənd ¦ȯr·dər 'dif·rəns }

second-order equation A differential equation where some term includes the second derivative of the unknown function and no derivative of higher order is present. { 'sek·ənd ¦ȯr·dər i'kwā·zhən }

second quadrant 1. The range of angles from 90 to 180°. **2.** In a plane with a system of cartesian coordinates, the region in which the x coordinate is negative and the y coordinate is positive. { 'sek·ənd 'kwä·drənt }

second species The class of sets G_0 such that all the sets G_n are nonempty, where, in general, G_n is the derived set of G_{n-1}. { ¦sek·ənd 'spē,shēz }

sector A portion of a circle bounded by two radii and an arc joining their end points. { 'sek·tər }

sectoral harmonic A spherical harmonic which is 0 on a set of equally spaced meridians of a sphere with center at the origin of spherical coordinates, dividing the sphere into sectors. { ¦sek·tə·rəl här'män·ik }

sectorgram *See* pie chart. { 'sek·tər,gram }

secular determinant For a square matrix A, the determinant of the matrix whose off-diagonal components are equal to those of A, and whose diagonal components are equal to the difference between those of A and a parameter λ; it is equal to the characteristic polynomial in λ of the linear transformation represented by A. { 'sek·yə·lər di'tər·mən·ənt }

segment 1. A segment of a line or curve is any connected piece. 2. A segment of a circle is a portion of the circle bounded by a chord and an arc subtended by the chord. 3. A segment of a totally ordered Abelian group G is a subset D of G such that if a is in D then so are all elements b satisfying $-a \leq b \leq a$. { 'seg·mənt }

Segrè characteristic A set of integers that are the orders of the Jordan submatrices of a classical canonical matrix, with integers that correspond to submatrices containing the same characteristic root being bracketed together. { 'se͵grā ͵kär·ik·tə͵ris·tik }

Seidel method A basic iterative procedure for solving a system of linear equations by reducing it to triangular form. Also known as Gauss-Seidel method. { 'zīd·əl ͵meth·əd }

self-adjoint operator A linear operator which is identical with its adjoint operator. { ¦self ə¦jȯint 'äp·ə͵rād·ər }

self-similarity The property whereby an object or mathematical function preserves its structure when multiplied by a certain scale factor. { ¦self ͵sim·ə'lar·əd·ē }

semiaxis A line segment that forms half of the axis of a geometric figure (such as an ellipse), having one end point at the center of symmetry of the figure. { ¦sem·ē 'ak·səs }

semiconjugate axis Either of the equal line segments into which the conjugate axis of a hyperbola is divided by the center of symmetry. { ͵sem·ē¦kän·jə·gət 'ak·səs }

semicubical parabola A plane curve whose equation in cartesian coordinates x and y is $y^2 = ax^3$, where a is some constant. Also known as isochrone. { ¦sem·i'kyü·bə·kəl pə'rab·ə·lə }

semigroup A set which is closed with respect to a given associative binary operation. { 'sem·i͵grüp }

semigroup theory The formal algebraic study of the structure of semigroups. { 'sem·i͵grüp ͵thē·ə·rē }

semilogarithmic coordinate paper Paper ruled with two sets of mutually perpendicular, parallel lines, one set being spaced according to the logarithms of consecutive numbers, and the other set uniformly spaced. { ¦sem·i¦läg·ə¦rith·mik kō¦ȯrd·ən·ət ͵pā·pər }

semimajor axis Either of the equal line segments into which the major axis of an ellipse is divided by the center of symmetry. { ͵sem·ē¦mā·jər 'ak·səs }

semimetric A real valued function $d(x,y)$ on pairs of points from a topological space which has all the same properties as a metric save that $d(x,y)$ may be zero even if x and y are distinct points. Also known as pseudometric. { ¦sem·i'me·trik }

semiminor axis Either of the equal line segments into which the minor axis of an ellipse is divided by the center of symmetry. { ͵sem·i¦mīn·ər 'ak·səs }

seminorm A scalar-valued function on a real or complex vector space satisfying the axioms of a norm, except that the seminorm of a nonzero vector may equal zero. { 'sem·ē͵nȯrm }

semiring of sets A collection S of sets that includes the empty set and the intersection of any two of its members, and is such that if A and B are members of S and A is a subset of B, then $B - A$ is the union of a finite number of disjoint members of S. { ˌsem·ē′riŋ əv ′setz }

semisimple module A module which is the sum of a family of simple modules. { ¦sem·iˌsim·pəl ′mä·jəl }

semisimple representation *See* completely reducible representation. { ¦sem·iˌsim·pəl ˌrep·ri·zen′ta·shən }

semisimple ring A ring in which 1 does not equal 0, and which is semisimple as a left module over itself. { ¦sem·iˌsim·pəl ′riŋ }

semitransverse axis Either of the equal line segments into which the transverse axis of a hyperbola is divided by the center of symmetry. { ¦sem·i¦tranzˌvərs ′ak·səs }

sentential calculus *See* propositional calculus. { sen′ten·chəl ′kal·kyə·ləs }

sentential connectives *See* propositional connectives. { sen¦ten·chəl kə′nek·tivz }

separable degree Let E be an algebraic extension of a field F, and let f be any embedding of F in a field L such that L is the algebraic closure of the image of F under f; the separable degree of E over F is the number of distinct embeddings of E in L which are extensions of f. { ′sep·rə·bəl di′grē }

separable element An element a is said to be separable over a field F if it is algebraic over F and if the extension field of F generated by a is a separable extension of F. { ′sep·rə·bəl ′el·ə·mənt }

separable extension A field extension K of a field F is separable if every element of K is a root of a separable polynomial whose coefficients are elements of F. { ′sep·rə·bəl ik′sten·chən }

separable polynomial A polynomial with no multiple roots. { ′sep·rə·bəl ˌpäl·i′nō·mē·əl }

separable space A topological space which has a countable subset that is dense. { ′sep·rə·bəl ′spās }

separated sets Sets A and B in a topological space are separated if both the closure of A intersected with B and the closure of B intersected with A are disjoint. { ′sep·əˌrād·əd ′sets }

separating transcendence base A transcendence base of a field E over a field F such that E is algebraic and separable over the field generated by F and the transcendence base. { ¦sep·əˌrād·iŋ tran′sen·dəns ˌbās }

separation axioms Properties of topological spaces such as Hansdorff, regular, and normal which reflect how points and closed sets may be enclosed in disjoint neighborhoods. { ˌsep·ə′rā·shən ′ak·sē·əmz }

separation of variables **1.** A technique where certain differential equations are rewritten in the form $f(x)dx = g(y)dy$ which is then solvable by integrating both sides of the equation. **2.** A method of solving partial differential equations in which the solution is written in the form of a product of functions, each of which depends on only one of the independent variables; the equation is then arranged so that each of the terms involves only one of the variables and its corresponding function, and each of these terms is then set equal to a constant, resulting in ordinary differential equations. Also known as product-solution method. { ˌsep·əˈrā·shən əv ˈver·ē·ə·bəlz }

septillion 1. The number 10^{24}. **2.** In British and German usage, the number 10^{42}. { sepˈtil·yən }

septinary number A number in which the quantity represented by each figure is based on a radix of 7. { ˈsep·təˌner·ē ˈnəm·bər }

sequence A listing of mathematical entities $x_1, x_2 ...$ which is indexed by the positive integers; more precisely, a function whose domain is an infinite subset of the positive integers. Also known as infinite sequence. { ˈsē·kwəns }

sequential compactness A topological space is sequentially compact if every sequence formed from its points has a convergent sequence contained in it. { siˈkwen·chəl kəmˈpak·nəs }

serially ordered set *See* linearly ordered set. { ¦sir·ē·ə·lē ¦ȯrd·ərd ˈset }

series An expression of the form $x_1 + x_2 + x_3 + \cdots$, where x_i are real or complex numbers. { ˈsir·ēz }

serpentine curve The curve given by the equation $x^2y + b^2y - a^2x = 0$, passing through and having symmetry about the origin while being asymptotic to the x axis in both directions. { ˈsər·pənˌtēn ˈkərv }

Serret-Frenet formulas *See* Frenet-Serret formulas. { səˈrā frəˈnā ˌfȯr·myə·ləz }

sesquillinear form A mapping $f(x,y)$ from $E \times F$ into R, where R is a commutative ring with an automorphism with period 2 and $E \times F$ is the cartesian product of two modules E and F over R, such that for each x in E the function which takes y into $f(x,y)$ is antilinear, and for each y in F the function which takes x into $f(x,y)$ is linear. { ¦ses·kwəˌlin·ē·ər ˈfȯrm }

set A collection of objects which has the property that, given any thing, it can be determined whether or not the thing is in the collection. { set }

set theory The study of the structure and size of sets from the viewpoint of the axioms imposed. { ˈset ˌthē·ə·rē }

sexadecimal *See* hexadecimal. { ¦sek·səˈdes·məl }

sexadecimal number system *See* hexadecimal number system. { ¦sek·səˈdes·məl ˈnəm·bər ˌsis·təm }

sexagesimal Pertaining to a multiplicity of 60 distinct alternative states or conditions or, simply, a positional numeration system to radix (or base) 60. { ¦sek·sə¦jez·ə·məl }

sexagesimal counting table A table for converting numbers using the 60 system into decimals, for example, minutes and seconds. { ¦sek·sə¦jez·ə·məl ˈkau̇nt·iŋ ˌtā·bəl }

sexagesimal measure of angles A system of angular units in which a complete revolution is divided into 360 degrees, a degree into 60 minutes, and a minute into 60 seconds. { ¦sek·sə¦jez·ə·məl ˈmezh·ər əv ˈaŋ·gəlz }

sextant A unit of plane angle, equal to 60° or $\pi/3$ radians. { ˈsek·stənt }

sextic Having the sixth degree or order. { 'sek·stik }

sextillion 1. The number 10^{21}. 2. In British and German usage, the number 10^{36}. { sek'stil·yən }

Shannon-McMillian-Breiman theorem Given an ergodic measure preserving transformation T on a probability space and a finite partition ζ of that space the limit as $n \to \infty$ of $1/n$ times the information function of the common refinement of $\zeta, T^{-1}\zeta, \ldots, T^{-n+1}\zeta$ converges almost everywhere and in the L_1 metric to the entropy of T given ζ. { 'shan·ən mik'mil·ən 'brī·mən ˌthir·əm }

Shannon's theorems These results are foundational to the mathematical study of information; mathematically they link the concept of entropy with the amount of efficient transmittal and reception of information. { 'shan·ənz ˌthir·əmz }

sheaf A fiber bundle with algebraic and topological structure usually associated to a differentiable manifold M which reflects the local behavior of differentiable functions on M. { shēf }

sheaf of planes All the planes passing through a given point. { 'shēf əv 'plānz }

sheet 1. A portion of a surface such that it is possible to travel continuously between any two points on it without leaving the surface. 2. A part of a Riemann surface such that any extension results in a multiple covering of some part of the complex plane over which the surface lies. { shēt }

sheffer stroke *See* NAND. { 'shef·ər ˌstrōk }

shifting theorem 1. If the Fourier transform of $f(t)$ is $F(x)$, then the Fourier transform of $f(t - a)$ is $\exp(iax)F(x)$. 2. If the Laplace transform of $f(x)$ is $F(y)$, then the Laplace transform of $f(x - a)$ is $\exp(-ay)F(y)$. { 'shift·iŋ ˌthir·əm }

short radius *See* apothem. { ¦shȯrt ¦rād·ē·əs }

side One of the line segments that bound a polygon. { sīd }

Sierpinski gasket A fractal which can be constructed by a recursive procedure; at each step a triangle is divided into four new triangles, only three of which are kept for further iterations. { sir'pin·skē ˌgas·kət }

Sierpinski set 1. A set of points S on a line such that both S and its complement contain at least one point in each uncountable set on the line that is a countable intersection of open sets. 2. A set of points in a plane that includes at least one point of each closed set of nonzero measure and does not include any subsets consisting of three collinear points. { sər'pin·skē ˌset }

sieve of Eratosthenes An iterative procedure which determines all the primes less than a given number. { 'siv əv ˌer·ə'täs·thəˌnēz }

sigma algebra A collection of subsets of a given set which contains the empty set and is closed under countable union and complementation of sets. Also known as sigma field. { 'sig·mə 'al·jə·brə }

sigma field *See* sigma algebra. { 'sig·mə ˌfēld }

sigma finite A measure is sigma finite on a space X if X is a countable disjoint union of sets each of which is measurable and has finite measure. { 'sig·mə 'fīˌnīt }

sigma ring A ring of sets where any countable union of its members is also a member. { 'sig·mə ˌriŋ }

sign 1. A symbol which indicates whether a quantity is greater than zero or less than zero; the signs are often the marks + and − respectively, but other arbitrarily selected symbols are used, especially in automatic data processing. **2.** A unit of plane angle, equal to 30° or $\pi/6$ radians. { sīn }

signed measure An extended real-valued function m defined on a sigma algebra of subsets of a set S such that (1) the value of m on the empty set is 0, (2) the value of m on a countable union of disjoint sets is the sum of its values on each set, and (3) m assumes at most one of the values $+\infty$ and $-\infty$. { ¦sīnd 'mezh·ər }

significance The arbitrary rank, priority, or order of relative magnitude assigned to a given position in a number. { sig'nif·i·kəns }

significant digit *See* significant figure. { sig'nif·i·kənt ˌdij·ət }

significant figure A prescribed decimal place which determines the amount of rounding off to be done; this is usually based upon the degree of accuracy in measurement. Also known as significant digit. { sig'nif·i·kənt ˌfig·yər }

signum The real function $\text{sgn}(x)$ defined for all x different from zero, where $\text{sgn}(x) = 1$ if $x > 0$ and $\text{sgn}(x) = -1$ if $x < 0$. { 'sig·nəm }

similar decimals Decimals that have the same number of decimal places. { ¦sim·ə·lər ¦des·məlz }

similar figures Two figures or bodies that are identical except for size; similar figures can be placed in perspective, so that straight lines joining corresponding parts of the two figures will pass through a common point. { 'sim·ə·lər 'fig·yərz }

similar fractions Two or more common fractions that have the same denominator. { ¦sim·ə·lər ¦frak·shənz }

similarity transformation 1. A transformation of a euclidean space obtained from such transformations as translations, rotations, and those which either shrink or expand the length of vectors. **2.** A mapping that associates with each linear transformation P on a vector space the linear transformation $R^{-1}PR$ that results when the coordinates of the space are subjected to a nonsingular linear transformation R. **3.** A mapping that associates with each square matrix P the matrix $Q = R^{-1}PR$, where R is a nonsingular matrix and R^{-1} is the inverse matrix of R; if P is the matrix representation of a linear transformation, then this definition is equivalent to the second definition. { ˌsim·ə'lar·əd·ē ˌtranz·fərˌmā·shən }

similarly placed conics Conics of the same type (both ellipses, both parabolas, or both hyperbolas) whose corresponding axes are parallel. { ¦sim·ə·lər·lē ¦plāst 'kän·iks }

similar matrices Two square matrices A and B related by the transformation $B = SAT$, where S and T are nonsingular matrices and T is the inverse matrix of S. { ¦sim·i·lər 'mā·triˌsēz }

similar terms Terms that contain the same unknown factors and the same powers of these factors. Also known as like terms. { ¦sim·ə·lər ¦tərms }

similar triangles Triangles whose corresponding angles are equal; the corresponding sides are then proportional in length. { ¦sim·ə·lər ¦trīˌaŋ·gəlz }

simple arc The image of a closed interval under a continuous, injective mapping from the interval into a plane. Also known as Jordan arc. { 'sim·pəl 'ärk }

simple character The character of an irreducible representation of a group. { 'sim·pəl 'kar·ik·tər }

simple closed curve A closed curve which never crosses itself. { 'sim·pəl 'klōzd 'kərv }

simple compression A transformation that compresses a configuration in a given direction, given by $x' = kx$, $y' = y$, $z' = z$, with $k < 1$, when the direction is that of the x axis. { ¦sim·pəl kəm¦presh·ən }

simple cusp *See* cusp of the first kind. { 'sim·pəl 'kəsp }

simple elongation A transformation that elongates a configuration in a given direction, given by $x' = kx$, $y' = y$, $z' = z$, with $k > 1$, when the direction is that of the x axis. { 'sim·pəl ˌē‚lȯŋ'gā·shən }

simple function **1.** For a region D of the complex plane, an analytic, injective function on D. Also known as schlicht function. **2.** *See* step function. { 'sim·pəl 'fəŋk·shən }

simple group A group G that is nontrivial and contains no normal subgroups other than the identity element and G itself. { ¦sim·pəl 'grüp }

simple integral An integral over only one variable. { 'sim·pəl 'int·ə·grəl }

simple order *See* linear order. { 'sim·pəl 'ȯr·dər }

simple path *See* path. { 'sim·pəl 'path }

simple ring A semisimple ring R such that for any two left ideals in R there is an isomorphism of R which maps one onto the other. { 'sim·pəl 'riŋ }

simple root A polynomial $f(x)$ has c as a simple root if $(x - c)$ is a factor but $(x - c)^2$ is not. { 'sim·pəl 'rüt }

simple shear A transformation that corresponds to a shearing motion in which a coordinate axis in the plane or a coordinate plane in space does not move, having the form $x' = x$, $y' = ax + y$, $z' = z$, where a is a constant, for a suitable choice of axes. { 'sim·pəl 'shir }

simple strain A one-dimensional strain or a simple shear. { ¦sim·pəl 'strān }

simplex An n-dimensional simplex in a euclidean space consists of $n + 1$ linearly independent points $p_0, p_1, ..., p_n$ together with all line segments $a_0p_0 + a_1p_1 + \cdots + a_np_n$ where the $a_i \geq 0$ and $a_0 + a_1 + \cdots + a_n = 1$; a triangle with its interior and a tetrahedron with its interior are examples. { 'sim‚pleks }

simplex method A finite iterative algorithm used in linear programming whereby successive solutions are obtained and tested for optimality. { 'sim‚pleks ¦meth·əd }

simplicial complex A set consisting of finitely many simplices where either two simplices are disjoint or intersect in a simplex which is a face common to each. { sim'plish·əl 'käm‚pleks }

simplicial graph A graph in which no line starts and ends at the same point, and in which no two lines have the same pair of end points. { sim′plish·əl ′graf }

simplicial homology A homology for a topological space where the nth group reflects how the space may be filled out by n-dimensional simplicial complexes and detects the presence of analogs of n-dimensional holes. { sim′plish·əl hə′mäl·ə·jē }

simplicial mapping A mapping of one simplicial complex into another in which the images of the simplexes of one complex are simplexes of the other complex. { sim′plish·əl ′map·iŋ }

simplicial subdivision A decomposition of the simplices composing a simplicial complex which results in a simplicial complex with a larger number of simplices. { sim′plish·əl ′səb·di‚vizh·ən }

simply connected region A region having no holes; all closed curves can be shrunk to a point without passing through points in the complement of the region. { ′sim·plē kə¦nek·təd ′rē·jən }

simply connected space A topological space whose fundamental group consists of only one element; equivalently, all closed curves can be shrunk to a point. { ′sim·plē kə¦nek·təd ′spās }

simply ordered set *See* linearly ordered set. { ¦sim·plē ¦ȯrd·ərd ′set }

simply normal number A number whose expansion with respect to a given base (not necessarily 10) is such that all the digits occur with equal frequency. { ¦sim·plē ¦nȯr·məl ′nəm·bər }

simply periodic function A periodic function $f(x)$ for which there is a period a such that every period of $f(x)$ is an integral multiple of a. Also known as singly periodic function. { ¦sim·plē ¦pir·ē‚äd·ik ′fəŋk·shən }

Simpson's rule Also known as parabolic rule. **1.** A basic approximation formula for definite integrals which states that the integral of a real-valued function f on an interval $[a,b]$ is approximated by $h[f(a) + 4f(g + h) + f(b)]/3$, where $h = (b - a)/2$; this is the area under a parabola which coincides with the graph of f at the abscissas a, $a + h$, and b. **2.** A method of approximating a definite integral over an interval which is equivalent to dividing the interval into equal subintervals and applying the formula in the first definition to each subinterval. { ′sim·sənz ‚rül }

simson *See* Simson line. { ′sim·sən }

Simson line The Simson line of a point P on the circumcircle of a triangle ABC is the line passing through the collinear points L, M, and N, where L, M, and N are the projections of P upon the sides BC, CA, and AB, respectively. Also known as simson. { ′sim·sən ‚līn }

simultaneous equations A collection of equations considered to be a set of joint conditions imposed on the variables involved. { ‚sī·məl′tā·nē·əs i′kwā·zhənz }

sin *A* *See* sine. { ′sīn ′ā }

sine The sine of an angle A in a right triangle with hypotenuse of length c given by the ratio a/c, where a is the length of the side opposite A; more generally, the sine function assigns to any real number A the ordinate of the point on the unit circle obtained by moving from (1,0) counterclockwise A units along the circle, or clockwise A units if A is less than 0. Denoted $\sin A$. { sīn }

sine curve The graph of $y = \sin x$, where x and y are cartesian coordinates. Also known as sinusoid. { 'sīn ˌkərv }

sine series A Fourier series containing only terms that are odd in the independent variable, that is, terms involving the sine function. { 'sīn ˌsirˌēz }

single cusp of the first kind *See* keratoid cusp. { ¦siŋ·gəl ˌkəsp əv thə 'fərst ˌkīnd }

single cusp of the second kind *See* ramphoid cusp. { ¦siŋ·gəl ˌkəsp əv thə 'sek·ənd ˌkīnd }

singleton A set that has only one element. { 'siŋ·gəl·tən }

singly periodic function *See* simply periodic function. { ¦siŋ·glē ˌpir·ēˌäd·ik 'fəŋk·shən }

singular integral *See* singular solution. { 'siŋ·gyə·lər ¦int·ə·grəl }

singular integral equation An integral equation where the integral appearing either has infinite limits of integration or the kernel function has points where it is infinite. { 'siŋ·gyə·lər ¦int·ə·grəl i'kwā·zhən }

singularity A point where a function of real or complex variables is not differentiable or analytic. Also known as singular point of a function. { ˌsiŋ·gyə'lar·əd·ē }

singular matrix A matrix which has no inverse; equivalently, its determinant is zero. { 'siŋ·gyə·lər 'mā·triks }

singular point **1.** For a differential equation, a point that is a singularity for at least one of the known functions appearing in the equation. **2.** A point on a curve at which the curve possesses no smoothly turning tangent, or crosses or touches itself, or has a cusp or isolated point. **3.** *See* singularity. { 'siŋ·gyə·lər 'pȯint }

singular solution For a differential equation, a solution that is not generic, that is, not obtainable from the general solution. Also known as singular integral. { 'siŋ·gyə·lər sə'lü·shən }

singular transformation A linear transformation which has no corresponding inverse transformation. { 'siŋ·gyə·lər tranz·fər'mā·shən }

singular values For a matrix A these are the positive square roots of the eigenvalues of A^*A, where A^* denotes the adjoint matrix of A. { 'siŋ·gyə·lər 'val·yüz }

sinistrorse curve *See* left-handed curve. { ¦sin·ə¦strȯrs 'kərv }

sinistrorsum *See* left-handed curve. { ˌsin·ə'strȯrs·əm }

sinusoid *See* sine curve. { 'sī·nəˌsȯid }

sinusoidal function The real or complex function $\sin(u)$ or any function with analogous continuous periodic behavior. { ˌsī·nə'sȯid·əl 'fəŋk·shən }

sinusoidal spiral A plane curve whose equation in polar coordinates (r, θ) is $r^n = a^n \cos n\theta$, where a is a constant and n is a rational number. { ˌsī·nə'sȯid·əl 'spī·rəl }

skew field A ring whose nonzero elements form a non-Abelian group with respect to the multiplicative operation. { 'skyü ˌfēld }

skew Hermitian matrix A square matrix which equals the negative of its adjoint. { 'skyü hər'mish·ən 'mā·triks }

skew lines Lines which do not lie in the same plane in euclidean three-dimensional space. { 'skyü ˌlīnz }

skew product A multiplicative operation or structure induced upon a cartesian product of sets, where each has some algebraic structure. { 'skyü ˌpräd·əkt }

skew quadrilateral A quadrilateral all four of whose vertices do not lie in a single plane. { ¦skyü ˌkwäd·rə'lad·ə·rəl }

skew surface A ruled surface that is not a developable surface. { 'skyü ˌsər·fəs }

skew-symmetric matrix *See* antisymmetric matrix. { 'skyü si¦me·trik 'mā·triks }

skew-symmetric tensor A tensor where interchanging two indices will only change the sign of the corresponding component. { 'skyü si¦me·trik 'ten·sər }

slide rule A mechanical device, composed of a ruler with sliding insert, marked with various number scales, which facilitates such calculations as division, multiplication, finding roots, and finding logarithms. { 'slīd ˌrül }

slope **1.** The slope of a line through the points (x_1,y_1) and (x_2,y_2) is the number $(y_2 - y_1)/(x_2 - x_1)$. **2.** The slope of a curve at a point p is the slope of the tangent line to the curve at p. { slōp }

slope angle The angle of inclination of a line in the plane, where this angle is measured from the positive x axis to the line in the counterclockwise direction. { 'slōp ˌaŋ·gəl }

slope-intercept form In a cartesian coordinate system, the equation of a straight line in the form $y = mx + b$, where m is the slope of the line and b is its intercept on the y axis. { ¦slōp 'in·tərˌsept ˌfȯrm }

Smarandache function A function η defined on the integers with the property that $\eta(n)$ is the smallest integer m such that $m!$ is divisible by n. { ˌsmär·ən'dä·chē ˌfənk·shən }

smoothing Approximating or perturbing a function by one which has a higher degree of differentiability. { 'smüth·iŋ }

smooth manifold A differentiable manifold whose local coordinate systems depend upon those of euclidean space in an infinitely differentiable manner. { 'smüth 'man·əˌfōld }

smooth map An infinitely differentiable function. { 'smüth 'map }

solenoidal A vector field has this property in a region if its divergence vanishes at every point of the region. { ¦säl·ə¦nȯid·əl }

solenoid group A compact Abelian, topological group that is one-dimensional and connected. { 'sä·ləˌnȯid ˌgrüp }

solid angle A surface formed by all rays joining a point to a closed curve. { 'säl·əd 'aŋ·gəl }

soliton A solution of a nonlinear differential equation that propogates with a characteristic constant shape. { 'säl·əˌtän }

solution set The set of values that satisfy a given equation. { sə′lü·shən ‚set }

solvable extension A finite extension E of a field F such that the Galois group of the smallest Galois extension of F containing E is a solvable group. { ′säl·və·bəl ik′sten·chən }

solvable group A group G which has subgroups $G_0, G_1, \ldots, G_n$, where $G_0 = G$, $G_n =$ the identity element alone, and each G_i is a normal subgroup of G_{i-1} with the quotient group G_{i-1}/G_i Abelian. { ′säl·və·bəl ′grüp }

solvmanifold A homogeneous space obtained by factoring a connected solvable Lie group by a closed subgroup. { sälv′man·ə‚fōld }

Sommerfeld-Watson transformation *See* Watson-Sommerfeld transformation. { ′zȯm·ər‚felt ′wät·sən ‚tranz·fər‚mā·shən }

space In context, usually a set with a topology on it or some other type of structure. { spās }

space coordinates A three-dimensional system of cartesian coordinates by which a point is located by three magnitudes indicating distance from three planes which intersect at a point. { ′spās kō′ȯrd·ən·əts }

space curve A curve in three-dimensional euclidean space; it may be a twisted curve or a plane curve. { ′spās ‚kərv }

space polar coordinates A system of coordinates by which a point is located in space by its distance from a fixed point called the pole, the colatitude or angle between the polar axis (a reference line through the pole) and the radius vector (a straight line connecting the pole and the point), and the longitude or angle between a reference plane containing the polar axis and a plane through the radius vector and polar axis. { ′spās ′pō·lər kō′ōrd·ən·əts }

span The span of a set of vectors is the set of all possible linear combinations of those vectors. { span }

spanning subgraph With reference to a graph G, a subgraph of G that contains all the vertices of G. { ′span·iŋ ′səb‚graf }

spanning tree A spanning tree of a graph G is a subgraph of G which is a tree and which includes all the vertices in G. { ′span·iŋ ‚trē }

sparse matrix A matrix most of whose entries are zeros. { ′spärs ′mā·triks }

sparseness The property of a nonlinear programming problem which has many variables, but whose objective and constraint functions each involve only relatively few variables. { ′spärs·nəs }

special functions The various families of solution functions corresponding to cases of the hypergeometric equation or functions used in the equation's study, such as the gamma function. { ′spesh·əl ′fəŋk·shənz }

special Jordan algebra A Jordan algebra that can be written as a symmetrized product over a matrix algebra. { ′spesh·əl ′jȯrd·ən ‚al·jə·brə }

special orthogonal group of dimension *n* The Lie group of special orthogonal transformations on an n-dimensional real inner product space. Symbolized SO_n; $SO(n)$. { ¦spesh·əl ȯr¦thäg·ən·əl ‚grüp əv di¦men·chən ′en }

special orthogonal transformation An orthogonal transformation whose matrix representation has determinant equal to 1. { ¦spesh·əl ȯr¦thäg·ən·əl ˌtranz·fər'mā·shən }

special unitary group of dimension *n* The Lie group of special unitary transformations on an n-dimensional inner product space over the complex numbers. Symbolized SU(n). { ¦spesh·əl ¦yü·nəˌter·ē ˌgrüp əv di¦men·chən 'en }

special unitary transformation A unitary transformation whose matrix representation has determinant equal to 1. { ¦spesh·əl ¦yü·nəˌter·ē ˌtranz·fər'mā·shən }

spectral density The density function for the spectral measure of a linear transformation on a Hilbert space. { 'spek·trəl 'den·səd·ē }

spectral factorization A process sometimes used in the study of control systems, in which a given rational function of the complex variable s is factored into the product of two functions, $F_R(s)$ and $F_L(s)$, each of which has all of its poles and zeros in the right and left half of the complex plane, respectively. { 'spek·trəl ˌfak·tə·rə'zā·shən }

spectral function In the theory of stationary stochastic processes, the function

$$F(y) = (2/\pi) \int_0^\infty \rho(x)(\sin xy/x)(dx),\ 0 \leq y \leq \infty$$

where $\rho(x)$ is the autocorrelation function of a stationary time series. { 'spek·trəl 'fəŋk·shən }

spectral measure A measure on the spectrum of an operator on a Hilbert space whose values are projection operators there; spectral theorems concerning linear operators often give an integral representation of the operator in terms of these projection valued measures. { 'spek·trəl 'mezh·ər }

spectral radius For the spectrum of an operator, this is the least upper bound of the set of all λ , where λ is in the spectrum. { 'spek·trəl 'rād·ē·əs }

spectral theorems Spectral theorems enable detailed study of various types of operators on Banach spaces by giving an integral or series representation of the operator in terms of its spectrum, eigenspaces, and simple projectionlike operators. { 'spek·trəl 'thir·əmz }

spectrum If T is a linear operator of a normed space X to itself and I is the identity transformation ($I(x) \equiv x$), the spectrum of T consists of all scalars λ for which either $T - \lambda I$ has no inverse or the range of $T - \lambda I$ is not dense in X. { 'spek·trəm }

speed-up theorem There is a computable function f with the property that for any algorithm A there is another algorithm B which computes f much faster than A. { 'spēd¦əp ˌthir·əm }

Sperner set A set S of subsets of a given set such that if A and B are in S, and A does not equal B, then neither A nor B is a subset of the other. { 'spər·nər ˌset }

sphere **1.** The set of all points in a euclidean space which are a fixed common distance from some given point; in euclidean three-dimensional space the Riemann sphere consists of all points (x,y,z) which satisfy the equation $x^2 + y^2 + z^2 = 1$. **2.** The set of points in a metric space whose distance from a fixed point is constant. { sfir }

spherical angle The figure formed by the intersection of two great circles on a sphere, and equal in size to the angle formed by the tangents to the great circles at the point of intersection. { ¦sfir·ə·kəl 'aŋ·gəl }

spherical Bessel functions Bessel functions whose order is half of an odd integer; they arise as the radial functions that result from solving Pockel's equation (or, equivalently, the time-independent Schrödinger equation for a free particle) by separation of variables in spherical coordinates. { ¦sfir·i·kəl 'bes·əl ˌfəŋk·shənz }

spherical coordinates A system of curvilinear coordinates in which the position of a point in space is designated by its distance r from the origin or pole, called the radius vector, the angle ϕ between the radius vector and a vertically directed polar axis, called the cone angle or colatitude, and the angle θ between the plane of ϕ and a fixed meridian plane through the polar axis, called the polar angle or longitude. Also known as spherical polar coordinates. { 'sfir·ə·kəl kō'ȯrd·ən·əts }

spherical curve A curve that lies entirely on the surface of a sphere. { 'sfir·ə·kəl 'kərv }

spherical cyclic curve *See* cyclic curve. { 'sfir·ə·kəl 'sī·klik 'kərv }

spherical degree A solid angle equal to one-ninetieth of a spherical right angle. { 'sfir·ə·kəl di'grē }

spherical distance The length of a great circle arc between two points on a sphere. { 'sfir·i·kəl 'dis·təns }

spherical excess The sum of the angles of a spherical triangle, minus 180°. { 'sfir·ə·kəl ek'ses }

spherical geometry The geometry of points on a sphere. { 'sfir·ə·kəl je'äm·ə·trē }

spherical harmonics Solutions of Laplace's equation in spherical coordinates. { 'sfir·ə·kəl här'män·iks }

spherical indicatrix of binormal to a curve All the end points of those radii from the sphere of radius one which are parallel to the positive direction of the binormal to a space curve. { 'sfir·ə·kəl ¦in·də¦kā·triks əv bī'nȯr·məl tü ə 'kərv }

spherical indicatrix of tangent to a curve Those points on the unit sphere traced out by a radius moving from point to point always parallel with the tangent to the curve. { 'sfir·ə·kəl ¦in·də¦kā·triks əv 'tan·jən tü ə 'kərv }

spherical polar coordinates *See* spherical coordinates. { 'sfir·ə·kəl 'pō·lər kō'ȯrd·ən·əts }

spherical polygon A part of a sphere that is bounded by arcs of great circles. { 'sfir·ə·kəl 'päl·əˌgän }

spherical pyramid A solid bounded by a spherical polygon and portions of planes passing through the sides of the polygon and the center of the sphere. { 'sfir·ə·kəl 'pir·əˌmid }

spherical radius For a circle on a sphere, the smaller of the spherical distances from one of the two poles of the circle to any point on the circle. { 'sfir·ə·kəl 'rād·ē·əs }

spherical sector The cap and cone formed by the intersection of a plane with a sphere, the cone extending from the plane to the center of the sphere and the cap extending from the plane to the surface of the sphere. { 'sfir·ə·kəl 'sek·tər }

spherical segment A solid that is bounded by a sphere and two parallel planes which intersect the sphere or are tangent to it. { 'sfir·ə·kəl 'seg·mənt }

spherical surface A surface whose total curvature has a constant positive value but that is not necessarily a sphere. { 'sfir·ə·kəl 'sər·fəs }

spherical surface harmonics Functions of the two angular coordinates of a spherical coordinate system which are solutions of the partial differential equation obtained by separation of variables of Laplace's equation in spherical coordinates. Also known as surface harmonics. { ¦sfir·ə·kəl ¦sər·fəs här'män·iks }

spherical triangle A three-sided surface on a sphere the sides of which are arcs of great circles. { 'sfir·ə·kəl 'trī‚aŋ·gəl }

spherical trigonometry The study of spherical triangles from the viewpoint of angle, length, and area. { 'sfir·ə·kəl ‚trig·ə'näm·ə·trē }

spherical wedge The portion of a sphere bounded by two semicircles and a lune (the surface of the sphere between the semicircles). { 'sfir·ə·kəl 'wej }

spheroid *See* ellipsoid of revolution. { 'sfir‚ȯid }

spheroidal excess The amount by which the sum of the three angles of a triangle on the surface of a spheroid exceeds 180°. { sfir'ȯid·əl ek‚ses }

spheroidal harmonics Solutions to Laplace's equation when phrased in ellipsoidal coordinates. { sfir'ȯid·əl här'män·iks }

spheroidal triangle The figure formed by three geodesic lines joining three points on a spheroid. Also known as geodetic triangle. { sfir'ȯid·əl 'trī‚aŋ·gəl }

spinode *See* cusp. { 'spi‚nōd }

spinor **1.** A vector with two complex components, which undergoes a unitary unimodular transformation when the three-dimensional coordinate system is rotated; it can represent the spin state of a particle of spin $1/2$. **2.** More generally, a spinor of order (or rank) n is an object with 2^n components which transform as products of components of n spinors of rank one. **3.** A quantity with four complex components which transforms linearly under a Lorentz transformation in such a way that if it is a solution of the Dirac equation in the original Lorentz frame it remains a solution of the Dirac equation in the transformed frame; it is formed from two spinors (definition 1). Also known as Dirac spinor. { 'spin·ər }

spin space The two-dimensional vector space over the complex numbers, whose unitary unimodular transformations are a two-dimensional double-valued representation of the three-dimensional rotation group; its vectors can represent the various spin states of a particle with spin $1/2$, and its unitary unimodular transformations can represent rotations of this particle. { 'spin ‚spās }

spiral A simple curve in the plane which continuously winds about itself either into some point or out from some point. { 'spī·rəl }

spiral of Archimedes The curve spiraling into the origin which in polar coordinates is given by the equation $r = a\theta$. Also known as Archimedes' spiral. { ¦spī·rəl əv ‚är·kə'mē·dēz }

spline A function used to approximate a specified function on an interval, consisting of pieces which are defined uniquely on a set of subintervals, usually as polynomials or some other simple form, and which match up with each other and the prescribed function at the end points of the subintervals with a sufficiently high degree of accuracy. { splīn }

sporadic simple group A simple group which cannot be classified in any known infinite family of simple groups. { spə¦rad·ik ¦sim·pəl ˌgrüp }

spur *See* trace. { spər }

square **1.** The square of a number r is the number r^2, that is, r times r. **2.** The plane figure with four equal sides and four interior right angles. { skwer }

square degree A unit of a solid angle equal to $(\pi/180)^2$ steradian, or approximately 3.04617×10^{-4} steradian. { 'skwer di'grē }

square grade A unit of solid angle equal to $(\pi/200)^2$ steradian, or approximately 2.46740×10^{-4} steradian. { 'skwer 'grād }

square matrix A matrix with the same number of rows and columns. { 'skwer 'mā·triks }

square number A number that is derived by squaring an integer. { 'skwer 'nəm·ber }

square root A square root of a real or complex number s is a number t for which $t^2 = s$. { 'skwer 'rüt }

squaring the circle For a circle with a specified radius, the problem of constructing a square that has the same area as the circle. { 'skwer·iŋ thə 'sər·kəl }

sr *See* steradian.

stability Stability theory of systems of differential equations deals with those solution functions possessing some particular property that still maintain the property after a perturbation. { stə'bil·əd·ē }

stability subgroup *See* stabilizer. { stə'bil·əd·ē 'səbˌgrüp }

stabilizer The stabilizer of a point x in a Riemann surface X, relative to a group G of conformal mappings of X onto itself, is the subgroup G_x of G consisting of elements g such that $g(x) = x$. Also known as stability subgroup. { 'stā·bəˌlīz·ər }

stable bundle The bundle E^s of a hyperbolic structure. { 'stā·bəl 'bən·dəl }

stable graph A graph from which an edge can be deleted to produce a subgraph whose group of automorphisms is a subgroup of the group of automorphisms of the original graph. { 'stā·bəl 'graf }

stable homeomorphism conjecture For dimension n, the assertion that each orientation-preserving homeomorphism of the real n space, R^n, into itself can be expressed as a composition of homeomorphisms, each of which is the identity on some nonempty open set in R^n. { 'stā·bəl ¦hō·mē·ō'mȯrˌfiz·əm kənˌjek·chər }

star For a member S of a family of sets, the collection of all sets in the family that contain S as a subset. { stär }

star algebra A real or complex algebra on which an involution is defined. { 'stär ˌal·jə·brə }

starlike region A region in the complex number plane such that the line segment joining any of its points to the origin lies entirely in the region. { 'stärˌlīk ˌrē·jən }

star-shaped set With respect to a point *P* of a euclidean space or vector space, a set such that if *Q* is a member of the set, then so is any point on the line segment *PQ*. { 'stär ˌshāpt ˌset }

star subalgebra A subalgebra of a star algebra which is mapped onto itself by the involution operation. { 'stär ¦səbˌal·jə·brə }

stationary phase A method used to find approximations to the integral of a rapidly oscillating function, based on the principle that this integral depends chiefly on that part of the range of integration near points at which the derivative of the trigonometric function involved vanishes. { 'stā·shəˌner·ē 'fāz }

stationary point **1.** A point on a curve at which the tangent is horizontal. **2.** For a function of several variables, a point at which all partial derivatives are 0. { 'stā·shəˌner·ē 'pȯint }

stationary stochastic process A stochastic process $x(t)$ is stationary if each of the joint probability distributions is unaffected by a change in the time parameter t. { 'stā·shəˌner·ē stō'kas·tik 'prä·səs }

statistics A discipline dealing with methods of obtaining data, analyzing and summarizing it, and drawing inferences from data samples by the use of probability theory. { stə'tis·tiks }

Steenrod algebra The cohomology groups of a topological space have additive operations on them, which can be added and multiplied so as to form the Steenrod algebra. { 'stenˌräd 'al·jə·brə }

Steenrod squares Operations which associate elements from different cohomology groups of a topological space and produce an element in another of the groups; these operations can be so added and multiplied as to produce the Steenrod algebra. { 'stenˌräd 'skwerz }

steepest descent method Certain functions can be approximated for large values by an asymptotic formula derived from a Taylor series expansion about a saddle point. Also known as saddle point method. { 'stēp·əst di'sent ˌmeth·əd }

Steiner triple system A balanced incomplete block design in which the number k of distinct elements in each block equals 3, and the number λ of blocks in which each combination of elements occurs together equals 1. { ¦stīn·ər ¦trip·əl 'sis·təm }

step function **1.** A function f defined on an interval $[a,b]$ so that $[a,b]$ can be partitioned into a finite number of subintervals on each of which f is a constant. Also known as simple function. **2.** More generally, a real function with finite range. { 'step ˌfəŋk·shən }

sterad *See* steradian. { 'stiˌrad }

steradian The unit of measurement for solid angles; it is equal to the solid angle subtended at the center of a sphere by a portion of the surface of the sphere whose area equals the square of the sphere's radius. Abbreviated sr; sterad. { stə'rād·ē·ən }

steregon The entire solid angle bounded by a sphere; equal to 4π steradians. { 'ster·ə‚gän }

stereographic projection The projection of the Riemann sphere onto the euclidean plane performed by emanating rays from the north pole of the sphere through a point on the sphere. { ¦ster·ē·ə¦graf·ik prə'jek·shən }

Stieltjes integral The Stieltjes integral of a real function $f(x)$ relative to a real function $g(x)$ of bounded variation on an interval $[a,b]$ is defined, analogously to the Riemann integral, as a limit of a sum of terms $f(a_i)\ [g(x_i) - g(x_{i-1})]$ taken as partitions of the interval shrink. Denoted

$$\int_a^b f(x)dg(x)$$

{ 'stēlt·yəs ‚int·ə·grəl }

Stieltjes transform A form of the Laplace transform of a function where the usual Riemann integral is replaced by a Stieltjes integral. { 'stēlt·yəs ‚tranz·fȯrm }

Stirling numbers The coefficients which occur in the Stirling interpolation formula for a difference operator. { 'stir·liŋ ‚nəm·bərz }

Stirling numbers of the second kind The numbers $S(n,r)$ giving the numbers of ways that n elements can be distributed among r indistinguishable cells so that no cell remains empty. { ¦stər·liŋ ‚nəm·bərz əv thə 'sek·ənd ‚kīnd }

Stirling's formula The expression $(n/e)^n\sqrt{2\pi n}$ is asymptotic to factorial n; that is, the limit as n goes to ∞ of their ratio is 1. { 'stir·liŋz ‚fȯr·myə·lə }

Stirling's series An asymptotic expansion for the logarithm of the gamma function, or an equivalent asymptotic expansion for the gamma function itself, from which Stirling's formula may be derived. { 'stər·liŋz ‚sir‚ēz }

stochastic Pertaining to random variables. { stō'kas·tik }

stochastic calculus The mathematical theory of stochastic integrals and differentials, and its application to the study of stochastic processes. { stō'kas·tik 'kal·kyə·ləs }

stochastic chain rule A generalization of the ordinary chain rule to stochastic processes; it states that the process $U_t = u(X_t^1, X_t^2, ..., X_t^n)$ satisfies

$$dU = \sum_i \partial_i u dX^i + {}^1\!/_2 \sum_{ij} \partial_i \partial_j u dX^i dX^j$$

with the conventions $(dt)^2 = 0$ and $dW^\alpha dW^\beta = \partial_{\alpha\beta} dt$, where the X^i are processes satisfying

$$dX_t^i = a_t^i dt + \sum_{\alpha=1}^{m} b_t^{i\alpha}\, dW_t^\alpha,\ i = 1, 2, \ldots, n;$$

$\{W_t^\alpha, t \geq 0\}$, $\alpha = 1, 2, \ldots, m$, are independent Wiener processes; the dW_t^α are the corresponding random disturbances occurring in the infinitesimal time interval dt; the a_t^i and $b_t^i\alpha$ are independent of future disturbances, and $u(x_1, x_2, ..., x_n)$ is a function whose derivatives $\partial_i u$ and $\partial_i\partial_j u$ are continuous. Also known as Itô's formula. { stō'kas·tik 'chān ‚rül }

stochastic differential An expression representing the random disturbances occurring in an infinitesimal time interval; it has the form dW_t, where $\{W_t, t \geq 0\}$ is a Wiener process. { stō'kas·tik ‚dif·ə'ren·chəl }

stochastic integral An integral used to construct the sample functions of a general diffusion process from those of a Wiener process; it has the form

$$\int_{W_0}^{W_s} a_t dW_t$$

where $\{W_t, t \geq 0\}$ is a Wiener process, dW_t represents the random disturbances occurring in an infinitesimal time interval dt, and a_t is independent of future disturbances. Also known as Itô's integral. { stō'kas·tik 'int·ə·grəl }

stochastic matrix A square matrix with nonnegative real entries such that the sum of the entries of each row is equal to 1. { stō'kas·tik 'mā·triks }

stochastic process A family of random variables, dependent upon a parameter which usually denotes time. Also known as random process. { stō'kas·tik 'prä·səs }

stochastic variable *See* random variable. { stō'kas·tik 'ver·ē·ə·bəl }

Stokes' integral theorem The analog of Green's theorem in n-dimensional euclidean space; that is, a line integral of $F_1(x_1,x_2,...,x_n)dx_1 + \cdots + F_n(x_1,x_2,...,x_n)dx_n$ over a closed curve equals an integral of an expression containing various partial derivatives of $F_1,...,F_n$ over a surface bounded by the curve. { ¦stōks 'int·ə·grəl ˌthir·əm }

Stokes phenomenon A change in the asymptotic representation of certain analytic functions that occurs in passing from one section of the complex plane to another. { 'stōks fəˌnäm·əˌnän }

Stone-Cech compactification The Stone-Cech compactification of a completely regular space X is a compact Hausdorff space $\beta(X)$ such that X is a dense subset of $\beta(X)$ and for any continuous function f from X to a compact space Y there is a unique continuous function from $\beta(X)$ to Y which is an extension of f. { ¦stōn ¦chek kəmˌpak·tə·fə'kā·shən }

Stone's representation theorem This theorem determines the nature of all unitary representations of locally compact Abelian groups. { 'stōnz ˌrep·rə·zən'tā·shən ˌthir·əm }

Stone's theorem Every Boolean ring is isomorphic to a ring of subsets of some set. { 'stōnz ˌthir·əm }

Stone-Weierstrass theorem If S is a collection of continuous real-valued functions on a compact space E, which contains the constant functions, and if for any pair of distinct points x and y in E there is a function f in S such that $f(x)$ is not equal to $f(y)$, then for any continuous real-valued function g on E there is a sequence of functions, each of which can be expressed as a polynomial in the functions of S with real coefficients, that converges uniformly to g. { 'stōn 'vī·ərˌsträs ˌthir·əm }

strategy In game theory a strategy is a specified collection of moves, which cover all possible situations, for the complete play of a given game. { 'strad·ə·jē }

strategy vector A vector characterizing a mixed strategy, whose components are the probability weights of the strategy. { 'strad·ə·jē ˌvek·tər }

strictly convex space A normal linear space such that, for any two vectors x and y, if $x + y = x + y$, then either $y = 0$ or $x = cy$, where c is a number. { ¦strik·lē ¦känˌveks 'spās }

strictly dominant strategy Relative to a given pure strategy for one player of a game, a second pure strategy for that player that has a greater payoff than the given strategy for any pure strategy of the opposing player. { ¦strik·lē ¦däm·ə·nənt 'strad·ə·jē }

strictly Hurwitz polynomial A polynomial whose roots all have strictly negative real parts. { 'strik·lē 'hər‚vits ‚päl·i'nō·mē·əl }

strongly continuous semigroup A semigroup of bounded linear operators on a Banach space B, together with a bijective mapping T from the positive real numbers onto the semigroup, such that $T(0)$ is the identity operator on B, $T(s + t) = T(s)T(t)$ for any two positive numbers s and t and, for each element x of B, $T(t)x$ is a continuous function of t. { ¦strȯŋ·lē kən¦tin·yə·wəs 'sem·i‚grüp }

strong topology The topology on a normed space obtained from the given norm; the basic open neighborhoods of a vector x are sets consisting of all those vectors y where the norm of $x - y$ is less than some number. { 'strȯŋ tə'päl·ə·jē }

strophoid **1.** A curve derived from a given curve C and two points, called the pole and the fixed point, consisting of the locus of points on a rotating line L passing through the pole whose distance from the intersection of L and C is equal to the distance of this intersection from the fixed point. **2.** The special case of the first definition in which C is a straight line and the fixed point lies on C. { 'strä‚fȯid }

structural stability Property of a differentiable flow on a compact manifold whose orbit structure is insensitive to small perturbations in the equations governing the flow or in the vector field generating the flow. { 'strək·chər·əl stə'bil·əd·ē }

structure constants A set of numbers that serve as coefficients in expressing the commutators of the elements of a Lie algebra. { 'strək·chər ‚kän·stəns }

Sturm-Liouville problem The general problem of solving a given linear differential equation of order $2n$ together with $2n$-boundary conditions. Also known as eigenvalue problem. { 'stərm lyü'vil ‚präb·ləm }

Sturm-Liouville system A given differential equation together with its boundary conditions having Sturm-Liouville problem form. { 'stərm lyü'vil ‚sis·təm }

Sturm sequence For a polynomial $p(x)$, this is the sequence of functions $f_0(x)$, $f_1(x)$,..., where $f_0(x) = p(x)$, $f_1(x) = p'(x)$, and $f_n(x)$ is the negative remainder that occurs by finding the greatest common divisor of $f_{n-2}(x)$ and $f_{n-1}(x)$ via the euclidean algorithm. { 'stərm ‚sēkwəns }

Sturm's theorem This gives a method to determine the number of real roots of a polynomial $p(x)$ which lie between two given values of x; the Sturm sequence of $p(x)$ provides the necessary information. { 'stərmz ‚thir·əm }

subadditive function A function f such that $f(x + y)$ is less than or equal to $f(x) + f(y)$ for all x and y in its domain. { ¦səb'ad·əd·iv ‚fəŋk·shən }

subalgebra **1.** A subset of an algebra which itself forms an algebra relative to the same operations. **2.** A subalgebra (of sets) is any algebra (of sets) contained in some given algebra. { ¦səb'al·jə·brə }

subbase for a topology A family S of subsets of a topological space X where by taking all finite intersections of sets from S and all unions of such intersections the entire topology of open sets of X is obtained. { 'səb‚bās fȯr ə tə'päl·ə·jē }

subdivision graph A graph which can be obtained from a given graph by breaking up each edge into one or more segments by inserting intermediate vertices between its two ends. { 'səb·di,vizh·ən ,graf }

subfactorial For an integer n, the number that is expressed as $n!\{(1/2!) - (1/3!) + (1/4!) - \cdots + [(-1)^n/n!]\}$. { ,səb,fak'tȯr·ē·əl }

subfield **1.** A subset of a field which itself forms a field relative to the same operations. **2.** A subfield (of sets) is any field (of sets) contained in some given field of sets. { 'səb,fēld }

subgraph A graph contained in a given graph which has as its vertices some subset of the vertices of the original. { 'səb,graf }

subgroup A subset N of a group G which is itself a group relative to the same operation. { 'səb,grüp }

subharmonic function A continuous function is subharmonic in a region R of the plane if its value at any point z_0 of R is less than or equal to its integral along a circle centered at z_0. { ¦səb·här'män·ik 'fəŋk·shən }

submodule A subset N of a module M over a ring R such that, if x and y are in N and a is in R, then $x + y$ and ax are in N, so that N is also a module over R. { 'səb,mä·jəl }

subnormal For a given point on a plane curve, the projection on the x axis of a rectangular coordinate system of the segment of the normal between the given point and the intersection of the normal with the x axis. { səb'nȯr·məl }

subnormal operator An operator A on a Hilbert space H is said to be subnormal if there exists a normal operator B on a Hilbert space K such that H is a subspace of K, the subspace H is invariant under the operator B, and the restriction of B to H coincides with A. { 'səb,nȯr·məl 'äp·ə,rād·ər }

subrange A subset of the range of values that a function may assume. { 'səb,rānj }

subring **1.** A subset I of a ring R where I is also a ring relative to the operations of R. **2.** A subring (of sets) is any ring (of sets) contained in some given ring (of sets). { 'səb,riŋ }

subscripted variable A symbolic name for an array of variables whose elements are identified by subscripts. { səb'skrip·təd 'ver·ē·ə·bəl }

subsequence A subsequence of a given sequence is any sequence all of whose entries appear in the original sequence and in the same manner of succession. { 'səb·sə·kwəns }

subset **1.** A subset A of a set B is a set all of whose elements are included in B. **2.** A fuzzy set A is a subset of a fuzzy set B if, for every element x, the value of the membership function of A at x is equal to or less than the value of the membership function of B at x. { 'səb,set }

subspace A subset of a space which, in the appropriate context, is a space in its own right. { 'səb,spās }

subtangent For a given point on a plane curve, the projection on the x axis of a rectangular coordinate system of the segment of the tangent between the point of tangency and the intersection of the tangent with the x axis. { ¦səb'tan·jənt }

subtend A line segment or an arc of a circle subtends an angle with vertex at a specified point if the end points of the line segment or arc lie on the sides of the angle. { səb′tend }

subtraction The addition of one quantity with the negative of another; in a system with an additive operation this is formally the sum of one element with the additive inverse of another. { səb′trak·shən }

subtraction sign The symbol −, used to indicate subtraction. Also known as minus sign. { səb′trak·shən ,sīn }

subtrahend A quantity which is to be subtracted from another given quantity. { ′səb·trə,hend }

subtree A subgraph of a tree which is itself a tree. { ′səb,trē }

successive approximations Any method of solving a problem in which an approximate solution is first calculated, this solution is then used in computing an improved approximation, and the process is repeated as many times as desired. { sək′ses·iv ə,präk·sə′mā·shənz }

successor For a vertex a in a directed graph, any vertex b for which there is an arc between a and b directed from a to b. { sək′ses·ər }

sum **1.** The addition of numbers or mathematical objects in context. **2.** The sum of an infinite series is the limit of the sequence consisting of all partial sums of the series. **3.** The sum $A + B$ of two matrices A and B, with the same number of rows and columns, is the matrix whose element c_{ij} in row i and column j is the sum of corresponding elements a_{ij} in A and b_{ij} in B. { səm }

summability method A method, such as Hölder summation or Cesaro summation, of attributing a sum to a divergent series by using some process to average the terms in the series. { ,səm·ə¦bil·əd·ē ,meth·əd }

summable function A function whose Lebesgue integral exists. { ¦səm·ə·bəl ′fəŋk·shən }

summation convention An abbreviated notation used particularly in tensor analysis and relativity theory, in which a product of tensors is to be summed over all possible values of any index which appears twice in the expression. { sə′mā·shən kən ,ven·chən }

sup *See* least upper bound.

superadditive function A function f such that $f(x + y)$ is greater than or equal to $f(x) + f(y)$ for all x and y in its domain. { ¦sü·pər,ad·əd·iv ′fəŋk·shən }

superharmonic function A continuous complex function f whose value at a point z_0 exceeds its average values computed by the integral of f around a circle centered at z_0. { ¦sü·pər·här′män·ik ′fəŋk·shən }

superposition The principle of superposition states that any given geometric figure in a euclidean space can be so moved about as not to change its size or shape. { ,sü·pər·pə′zish·ən }

superset A set whose elements include all the elements of a given set. { ′sü·pər ,set }

supplemental chords Two chords joining a point on the circumference of a circle to the ends of a diameter of the circle.

supplementary angle One angle is supplementary to another angle if their sum is 180°. { ¦səp·lə¦men·trē 'aŋ·gəl }

support The support of a real-valued function *f* on a topological space is the closure of the set of points where *f* is not zero. { sə'pȯrt }

support function Relative to a convex body in a real inner product space, a function whose value at a point *P* is the maximum of the inner product of *P* and *Q* for *Q* in the convex body. { sə'pȯrt ˌfəŋk·shən }

supremum *See* least upper bound. { su'prē·məm }

surd A sum of one or more roots of rational numbers, some or all of which are themselves irrational numbers. { sərd }

surface A subset of three-space consisting of those points whose cartesian coordinates x, y, and z satisfy equations of the form $x = f(u,v)$, $y = g(u,v)$, $z = h(u,v)$, where f, g, and h are differentiable real-valued functions of two parameters u and v which take real values and vary freely in some domain. { 'sər·fəs }

surface harmonics *See* spherical surface harmonics. { 'sər·fəs härˌmän·iks }

surface integral The integral of a function of several variables with respect to surface area over a surface in the domain of the function. { 'sər·fəs 'int·ə·grəl }

surface of center The locus of points that are one of the two centers of principal curvature at some point on a given surface. { ¦sər·fəs əv 'sen·tər }

surface of Joachimsthal A surface such that all the members of one of its two families of lines of curvature are plane curves and their planes all pass through a common axis. { ¦sər·fəs əv yō'äk·əmzˌtäl }

surface of Liouville A surface that can be assigned parameters u and v such that a linear element ds on the surface is given by $ds^2 = [f(u) + g(v)][du^2 + dv^2]$, where f and g are functions of u and v. { ¦sər·fəs əv 'lyüˌvēl }

surface of Monge A surface generated by a plane curve as its plane rolls without slipping over a developable surface. { ¦sər·fəs əv 'mȯnzh }

surface of negative curvature A surface whose Gaussian curvature is negative at every point. { ¦sər·fəs əv ¦neg·əd·iv 'kər·və·chər }

surface of positive curvature A surface whose Gaussian curvature is positive at every point. { ¦sər·fəs əv 'päs·əd·iv ˌkər·və·chər }

surface of revolution A surface realized by rotating a planar curve about some axis in its plane. { ¦sər·fəs əv ˌrev·ə'lü·shən }

surface of translation A surface that can be generated from two curves by translating either one of them parallel to itself in such a way that each of its points describes a curve that is a translation of the other curve. Also known as translation surface. { ¦sər·fəs əv tranz'lā·shən }

surface of Voss A surface that has a conjugate system of geodesics. { ¦sər·fəs əv 'vȯs }

surface of zero curvature A surface whose Gaussian curvature is zero at every point. { ¦sər·fəs əv 'zir·ō ˌkər·və·chər }

surface patch A surface or a portion of a surface that is bounded by a closed curve. { 'sər·fəs ˌpach }

surjection A mapping f from a set A to a set B such that for every element b of B there is an element a of A such that $f(a) = b$. Also known as surjective mapping. { sər'jek·shən }

surjective mapping *See* surjection. { sər'jek·tiv 'map·iŋ }

swastika A plane curve whose equation in cartesian coordinates x and y is $y^4 - x^4 = xy$. { 'swäs·tə·kə }

syllogism A statement together with a conclusion; this usually has the form "if p then q." { 'sil·əˌjiz·əm }

Sylow subgroup A subgroup H of a given group G such that the order of H is p^n, where p is a prime and n is an integer, and p^n is the highest power of p dividing the order of G. { ¦sīˌlō 'səbˌgrüp }

Sylvester's theorem If A is a matrix with distinct eigenvalues $\lambda_1, \ldots, \lambda_n$, then any analytic function $f(A)$ can be realized from the λ_i, $f(\lambda_i)$, and the matrices $A - \lambda_i I$, where I is the identity matrix. { sil'ves·tərz ˌthir·əm }

symbolic logic The formal study of symbolism and its use in the foundations of mathematical logic. { sim'bäl·ik 'läj·ik }

symmetric design A balanced incomplete block design in which the number b of blocks equals the number v of elements arranged among the blocks. { si¦me·trik di'zīn }

symmetric difference The symmetric difference of two sets consists of all points in one or the other of the sets but not in both. { sə'me·trik 'dif·rəns }

symmetric form A bilinear form f which is unchanged under interchange of its independent variables; that is, $f(x,y) = f(y,x)$ for all values of the independent variables x and y. { si¦me·trik 'fȯrm }

symmetric function A function whose value is unchanged for any permutation of its variables. { sə'me·trik 'fəŋk·shən }

symmetric group The group consisting of all permutations of a finite set of symbols. { sə'me·trik 'grüp }

symmetric matrix A matrix which equals its transpose. { sə'me·trik 'mā·triks }

symmetric relation The property of a relation on a set that requires y to be related to x whenever x is related to y. { sə'me·trik ri'lā·shən }

symmetric space A differentiable manifold which has a differentiable multiplication operation that behaves similarly to the multiplication of a complex number and its conjugate. { sə'me·trik 'spās }

symmetric spherical triangles Spherical triangles whose corresponding angles and corresponding sides are equal but appear in opposite order as viewed from the center of the sphere. { sə¦me·trik ¦sfir·ə·kəl 'trīˌaŋ·gəlz }

symmetric tensor A tensor that is left unchanged by the interchange of two contravariant (or covariant) indices. { sə¦me·trik 'ten·sər }

symmetric transformation A transformation T defined on a Hilbert space such that the inner products (x,Ty) and (Tx,y) are equal for any vectors x and y in the domain of T. { sə¦me·trik ˌtranz·fər′mā·shən }

symmetry 1. A geometric object G has this property relative to some configuration S of its points if S determines two pieces of G which can be reflected onto each other through S. **2.** A rigid motion of a geometric figure that maps the figure onto itself. { ′sim·əˌtrē }

symmetry group A group composed of all rigid motions or similarity transformations of some geometric object onto itself. { ′sim·əˌtrē ˌgrüp }

symmetry plane *See* plane of mirror symmetry. { ′sim·əˌtrē ˌplān }

symmetry principle The centroid of a geometrical figure (line, area, or volume) is at a point on a line or plane of symmetry of the figure. { ′sim·əˌtrē ˌprin·sə·pəl }

symmetry transformation A rigid motion sending a geometric object onto itself; examples are rotations and, for the case of a polygon, permutations of the vertices. Also known as symmetry function. { ′sim·əˌtrē ˌtranz·fər′mā·shən }

symplectic group of dimension *n* The Lie group of symplectic transformations on an n-dimensional vector space over the quaternions. Symbolized Sp(n). { sim¦plek·tik ˌgrüp əv di¦men·chən ′en }

symplectic transformation A linear transformation of a vector space over the quaternions that leaves the lengths of vectors unchanged. { sim¦plek·tik ˌtranz·fər′mā·shən }

synclastic Property of a surface or portion of a surface for which the centers of curvature of the principal sections at each point lie on the same side of the surface. { sin′klas·tik }

synthetic division A long division process for dividing a polynomial $p(x)$ by a polynomial $(x - a)$ where only the coefficients of these polynomials are used. { sin′thed·ik də′vizh·ən }

system of distinct representatives A family of subsets S_i of a given finite set S such that the family has as many members as there are elements in S, and such that it is possible to assign each element x_i of S to a distinct subset S_i with x_i in S_i. { ¦sis·təm əv di¦stinkt ˌrep·rə′zen·tə′tivz }

T

T *See* tera-.

table An array or listing of computed quantities. { 'tā·bəl }

tabular interpolation Method of finding from a table the values of the dependent variable for intermediate values of the independent variable. { 'tab·yə·lər in‚tər·pə'lā·shən }

tacnode *See* double cusp. { 'tak‚nōd }

tail For a stochastic process represented by $x(t_1)$, $x(t_2)$,..., the process obtained by deleting the first n terms, for some n. { tāl }

Talbot's curve The negative pedal of an ellipse, with eccentricity greater than $\sqrt{2/2}$, with respect to its center. { 'tal·bəts ‚kərv }

tan *See* tangent. { tan }

tangent **1.** A line is tangent to a curve at a fixed point P if it is the limiting position of a line passing through P and a variable point on the curve Q, as Q approaches P. **2.** The function which is the quotient of the sine function by the cosine function. Abbreviated tan. **3.** The tangent of an angle is the ratio of its sine and cosine. Abbreviated tan. { 'tan·jənt }

tangent bundle The fiber bundle $T(M)$ associated to a differentiable manifold M which is composed of the points of M together with all their tangent vectors. Also known as tangent space. { 'tan·jənt ‚bənd·əl }

tangent cone A cone each of whose elements is tangent to a given quadric surface. { 'tan·jənt ‚kōn }

tangential component A component of a given vector acting at right angles to a given radius of a given circle. { tan'jen·chəl kəm'pō·nənt }

tangential coordinates For a surface, a set of four coordinates, three of which are the direction cosines of the normal to the surface, while the fourth is the algebraic distance from the origin to the plane tangent to the surface. { tan'jen·chəl kō'ȯrd·ən·əts }

tangential curvature *See* geodesic curvature. { tan'jen·chəl 'kər·və·chər }

tangential developable *See* tangent surface. { tan'jen·chəl di'vel·əp·ə·bəl }

tangential polar equation An equation of a curve exprepssed in terms of the distance of a point P on the curve from a reference point O and the perpendicular distance from O to the tangent to the curve at P. { tan¦jen·chəl ¦pō·lər i'kwā·zhən }

tangent plane The tangent plane to a surface at a point is the plane having every line in it tangent to some curve on the surface at that point. { 'tan·jənt 'plān }

tangent space **1.** The vector space of all tangent vectors at a given point of a differentiable manifold. **2.** *See* tangent bundle. { 'tan·jənt ˌspās }

tangent surface The ruled surface generated by the tangents to a specified space curve. Also known as tangential developable. { 'tan·jənt ˌsər·fəs }

tangent vector A tangent vector at a point of a differentiable manifold is any vector tangent to a differentiable curve in the manifold at this point; alternatively, a member of the tangent plane to the manifold at the point. { 'tan·jənt ˌvek·tər }

tanh *See* hyperbolic tangent. { ¦tan'āch }

Taylor series The Taylor series corresponding to a function $f(x)$ at a point x_0 is the infinite series whose nth term is $(1/n!)\cdot f^{(n)}(x_0)(x - x_0)^n$, where $f^{(n)}(x)$ denotes the nth derivative of $f(x)$. { 'tā·lər ˌsir·ēz }

Taylor's theorem The theorem that under certain conditions a real or complex function can be represented, in a neighborhood of a point where it is infinitely differentiable, as a power series whose coefficients involve the various order derivatives evaluated at that point. { 'tā·lərz ˌthir·əm }

telegrapher's equation The partial differential equation $(\partial^2 f/\partial x^2) = a^2(\partial^2 f/\partial y^2) + b(\partial f/\partial y) + cf$, where a, b, and c are constants; appears in the study of atomic phenomena. { tə'leg·rə·fərz iˌkwā·zhən }

telescopic series The series whose nth term is $1/[(k + n - 1)(k + n)] = [1/(k + n - 1)] - [1/(k + n)]$, where k is not zero or a negative integer, and whose sum is $1/k$. { ¦tel·əˌskäp·ik 'sirˌēz }

ten's complement In decimal arithmetic, the unique numeral that can be added to a given N-digit numeral to form a sum equal to 10^N (that is, a one followed by N zeros). { 'tenz 'käm·plə·mənt }

tensor **1.** An object relative to a locally euclidean space which possesses a specified system of components for every coordinate system and which changes under a transformation of coordinates. **2.** A multilinear function on the cartesian product of several copies of a vector space and the dual of the vector space to the field of scalars on the vector space. { 'ten·sər }

tensor analysis The abstract study of mathematical objects having components which express properties similar to those of a geometric tensor; this study is fundamental to Riemannian geometry and the structure of euclidean spaces. Also known as tensor calculus. { 'ten·sər əˌnal·ə·səs }

tensor calculus *See* tensor analysis. { 'ten·sər ˌkal·kyə·ləs }

tensor contraction For a tensor having an upper and a lower index, summation over the components in which these indexes have the same value, in order to obtain a new tensor two lower in rank. { 'ten·sər kən'trak·shən }

tensor differentiation An operation on a tensor in which a term involving a Christoffel symbol is subtracted from the ordinary derivative, to obtain another tensor of one higher rank. { 'ten·sər ˌdif·əˌren·chē'ā·shən }

tensor field A tensor or collection of tensors defined in some open subset of a Riemann space. { 'ten·sər ˌfēld }

tensorial set Any collection of quantities that are associated with a system of spatial coordinates and which undergo a linear transformation when this system rotates; examples are the components of a tensor and the eigenfunctions of a quantum mechanical operator. { ten'sȯr·ē·əl 'set }

tensor product 1. The product of two tensors is the tensor whose components are obtained by multiplying those of the given tensors. 2. In algebra, a multiplicative operation performed between modules. { 'ten·sər ˌpräd·əkt }

tensor quantity A quantity mathematically represented by a tensor or possessing properties analogous to a tensor. { 'ten·sər ˌkwän·əd·ē }

tensor space A fiber bundle composed of the points of a Riemannian manifold and tensor fields. { 'ten·sər ˌspās }

tera- A prefix representing 10^{12}, which is equivalent to 1,000,000,000,000 or a million million. Abbreviated T. { 'ter·ə }

term 1. For an expression, any one of several quantities whose sum is the expression. 2. For a fraction, either the numerator or the denominator. { tərm }

terminal line One of the two rays that form an angle and may be regarded as having been rotated about a fixed point on another line (the initial line) to form the angle. { 'tərm·ən·əl ˌlīn }

terminal vertex A vertex in a rooted tree that has no successor. Also known as leaf. { 'tər·mən·əl 'vərˌteks }

terminating continued fraction A continued fraction that has a finite number of terms. { ¦tər·məˌnād·iŋ kən¦tin·yüd 'frak·shən }

terminating decimal A decimal that has only a finite number of nonzero digits to the right of the decimal point. { ˌtər·mə¦nād·iŋ 'des·məl }

ternary expansion The numerical representation of a real number relative to the base 3, the digits determined by how the given number can be written in terms of powers of 3. { 'tər·nə·rē ik'span·chən }

ternary notation A system of notation using the base of 3 and the characters 0, 1, and 2. { 'tər·nə·rē nō'tā·shən }

tesselation A covering of a plane without gaps or overlappings by polygons, all of which have the same size and shape. { ˌtes·ə'lā·shən }

tesseral harmonic A spherical harmonic which is 0 on both a set of equally spaced meridians and a set of parallels of latitude of a sphere with center at the origin of spherical coordinates, dividing the sphere into rectangular and triangular regions. { ¦tes·ə·rəl här'män·ik }

test function An infinitely differentiable function of several real variables used in studying solutions of partial differential equations. { 'test ˌfəŋk·shən }

tetradic An operator that transforms one dyadic into another. { tə'trad·ik }

tetrahedral angle A polyhedral angle with four faces. { ˌte·trə¦hē·drəl 'aŋ·gəl }

tetrahedron A four-sided polyhedron. { ˌte·trəˈhē·drən }

theorem A proven mathematical statement. { ˈthir·əm }

theory The collection of theorems and principles associated with some mathematical object or concept. { ˈthē·ə·rē }

theory of equations The study of polynomial equations from the viewpoint of solution methods, relations among roots, and connections between coefficients and roots. { ˈthē·ə·rē əv iˈkwā·zhənz }

theta functions Complex functions used in the study of Riemann surfaces and of elliptic functions and elliptic integrals; they are:

$$\theta_1(z) = 2 \sum_{n=0}^{\infty} (-1)^n q^{(n+1/2)^2} \sin (2n + 1)z$$

$$\theta_2(z) = 2 \sum_{n=0}^{\infty} q^{(n+1/2)^2} \cos (2n + 1)z$$

$$\theta_3(z) = 1 + 2 \sum_{n=1}^{\infty} q^{n^2} \cos 2nz$$

$$\theta_4(z) = 1 + 2 \sum_{n=1}^{\infty} (-1)^n q^{n^2} \cos 2nz$$

where $q = \exp \pi i \tau$, and τ is a constant complex number with positive imaginary part. { ˈthād·ə ˌfəŋk·shənz }

theory of games *See* game theory. { ˈthē·ə·rē əv ˈgāmz }

third proportional For numbers a and b, a number x such that $a/b = b/x$. { ˈthərd prəˈpȯr·shən·əl }

third quadrant **1.** The range of angles from 180 to 270°. **2.** In a plane with a system of cartesian coordinates, the region in which the x and y coordinates are both negative. { ˈthərd ˈkwä·drənt }

three-eighths rule **1.** An approximation formula for definite integrals which states that the integral of a real-valued function f on an interval $[a,b]$ is approximated by $(3/8)h[f(a) + 3f(a + h) + 3f(a + 2h) + f(b)]$, where $h = (b - a)/3$; this is the integral of a third-degree polynomial whose value equals that of f at a, $a + h$, $a + 2h$, and b. **2.** A method of approximating a definite integral over an interval which is equivalent to dividing the interval into equal subintervals and applying the formula in the first definition to each subinterval. { ¦thrē ˈāths ˌrül }

three-index symbols *See* Christoffel symbols. { ˈthrē ¦inˌdeks ˈsim·bəlz }

three-space A vector space over the real numbers whose basis has three vectors. { ˈthrē ˌspās }

threshold A logic operator such that, if $P, Q, R, S, \ldots$ are statements, then the threshold will be true if at least N statements are true, false otherwise. { ˈthreshˌhōld }

Tietze extension theorem A topological space X is normal if and only if every continuous function of a closed subset to [0,1] has a continuous extension to all of X. { 'tēt·sə ik'sten·chən ˌthir·əm }

time-series analysis The general study of mathematical systems or processes analogous to that of data taken at time intervals. { 'tīm ˌsir·ēz əˌnal·ə·səs }

times sign *See* multiplication sign. { 'tīmz ˌsīn }

Titchmarsh's theorem The proposition that, if $f(x)$ and $g(x)$ are continuous functions on the positive real numbers and are not identically equal to 0, then their convolution is not identically 0. { 'tichˌmärsh·əz ˌthir·əm }

topological dynamics The study and application of transformations, or groups of such transformations (particularly topological transformation groups), defined on a topological space (usually compact), with particular regard to properties of interest in the qualitative theory of differential equations. { ¦täp·ə¦läj·ə·kəl dī'nam·iks }

topological groups Groups which also have a topology with the property that the group operation and the inverse operation determine continuous functions. { ¦täp·ə¦läj·ə·kəl 'grüps }

topological *K* theory *See* K theory. { ¦täp·ə¦läj·ə·kəl 'kā ˌthē·ə·rē }

topological linear space *See* topological vector space. { ¦täp·ə¦läj·ə·kəl 'lin·ē·ər ˌspās }

topologically closed set *See* closed set. { ¦täp·ə¦läj·ə·klē ¦klōzd 'set }

topological mapping *See* homeomorphism. { ¦täp·ə¦läj·ə·kəl 'map·iŋ }

topological product The topological space obtained from taking the cartesian product of topological spaces. { ¦täp·ə¦läj·ə·kəl 'präd·əkt }

topological property A property that holds true for any topological space homeomorphic to one possessing the property. { ¦täp·ə¦läj·jə·kəl 'präp·ərd·ē }

topological space A set endowed with a topology. { ¦täp·ə¦läj·ə·kəl 'spās }

topological vector space A vector space which has a topology with the property that vector addition and scalar multiplication are continuous functions. Also known as linear topological space; topological linear space. { ˌtäp·ə¦läj·ə·kəl 'vek·tər ˌspās }

topology **1.** A collection of subsets of a set X, which includes X and the empty set, and has the property that any union or finite intersection of its members is also a member. **2.** The generalized study of properties of spaces invariant under deformations and stretchings. { tə'päl·ə·jē }

toric surface A surface generated by rotating an arc of a circle about a line that lies in the plane of the circle but does not pass through its center. Also known as toroidal surface. { 'tȯr·ik ˌsər·fəs }

toroidal coordinate system A three-dimensional coordinate system whose coordinate surfaces are the toruses and spheres generated by rotating the families of circles defining a two-dimensional bipolar coordinate system about the perpendicular bisector of the line joining the common points of intersection of one of the families, together with the planes passing through the axis of rotation. { tə¦rȯid·əl kō'ȯrd·ən·ət ˌsis·təm }

toroidal surface *See* toric surface. { təˈrȯid·əl ˌsər·fəs }

torsion The rate of change of the positive direction of the binormal of a space curve with respect to arc length along the curve; its sign is defined as positive if it is in the same direction as the principal normal, and negative if it is in the opposite direction. { ˈtȯr·shən }

torsion coefficients For a finitely generated abelian group G, the orders of the finite cyclic groups such that G is the direct sum of these groups and infinite cyclic groups. { ˈtȯr·shən ˌkō·əˌfish·əns }

torsion element **1.** A torsion element of an Abelian group G is an element of G with finite period. **2.** A torsion element of a module M over an entire, principal ring R is an element x in M for which there exists an element a in R such that $a \neq 0$ and $ax = 0$. { ˈtȯr·shən ˌel·ə·mənt }

torsion-free group A group whose only torsion element is the unit element. { ¦tȯr·shən ˌfrē ˌgrüp }

torsion group **1.** A group whose elements all have finite period. **2.** For a topological space X, one of a sequence of finite groups $G_n(X)$ such that the homology group $H_n(X)$ is the direct sum of $G_n(X)$ and a number of infinite cyclic groups. { ˈtȯr·shən ˌgrüp }

torsion module A module M over an entire principal ring R is said to be a torsion module if for any element x in M there exists an element a in R such that $a \neq 0$ and $ax = 0$. { ˈtȯr·shən ˌmä·jül }

torsion subgroup The torsion subgroup of an Abelian group G is the subset of all torsion elements of G. { ¦tȯr·shən ˈsəbˌgrüp }

torsion submodule The torsion submodule of a module E over an entire principal ring is the submodule consisting of all torsion elements of E. { ¦tȯr·shən ˈsəbˌmä·jül }

torus **1.** The surface of a doughnut-shaped object. **2.** The topological space obtained by identifying the opposite sides of a rectangle. **3.** The group which is the product of two circles. { ˈtȯr·əs }

total curvature *See* Gaussian curvature. { ˈtōd·əl ˈkər·və·chər }

total differential The total differential of a function of several variables, $f(x_1,x_2,\ldots,x_n)$, is the function given by the sum of terms $(\partial f/\partial x_i)dx_i$ as i runs from 1 to n. Also known as differential. { ˈtōd·əl ˌdif·əˈren·chəl }

totally bounded set *See* precompact set. { ˈtōd·əl·ē ˈbau̇n·dəd ˈset }

totally disconnected A topological space has this property if the largest connected subset containing any given point is only the point itself. { ˈtōd·əl·ē ˌdis·kəˈnek·təd }

totally imaginary field An extension field F of the field of rational numbers such that no embedding of F in the complex numbers is contained in the real numbers. { ¦tōd·əl·ē iˌmaj·əˌner·ē ˈfēld }

total order **1.** The total order of an analytic function in a domain D is the algebraic sum of its orders at all poles and zeros in D. **2.** *See* linear order. { ˈtōd·əl ˈȯr·dər }

total space The topological space E in the bundle (E,p,B). { ¦tōd·əl ˈspās }

total variation For a real function defined on an interval, the least upper bound of the function's variation relative to all possible partitions of the interval. { 'tōd·əl ˌver·ē'ā·shən }

totitive An integer that is less than a given integer and relatively prime to it. { 'tōd·əˌtiv }

tournament A graph in which there is one line between every pair of points and no loops, and in which a unique direction is assigned to every line. { 'tu̇r·nə·mənt }

tower For a set S with a given algebraic structure, this is a set of subsets, $S_0 = S, S_1, S_2, \ldots, S_n$, such that S_{i+1} is a subset of S_i, $i = 1, 2, \ldots, n - 1$, and each S_i is closed under all possible operations in the algebraic structure of S. { tau̇·ər }

trace 1. The trace of a matrix is the sum of the entries along its principal diagonal. Designated Tr. Also known as spur. **2.** The trace of a linear transformation on a finite-dimensional vector space is the trace (in the sense of the first definition) of the matrix associated with it. **3.** One of the curves along which a given surface cuts a coordinate plane. **4.** A point at which a given straight line in space passes through a coordinate plane. Also known as piercing point. **5.** The projection of a given straight line in space on a coordinate plane. { trās }

tractrix A curve in the plane where every tangent to it has the same length. Also known as equitangential curve. { 'trakˌtriks }

trailing zero Any zero following the last nonzero integer of a number. { 'trāl·iŋ 'zir·ō }

trajectory A curve that intersects all the members of a given family of curves at the same angle. { trə'jek·trē }

transcendence base A transcendence base of a field E over a subfield F is a subset S of E which is algebraically independent over F and is not a proper subset of any other subset S' which is algebraically independent over F. { tran'sen·dəns ˌbās }

transcendence degree The transcendence degree of a field E of a subfield F is the number of elements in a transcendence base of E over F. Also known as transcendence dimension. { tran'sen·dəns diˌgrē }

transcendence dimension *See* transcendence degree. { tran'sen·dəns diˌmen·chən }

transcendental curve The graph of a transcendental function. { ˌtran·sən¦den·təl 'kərv }

transcendental element An element of a field K is transcendental relative to a subfield F if it satisfies no polynomial whose coefficients come from F. { ¦tranˌsen¦dent·əl 'el·ə·mənt }

transcendental field extension A field extension K of F where the elements of K not in F are all transcendental relative to F. { ¦tranˌsen¦dent·əl 'fēld ikˌsten·chən }

transcendental functions Functions which cannot be given by any algebraic expression involving only their variables and constants. { ¦tranˌsen¦dent·əl 'fəŋk·shənz }

transcendental number An irrational number that is the root of no polynomial with rational-number coefficients. { ¦tranˌsen¦dent·əl 'nəm·bər }

transcendental term In an expression, a term that cannot be expressed solely by numbers and algebraic symbols. { ˌtran·sənˈden·təl ˈtərm }

transfinite induction A reasoning process by which if a theorem holds true for the first element of a well-ordered set N and is true for an element n whenever it holds for all predecessors of n, then the theorem is true for all members of N. { tranzˈfīˌnīt inˈdək·shən }

transfinite number Any ordinal or cardinal number equal to or exceeding aleph null. { tranzˈfīˌnīt ˈnəm·bər }

transform **1.** An expression, commonly used in harmonic analysis, formed from a given function f by taking an integral of $f \cdot g$, where g is a member of an orthogonal family of functions. **2.** The value of a transformation at some point. **3.** A matrix B related to a given matrix A by $B = C^{-1}AC$, where C is a nonsingular matrix. **4.** *See* conjugate. { tranzˈfȯrm (verb) *or* ˈtranzˌfȯrm (noun) }

transformation A function, usually between vector spaces. { ˌtranz·fərˈmā·shən }

transformation group **1.** A collection of transformations which forms a group with composition as the operation. **2.** A dynamical system or, more generally, a topological group G together with a topological space X where each g in G gives rise to a homeomorphism of X in a continuous manner with respect to the algebraic structure of G. { ˌtranz·fərˈmā·shən ˌgrüp }

transformation methods A category of numerical methods for finding the eigenvalues of a matrix, in which a series of orthogonal transformations are used to reduce the matrix to some simpler matrix, usually a triple-diagonal one, before an attempt is made to find the eigenvalues. { ˌtranz·fərˈmā·shən ˌmeth·ədz }

transformation of similitude *See* homothetic transformation. { ˌtranz·fərˈmā·shən əv siˈmil·əˌtüd }

transition probability Conditional probability concerning a discrete Markov chain giving the probabilities of change from one state to another. { tranˈzish·ən ˌpräb·əˈbil·əd·ē }

transitive group A group of permutations of a finite set such that for any two elements in the set there exists an element of the group which takes one into the other. { ˈtran·səd·iv ˌgrüp }

transitive relation A relation $<$ on a set such that if $a < b$ and $b < c$, then $a < c$. { ˈtran·səd·iv riˈlā·shən }

translation **1.** A function changing the coordinates of a point in a euclidean space into new coordinates relative to axes parallel to the original. **2.** A function on a group to itself given by operating on each element by some one fixed element. **3.** Let E be a finitely generated extension of a field k, F be an extension of k, and both E and F be contained in a common field; the translation of E to F is the extension EF of F, where EF is the compositum of E and F. Also known as lifting. { tranˈslā·shən }

translation surface *See* surface of translation. { tranˈslā·shən ˌsər·fəs }

transpose The matrix obtained from a given matrix by interchanging its rows and columns. { ˈtranzˌpōz }

transposition A permutation of a set of symbols which exchanges exactly two while leaving all others unaffected. { ˌtranz·pəˈzish·ən }

transversal 1. A line intersecting a given family of lines. 2. A curve orthogonal to a hypersurface. 3. If π is a given map of a set X onto a set Y, a transversal for π is a subset T of X with the property that T contains exactly one point of $\pi^{-1}(y)$ for each $y \in Y$. { trans'vər·səl }

transverse axis The portion of a line passing through the foci of a hyperbola that lies between the two branches of the hyperbola. { trans¦vərs 'ak·səs }

trapezium A quadrilateral where no sides are parallel. { trə'pē·zē·əm }

trapezoid A quadrilateral having two parallel sides. { 'trap·ə,zȯid }

trapezoidal integration A numerical approximation of an integral by means of the trapezoidal rule. { ¦trap·ə¦zȯid·əl ,int·ə'grā·shən }

trapezoidal rule The rule that the integral from a to b of a real function $f(x)$ is approximated by

$$\frac{b-a}{2n}\left[f(a)+\sum_{j=1}^{n-1}2f(x_j)+f(b)\right]$$

where $x_0 = a$, $x_j = x_{j-1} + (b - a)/n$ for $j = 1, 2, ..., n - 1$. { ¦trap·ə¦zȯid·əl 'rül }

traveling salesman problem The problem of performing successively a number of tasks, represented by vertices of a graph, with the least expenditure on transitions from one task to another, represented by edges of the graph with journey costs attached. { ¦trav·əl·iŋ 'sālz·mən ,präb·ləm }

tree A connected graph contained in a given connected graph having all the vertices of the original but without any closed circuit. { trē }

trefoil A multifoil consisting of three congruent arcs of a circle arranged around an equilateral triangle. { 'trē,fȯil }

triangle The figure realized by connecting three noncollinear points by line segments. { 'trī,aŋ·gəl }

triangle inequality For real or complex numbers or vectors in a normed space x and y, the absolute value or norm of $x + y$ is less than or equal to the sum of the absolute values or norms of x and y. { 'trī,aŋ·gəl ,in·i'kwäl·ə·dē }

triangle of vectors A triangle, two of whose sides represent vectors to be added, while the third represents the sum of these two vectors. { 'trī,aŋ·gəl əv 'vek·tərz }

triangulable space A topological space that is homeomorphic to a simplicial complex. { trī¦aŋ·gyə·lə·bəl 'spās }

triangular matrix A matrix where either all entries above or all entries below the principal diagonal are zero. { trī'aŋ·gyə·lər 'mā·triks }

triangular numbers The numbers 1, 3, 6, 10, . . . , which are the numbers of dots in successive triangular arrays, and are given by the expression $(n + 1)(n/2)$, where $n = 1, 2, 3, \ldots$. { trī¦aŋ·gyə·lər 'nəm·bərz }

triangulation A decomposition of a topological manifold into subsets homeomorphic with a polyhedron in some euclidean space. { trī,aŋ·gyə'lā·shən }

triangulation problem The problem of whether each topological n manifold admits a piecewise linear structure. { trī,aŋ·gyə'lā·shən ,präb·ləm }

trident of Newton The curve in the plane given by the equation $xy = ax^3 + bx^2 + cx + d$, where $a \neq 0$; this cuts the x axis in one or three points and is asymptotic to the y axis if $d \neq 0$. { 'trīd·ənt əv 'nüt·ən }

tridiagonal matrix A square matrix in which all entries other than those on the principal diagonal and the two adjacent diagonals are zero. { ¦trī·dī'ag·ən·əl 'mā·triks }

trigonometric cofunctions Trigonometric functions that are equal when their arguments are complementary angles, such as sine and cosine, tangent and cotangent, and secant and cosecant. { ¦trig·ə·nə¦me·trik ˌkō¦fəŋk·shənz }

trigonometric functions The real-valued functions such as $\sin(x)$, $\tan(x)$, and $\cos(x)$ obtained from studying certain ratios of the sides of a right triangle. Also known as circular functions. { ¦trig·ə·nə¦me·trik 'fəŋk·shənz }

trigonometric polynomial A finite series of functions of the form $a_n \cos nx + b_n \sin nx$; occasionally used synonymously with trigonometric series. { ¦trig·ə·nə¦me·trik ˌpäl·i'nō·mē·əl }

trigonometric series An infinite series of functions with nth term of the form $a_n \cos nx + b_n \sin nx$. { ¦trig·ə·nə¦me·trik 'sir·ēz }

trigonometric substitutions The substitutions $x = a \sin u$, $x = a \tan u$, and $x = a \sec u$, which are used to rationalize expressions of the form $\sqrt{a^2 - x^2}$, $\sqrt{x^2 + a^2}$, and $\sqrt{x^2 - a^2}$, respectively, when they appear in integrals. { ¦trig·ə·nə¦me·trik ˌsəb·stə'tü·shənz }

trigonometry The study of triangles and the trigonometric functions. { ˌtrig·ə'näm·ə·trē }

trihedral Any figure obtained from three noncoplanar lines intersecting in a common point. { trī'hē·drəl }

trihedral angle A polyhedral angle with three faces. { trī¦hē·drəl 'aŋ·gəl }

trillion **1.** The number 10^{12}. **2.** In British and German usage, the number 10^{18}. { 'tril·yən }

trinomial A polynomial comprising three terms. { trī'nō·mē·əl }

trinomial surd A sum of three roots of rational numbers, at least two of which are irrational numbers that cannot be combined without evaluating them. { trī'nō·mē·əl 'sərd }

triple-diagonal matrix *See* continuant matrix. { ¦trip·əl di¦ag·ən·əl 'mā·triks }

triple scalar product *See* scalar triple product. { 'trip·əl 'skā·lər ˌpräd·əkt }

triple vector product The triple vector product of vectors **a, b,** and **c** is the cross product of a with the cross product of b and c; written **a × (b × c).** { ¦trip·əl ¦vek·tər ˌpräd·əkt }

trirectangular spherical triangle A spherical triangle with three right angles. { ˌtrī·rek¦taŋ·gyə·lər ¦sfir·ə·kəl 'trīˌaŋ·gəl }

trisection The problem of dividing an angle into three equal parts, which is impossible to do with straight edge and compass alone. { trī'sek·shən }

trisectrix The planar curve given by $x^3 + xy^2 + ay^2 - 3ax^2 = 0$ which is symmetric about the x axis and asymptotic to the line $x = -a$; this is useful in studying the trisection of an angle problem. Also known as trisectrix of Maclaurin. { trī'sek·triks }

trisectrix of Catalan *See* Tschirnhausen's cubic. { trī'sek·triks əv 'kad·ə‚lan }

trisectrix of Maclaurin *See* trisectrix. { trī'sek·triks əv mə'klȯr·ən }

trit A digit in a balanced ternary system, that is, a balanced digit system with base 3. { trit }

trivial solution A solution of a set of homogeneous linear equations in which all the variables have the value zero. { ¦triv·ē·əl sə'lü·shən }

trochoid The path in the plane obtained from a point on the radius of a circle or the extension of the radius as the circle rolls along a fixed straight line. { 'trō‚kȯid }

true complement *See* radix complement. { 'trü 'käm·plə·mənt }

truncate **1.** To drop digits at the end of a numerical value; the number 3.14159265 is truncated to five figures in 3.1415, whereas it would be 3.1416 if rounded off to five figures. **2.** To approximate the sum of an infinite series by the sum of a finite number of its terms. **3.** To terminate an infinite sequence of successively better approximations of a quantity after a finite number of such approximations. **4.** To construct from a geometric solid another solid consisting of those portions of the original solid that lie between two planes. { 'trəŋ‚kāt }

truncated cone The portion of a cone between two nonparallel planes whose line of intersection lies outside the cone. { ¦trəŋ·kād·əd 'kōn }

truncated prism The part of a prism that lies between two nonparallel planes that cut the prism and intersect outside the prism. { ¦trəŋ·kād·əd 'priz·əm }

truncated pyramid The part of a pyramid between the base and a plane that is not parallel to the base. { ¦trəŋ·kād·əd 'pir·ə‚mid }

truncation **1.** Approximating the sum of an infinite series by the sum of a finite number of its terms. **2.** *See* rounding. { trəŋ'kā·shən }

truncation error **1.** The computation error resulting from use of only a finite number of terms of an infinite series. **2.** The error resulting from the approximation of a derivative or differential by a finite difference. { trəŋ'kā·shən ‚er·ər }

truth set A set containing all the elements that make a given statement of relationships true when they are substituted in this statement. { 'trüth ‚set }

truth table A table listing statements concerning an event and their respective truth values. { 'trüth ‚tā·bəl }

truth value The result of a logical proposition; either "true" or "false" in classical logic. { 'trüth ‚val·yü }

Tschirnhausen's cubic A plane curve consisting of the envelope of the line through a variable point P on a parabola which is perpendicular to the line from the focus of the parabola to P. Also known as l'Hôpital's cubic; trisectrix of Catalan. { 'chərn‚hau̇z·ənz 'kyü·bik }

T_0 space A topological space where, for each pair of points, at least one has a neighborhood not containing the other. { ˌtē ˈzir·ō ¦spās }

T_1 space A topological space where, for each pair of distinct points, each one has a neighborhood not containing the other. { ˈtē ˌwən ¦spās }

T_2 space *See* Hausdorff space. { ˈtē ˌtü ¦spās }

T_3 space A regular topological space that is also a T_1 space. { ˈtē ˌthrē ¦spās }

T_4 space A normal space that is also a T_1 space. { ˈtē ˌfȯr ¦spās }

Tukey lemma The proposition that any nonempty family of finite character has a maximal member. { ˈtü·kēˌlem·ə }

Turing computable function A function that can be computed on a Turing machine. { ˈtu̇r·iŋ kəmˈpad·ə·bəl ˈfəŋk·shən }

Turing's thesis *See* Church's thesis. { ˈtu̇r·iŋz ˌthē·səs }

turn *See* circle. { ˈtərn }

turning value A relative maximum or relative minimum of a function. { ˈtərn·iŋ ˌval·yü }

twisted curve A curve that does not lie wholly in any one plane. { ¦twis·təd ˈkərv }

two-cycle The repetition of numbers generated by a mapping on every second iteration of the mapping. { ¦tü ˈsī·kəl }

two-person game A game consisting of exactly two players with competing interests. { ˈtü ¦pər·sən ˈgām }

two's complement A number derived from a given n-bit number by requiring the two numbers to sum to a value of 2^n. { ˈtüz ˈkäm·plə·mənt }

two-sided ideal A two-sided ideal I is a sub-ring of a ring R where the products xy and yx are always in I for every x in R and y in I. { ˈtü ¦sīd·əd īˈdēl }

two-sphere The surface of a ball; the two-dimensional sphere in euclidean three-dimensional space obtained from all points whose distance from the origin is one. { ˈtü ˌsfir }

two-valued logic A system of logic where each statement has two possible values or states, truth or falsehood. { ˈtü ¦val·yüd ˈläj·ik }

two-valued variable A variable which assumes values in a set containing exactly two elements, often symbolized as 0 and 1. { ˈtü ¦val·yüd ˈver·ē·ə·bəl }

Tychonoff space *See* completely regular space. { tīˈkäˌnȯf ˌspās }

Tychonoff theorem A product of topological spaces is compact if and only if each individual space is compact. { tīˈkäˌnȯf ˌthir·əm }

U

ultrafilter A filter base which has no properly subordinated filter base. { ¦əl·trə′fil·tər }

ultraspherical polynomials *See* Gegenbauer polynomials. { ¦əl·trə′sfer·ə·kəl ˌpäl·i′nō·mē·əlz }

umbilic *See* umbilical point. { u̇m′bil·ik }

umbilical point A point on a surface at which the normal curvature is the same in all directions. Also known as navel point; umbilic. { əm′bil·ə·kəl ˌpȯint }

unary operation An operation in which only a single operand is required to produce a unique result; some examples are negation, complementation, square root, transpose, inverse, and conjugate. { ′yü·nə·rē ˌäp·əˌrā·shən }

unavoidable set of configurations A set of graphs such that any planar graph has at least one member of the set as a subgraph. { ¦ən·ə′vȯid·ə·bəl ′set əv kənˌfig·yə′rā·shənz }

unbounded manifold A manifold with no boundary. { ¦ən′bau̇n·dəd ′man·əˌfōld }

unbounded set of real numbers A set with the property that if R is any positive real number, there is a number in the set which is smaller than $-R$ or a number larger than R. { ¦ən′bau̇n·dəd ′set əv ′rēl ′nəm·bərz }

unconditional convergence A convergent series converges unconditionally if every series obtained by rearranging its terms also converges; equivalent to absolute convergence. { ¦ən·kən′dish·ən·əl kən′vər·jəns }

unconditional inequality An inequality which holds true for all values of the variables involved, or which contains no variables; for example, $y + 2 > y$, or $4 > 3$. Also known as absolute inequality. { ¦ən·kən′dish·ən·əl ˌin·i′kwäl·əd·ē }

unconstrained optimization problem A nonlinear programming problem in which there are no constraint functions. { ¦ən·kən′strānd ˌäp·tə·mə′zā·shən ˌpräb·ləm }

uncountable set An infinite set which cannot be put in one-to-one correspondence with the set of integers; for example, the set of real numbers. { ¦ən′kau̇nt·ə·bəl ′set }

undetermined multipliers *See* Lagrangian multipliers. { ¦ən·di′tər·mənd ′məl·təˌplī·ərz }

ungula A solid bounded by a portion of a circular cylindrical surface and portions of two planes, one of which is perpendicular to the generators of the cylindrical surface. { 'əŋ·gyə·lə }

uniform bound A number M such that $|f_n(x)|$ M for every x and for every function in a given sequence of functions $\{f_n(x)\}$. { 'yü·nə‚fȯrm 'bau̇nd }

uniform boundedness principle A family of pointwise bounded, real-valued continuous functions on a complete metric space X is uniformly bounded on some open subset of X. { 'yü·nə‚fȯrm 'bau̇n·dəd·nəs ‚prin·sə·pəl }

uniform continuity A property of a function f on a set, namely: given any $\epsilon > 0$ there is a $\delta > 0$ such that $|f(x_1) - f(x_2)|$ ϵ provided $|x_1 - x_2|$ δ for any pair x_1, x_2 in the set. { 'yü·nə‚fȯrm känt·ən'ü·əd·ē }

uniform convergence A sequence of functions $\{f_n(x)\}$ converges uniformly on E to $f(x)$ if given $\epsilon > 0$ there is an N such that $|f_n(x) - f(x)| < \epsilon$ for all x in E provided $n > N$. { 'yü·nə‚fȯrm kən'vər·jəns }

uniformly convex space A normed vector space such that for any number $\epsilon > 0$ there is a number $\delta > 0$ such that, for any two vectors x and y, if $\|x\| \leqq 1 + \delta$, $\|y\| \leqq 1 + \delta$, and $\|x + y\| > 2$, then $\|x - y\| < \epsilon$. { ¦yü·nə‚fȯrm·lē ¦kän‚veks 'spās }

uniformly summable series For a given summability method and for a given interval, a series for which the sequence that defines the sum converges uniformly on the interval. { ¦yü·nə‚fȯrm·lē ¦səm·ə·bəl 'sir‚ēz }

uniform scale A scale in which equal distances correspond to equal numerical values. Also known as linear scale. { ¦yü·nə‚fȯrm 'skāl }

uniform space A topological space X whose topology is derived from a family of subsets of $X \times X$, called a uniformity; intuitively, this gives a notion of "nearness" which is uniform throughout the space. { 'yü·nə‚fȯrm 'spās }

unilateral surface A one-sided surface; equivalently, any nonorientable two-dimensional manifold such as the Möbius strip and the Klein bottle. { ¦yü·nə'lad·ə·rəl 'sər·fəs }

unimodular matrix A unimodulus matrix with integer entries. { ¦yü·nə'mäj·ə·lər 'mā·triks }

unimodulus matrix A square matrix whose determinant is 1. { ¦yü·nə'mäj·ə·ləs 'mā·triks }

union **1.** A union of a given family of sets is a set consisting of those elements that are members of at least one set in the family. **2.** For two fuzzy sets A and B, the fuzzy set whose membership function has a value at any element x that is the maximum of the values of the membership functions of A and B at x. **3.** The union of two Boolean matrices A and B, with the same number of rows and columns, is the Boolean matrix whose element c_{ij} in row i and column j is the union of corresponding elements a_{ij} in A and b_{ij} in B. { 'yün·yən }

unique factorization theorem A positive integer may be expressed in precisely one way as a product of prime numbers. { yu̇'nēk ‚fak·tə·rə'zā·shən ‚thir·əm }

unit An element of a ring with identity that has both a left inverse and a right inverse. { 'yü·nət }

unitary group The group of unitary transformations on a k-dimensional complex vector space. Usually denoted U(k). { 'yü·nə‚ter·ē 'grüp }

unitary matrix A matrix whose inverse is equal to the complex conjugate of its transpose. { 'yü·nə‚ter·ē 'mā·triks }

unitary space *See* inner product space. { 'yü·nə‚ter·ē 'spās }

unitary transformation A linear transformation on a vector space which preserves inner products and norms; alternatively, a linear operator whose adjoint is equal to its inverse. { 'yü·nə‚ter·ē ‚tranz·fər'mā·shən }

unit ball The set of all points in euclidean n-space whose distance from the origin is at most 1. { 'yü·nət 'bȯl }

unit binormal A unit vector in the same direction as the binormal to a point on a surface or space curve. { 'yü·nət bī'nȯr·məl }

unit circle The locus of points in the plane which are precisely one unit from the origin. { 'yü·nət 'sər·kəl }

unit conversion factor *See* conversion factor. { ¦yü·nət kən'vər·zhən ‚fak·tər }

unit element An element in a ring which acts as a multiplicative identity. { 'yü·nət 'el·ə·mənt }

unit fraction A common fraction whose numerator is unity. { 'yü·nət ‚frak·shən }

unit impulse *See* delta function. { 'yü·nət 'im‚pəls }

unit normal A unit vector in the direction of the principal normal to a surface or space curve. { 'yü·nət 'nȯr·məl }

unit operator The identity operator. { 'yü·nət 'äp·ə‚rād·ər }

unit sphere The set of points in three-space (more generally n-space) which are precisely one unit distance from the origin. { 'yü·nət 'sfir }

unit tangent A unit vector in the tangent plane at a point of a surface. { 'yü·nət 'tan·jənt }

unit vector A vector whose length is one unit. { 'yü·nət 'vek·tər }

univalent function *See* injection. { ¦yü·nə¦vā·lənt 'fəŋk·shən }

universal algebra The study of algebraic systems such as groups, rings, modules, and fields and the examination of what families of theorems are analogous in each system. { ¦yü·nə¦vər·səl 'al·jə·brə }

universal element An element of a Boolean algebra that includes every element of the algebra. { ¦yü·nə¦vər·səl 'el·ə·mənt }

universally attracting object An object O in a category C such that there exists a unique morphism of each object of C into O. { ¦yü·nə¦vər·sə·lē ə'trak·tiŋ ‚äb ‚jekt }

universally repelling object An object O of a category C such that there exists a unique morphism of O into each object of C. { ¦yü·nə¦vər·sə·lē ri'pel·iŋ ˌäb ˌjekt }

universal object An object which is universally attracting or universally repelling. { ¦yü·nə'vər·səl ˌäbˌjekt }

universal quantifier A logical relation, often symbolized $\forall$, that may be expressed by the phrase "for all" or "for every"; if P is a predicate, the statement $(\forall x)P(x)$ is true if $P(x)$ is true for all values of x in the domain of P, and is false otherwise. { ¦yü·nə¦vər·səl 'kwän·təˌfī·ər }

universal set A set that contains all the elements of concern in the study of a particular problem. { ¦yü·nə¦vər·səl 'set }

unsigned integer A whole number that is equal to or greater than zero and does not carry a positive or negative sign. { ən'sīnd 'int·ə·jər }

unsigned real number A number that does not carry a sign indicating whether it is positive or negative, and that is therefore assumed to be positive. { ən'sīnd 'rēl 'nəm·bər }

unstable graph A graph from which it is not possible to delete an edge to produce a subgraph whose group of automorphisms is a subgroup of the group of automorphisms of the original graph. { ¦ənˌstā·bəl 'graf }

upper bound If S is a subset of an ordered set A, an upper bound b for S in A is an element b of A such that $x \leq b$ for all x belonging to A. { ¦əp·ər 'baund }

upper integral The upper Riemann integral for a real-valued function $f(x)$ on an interval is computed to be the infimum of all finite sums over all partitions of the interval, the sums having terms given by $(x_i - x_{i-1})y_i$, where the x_i are from a partition, and y_i is the largest value of $f(x)$ over the interval from x_{i-1} to x_i. { 'əp·ər 'int·ə·grəl }

upper semicontinuous decomposition A partition of a topological space with the property that for every member D of the partition and for every open set U containing D there is an open set V containing D which is contained in U and is the union of members of the partition. { 'əp·ər ¦sem·i·kən'tin·yə·wəs dēˌkäm·pə'zish·ən }

upper semicontinuous function A real-valued function $f(x)$ is upper semicontinuous at a point x_0 if for any small positive ϵ, $f(x)$ always is less than $f(x_0) + \epsilon$ for all x in some neighborhood of x_0. { 'əp·ər ¦sem·i·kən'tin·yə·wəs 'fəŋk·shən }

Urysohn lemma If A and B are disjoint, closed sets in a normal space X, there is a real-valued function f such that $0 \leq f(x) \leq 1$ for all $x \in X$, and $f(A) = 0$ and $f(B) = 1$. { 'ür·ēˌzōn ˌlem·ə }

Urysohn theorem The theorem that a regular T_1 space whose topology has a countable base is metrizable. { 'ür·ēˌzōn ˌthir·əm }

V

valence The number of lines incident on a specified point of a graph. { 'vā·ləns }

validity Correctness; especially the degree of closeness by which iterated results approach the correct result. { və'lid·əd·ē }

valuation A scalar function of a field which has properties similar to those of absolute value. { ˌval·yə'wā·shən }

value 1. The value of a function f at an element x is the element y which f associates with x; that is, $y = f(x)$. 2. The expected payoff of a matrix game when each player follows an optimal strategy. { 'val·yü }

value group For a discrete valuation v on a field K, this is the group formed by the elements $v(x)$ corresponding to nonzero elements x in K. { 'val·yü ˌgrüp }

Vandermonde determinant The determinant of the $n \times n$ matrix whose ith row appears as $1, x_i, x_i^2, \ldots, x_i^{n-1}$ where the x_i^k appear as variables in a given polynomial equation; this provides information about the roots. { 'van·dərˌmōnd diˌtər·mə·nənt }

Vandermonde's theorem A theorem stating that a binomial $(x + y)^a$, where a is an exponent involving the variables x and y, can be stated in terms of a sum of expressions $x^c y^d$, where the exponents c and d involve the variables x and y also. { 'van·dərˌmōndz ˌthir·əm }

variable A symbol which is used to represent some undetermined element from a given set, usually the domain of a function. { 'ver·ē·ə·bəl }

variate *See* random variable. { 'ver·ē·ət }

variational principle A technique for solving boundary value problems that is applicable when the given problem can be rephrased as a minimization problem. { ˌver·ē'ā·shən·əl ˌprin·sə·pəl }

vector 1. An element of a vector space. 2. A matrix consisting of a single row or a single column of entries. { 'vek·tər }

vector analysis The formal study of vectors. { 'vek·tər əˌnal·ə·səs }

vector bundle A locally trivial bundle whose fibers are isomorphic vector spaces. { 'vek·tər ˌbən·dəl }

vector equation An equation involving vectors. { 'vek·tər iˌkwā·zhən }

vector field 1. The field of vectors arising from considering a system of differential equations on a differentiable manifold. **2.** A function whose range is in a vector space. { 'vek·tər ˌfēld }

vector space A system of mathematical objects which have an additive operation producing a group structure and which can be multiplied by elements from a field in a manner similar to contraction or magnification of directed line segments in euclidean space. Also known as linear space. { 'vek·tər ˌspās }

vector sum For a set of located vectors in euclidean space, $\mathbf{v}_1$, $\mathbf{v}_2$, . . . , $\mathbf{v}_n$, this is the vector whose initial point is the initial point of $\mathbf{v}_1$ and whose terminal point is the terminal point of $\mathbf{v}_n$, when the vectors are laid end to end so that the terminal point of one vector $\mathbf{v}_i$ is the initial point of the next vector $\mathbf{v}_{i+1}$. Also known as resultant. { 'vek·tər ˌsəm }

Venn diagram A pictorial representation of set theoretic operations such as union, intersection, and complementation of sets. { 'ven ˌdī·əˌgram }

vers *See* versed sine.

versed cosine *See* coversed sine. { 'vərst 'kōˌsīn }

versed sine The versed sine of A is $1 -$ cosine A. Denoted vers. Also known as versine. { 'vərst 'sīn }

versiera *See* witch of Agnesi. { ˌvər·sē'er·ə }

versine *See* versed sine. { 'vərˌsīn }

vertex **1.** For a polygon or polyhedron, any of those finitely many points which together with line segments or plane pieces determine the figure or solid. **2.** The common point at which the two sides of an angle intersect. **3.** The fixed point through which pass all the elements of a cone or conical surface. **4.** An intersection of a conic with one of its axes of symmetry. { 'vərˌteks }

vertex angle In a triangle, the angle opposite the base. { 'vərˌteks ˌaŋ·gəl }

vertical angles The two angles produced by a pair of intersecting lines and lying on opposite sides of the point of intersection. { 'vərd·ə·kəl 'aŋ·gəlz }

Vitali set A set of real numbers such that the difference of any two members of the set is an irrational number and any real number is the sum of a rational number and a member of the set. { vē'täl·ē ˌset }

vol *See* volume.

Volterra equations Given functions $f(x)$ and $K(x,y)$, these are two types of equations with unknown function y:

$$f(x) = \int_a^x K(x,t)y(t)dt$$

$$y(x) = f(x) + \lambda \int_a^x K(x,t)y(t)dt$$

{ vol'ter·ə iˌkwā·shənz }

volume A measure of the size of a body or definite region in three-dimensional space;

it is equal to the least upper bound of the sum of the volumes of nonoverlapping cubes that can be fitted inside the body or region, where the volume of a cube is the cube of the length of one of its sides. Abbreviated vol. { 'väl·yəm }

volume integral An integral of a function of several variables with respect to volume measure taken over a three-dimensional subset of the domain of the function. { 'väl·yəm 'int·ə·grəl }

von Neumann algebra A subalgebra A of the algebra $B(H)$ of bounded linear operators on a complex Hilbert space, such that the adjoint operator of any operator in A is also in A, and A is closed in the strong operator topology in $B(H)$. Also known as ring of operators; W* algebra. { fȯn ¦nȯi·män 'al·jə·brə }

W

W* algebra *See* von Neumann algebra. { ¦dəb·əl‚yü stär 'al·jə·brə }

walk In graph theory, a set of vertices (v_0, v_1, . . . , v_n) in a graph, such that v_i and v_{i+1} are joined by a common edge for $i = 0, 1, \ldots, n - 1$. Also known as path. { wȯk }

Wallis formulas Formulas that determine the values of the definite integrals from 0 to $\pi/2$ of the functions $\sin^n(x)$, $\cos^n(x)$, and $\cos^m(x)\sin^n(x)$ for positive integers m and n. Also known as Wallis theorem. { 'wäl·əs ‚fȯr·myə·ləz }

Wallis product An infinite product representation of $\pi/2$, namely,

$$\frac{\pi}{2} = \frac{2}{1}\frac{2}{3}\frac{4}{3}\frac{4}{5}\cdots\frac{2n}{2n-1}\frac{2n}{2n+1}$$

{ 'wäl·əs ‚präd·əkt }

Wallis theorem *See* Wallis formulas. { 'wäl·əs ‚thir·əm }

Watson-Sommerfeld transformation A procedure for transforming a series whose lth term is the product of the lth Legendre polynomial and a coefficient, a_l, having certain properties, into the sum of a contour integral of $a(l)$ and terms involving poles of $a(l)$, where $a(l)$ is a meromorphic function such that $a(l)$ equals a_l at integral values of l; used in studying rainbows, propagation of radio waves around the earth, scattering from various potentials, and scattering of elementary particles. Also known as Sommerfeld-Watson transformation. { 'wät·sən 'zȯm·ər‚felt ‚tranz·fər'mā·shən }

Watt's curve The curve traced out by the midpoint of a line segment whose end points move along two circles of equal radius. { 'wäts ‚kərv }

weak convergence A sequence of elements x_1, x_2,... from a topological vector space X converges weakly if the sequence $f(x_1)$, $f(x_2)$,... converges for every continuous linear functional f on X. { 'wēk kən'vər·jəns }

weakly complete space A topological vector space in which an element x is associated with any weakly convergent sequence of elements x_n such that the limit of $f(x_n)$ equals $f(x)$ for any continuous linear functional f. { ¦wēk·lē kəm¦plēt 'spās }

weak topology A topology on a topological vector space X whose open neighborhoods around a point x are obtained from those points y of X for which every $f_i(x)$ is close to $f_i(y)$, f_i appearing in a finite list of linear functionals. { 'wēk tə'päl·ə·jē }

Weber differential equation A special case of the confluent hypergeometric equation that has as solution a confluent hypergeometric series. Also known as Weber-Hermite equation. { 'vā·bər ˌdif·ə'ren·chəl i'kwā·zhən }

Weber-Hermite equation *See* Weber differential equation. { 'vā·bər er'mēt iˌkwā·zhən }

wedge A polyhedron whose base is a rectangle and whose lateral faces consist of two equilateral triangles and two trapezoids. { wej }

wedge product A product defined on forms such that a wedge product of a p-form and a q-form results in a $p + q$ form. { 'wej ˌpräd·əkt }

Weierstrass' approximation theorem A continuous real-valued function on a closed interval can be uniformly approximated by polynomials. { 'vī·ərˌshträs əˌpräk·sə'mā·shən ˌthir·əm }

Weierstrass functions Used in the calculus of variations, these determine functions satisfying the Euler-Lagrange equation and Jacobi's condition while maximizing a given definite integral. { 'vī·ərˌshträs ˌfəŋk·shənz }

Weierstrassian elliptic function A function that plays a central role in the theory of elliptic functions; for z, g_2 and g_3 real or complex numbers, let y be that number such that

$$z = \int_y^\infty \frac{dt}{\sqrt{4t^3 - g_2 t - g_3}};$$

the Weierstrassian elliptic function of z with parameters g_2 and g_3 is $p\,(z; g_2, g_3) = y$. { ¦vī·ər¦shträs·ē·ən i'lip·tik 'fəŋk·shən }

Weierstrass M test An infinite series of numbers will converge or functions will converge uniformly if each term is dominated in absolute value by a nonnegative constant M_n, where these M_n form a convergent series. Also known as Weierstrass' test for convergence. { 'vī·ərˌshträs 'em ˌtest }

Weierstrass' test for convergence *See* Weierstrass test. { 'vī·ərˌshträs 'test fər kən'vər·jəns }

Weierstrass transform This transform of a real function $f(y)$ is the function given by the integral from $-\infty$ to ∞ of $(4\pi t)^{-1/2}\exp[-(x - y)^2/4]\,f(y)dy$; this is used in studying the heat equation. { 'vī·ərˌshträs 'tranzˌfȯrm }

weight function Two real valued functions f and g are orthogonal relative to a weight function σ on an interval if the integral over the interval of $f \cdot g \cdot \sigma$ vanishes. { 'wāt ˌfəŋk·shən }

Weingarten formulas Equations concerning the normals to a surface at a point. { 'wīnˌgart·ən ˌfor·myə·ləs }

Weingarten surface A surface such that either of the principal radii is uniquely determined by the other. { 'wīnˌgärt·ən ˌsər·fəs }

well-formed formula A finite sequence or string of symbols that is grammatically or syntactically correct for a given set of grammatical or syntactical rules. { 'wel ¦fȯrmd ˌfor·myə·lə }

well-ordered set A linearly ordered set where every subset has a least element. { 'wel ¦ȯr·dərd 'set }

well-ordering principle The proposition that every set can be endowed with an order so that it becomes a well-ordered set; this is equivalent to the axiom of choice. { 'wel ¦ȯr·dər·iŋ 'prin·sə·pəl }

well-posed problem A problem that has a unique solution which depends continuously on the initial data. { 'wel ¦pōzd 'präb·ləm }

Wheweel equation An equation which relates the arc length along a plane curve to the angle of inclination of the tangent to the curve. { 'wā‚wēl i‚kwā·zhən }

white stochastic process A stochastic process such that there is no correlation between any of its components at different times, including autocorrelations. { 'wīt stō'kas·tik 'prä‚səs }

Whitney sum A tangent bundle TX over a differentiable manifold X is a Whitney sum of continuous bundles A and B over X if for each x the fibers of A and B at x are complementary subspaces of the tangent space at x. { 'wit·nē ‚səm }

Whittaker differential equation A special form of Gauss' hypergeometric equation with solutions as special cases of the confluent hypergeometric series. { 'wid·ə·kər ‚dif·ə¦ren·chəl i'kwā·zhən }

Wiener-Hopf equations Integral equations arising in the study of random walks and harmonic analysis; they are

$$g(x) = \int_0^\infty K(\,|\,x - t\,|\,)\,f(t)\,dt$$

$$f(x) = \int_0^\infty K(\,|\,x + t\,|\,)\,f(t)\,dt + g(x)$$

where g and K are known functions on the positive real numbers and f is the unknown function. { 'vē·nər 'hȯpf i‚kwā·zhənz }

Wiener-Hopf technique A method used in solving certain integral equations, boundary-value problems, and other problems, which involves writing a function that is holomorphic in a vertical strip of the complex z plane as the product of two functions, one of which is holomorphic both in the strip and everywhere to the right of the strip, while the other is holomorphic in the strip and everywhere to the left of the strip. { ¦vēn·ər 'hȯpf ‚tek‚nēk }

Wiener-Khintchine theorem The theorem that determines the form of the correlation function of a given stationary stochastic process. { 'vē·nər kin'chēn ‚thir·əm }

Wiener process A stochastic process with normal density at each stage, arising from the study of Brownian motion, which represents the limit of a sequence of experiments. Also known as Gaussian noise. { 'vē·nər ‚prä·səs }

Wilson's theorem The number $(n - 1)! + 1$ is divisible by n if and only if n is a prime. { 'wil·sənz ‚thir·əm }

winding number The number of times a given closed curve winds in the counterclockwise direction about a designated point in the plane. { 'wīnd·iŋ ‚nəm·bər }

witch of Agnesi The curve, symmetric about the y axis and asymptotic in both directions to the x axis, given by $x^2y = 4a^2(2a - y)$. Also known as versiera. { 'wich əv än'nyā·zē }

Witt-Grothendieck group The Grothendieck group of the monoid consisting of isometry classes of nondegenerate symmetric forms on vector spaces over a given field, where the product of two such forms is given by their orthogonal sum. { ¦wit ′grōt·ən‚dēk ‚grüp }

Witt group The group of isometry classes of symmetric forms on vector spaces over a given field, where the product of two such forms is given by their orthogonal sum. { ′wit ‚grüp }

Witt's theorem If F and F' are subspaces of a vector space E with a nondegenerate, symmetric form g, then an isometry of g from F onto F' can be extended to an isometry of g from E onto itself. { ′wits ‚thir·əm }

Wronskian An $n \times n$ matrix whose ith row is a list of the $(i - 1)$st derivatives of a set of functions $f_1, \ldots, f_n$; ordinarily used to determine linear independence of solutions of linear homogeneous differential equations. { ′vrän·skē·ən }

x axis 1. A horizontal axis in a system of rectangular coordinates. **2.** That line on which distances to the right or left (east or west) of the reference line are marked, especially on a map, chart, or graph. { 'eks ˌak·səs }

x coordinate One of the coordinates of a point in a two- or three-dimensional cartesian coordinate system, equal to the directed distance of a point from the y axis in a two-dimensional system, or from the plane of the y and z axes in a three-dimensional system, measured along a line parallel to the x axis. { 'eks kō'ȯrd·ən·ət }

Y

y axis **1.** A vertical axis in a system of rectangular coordinates. **2.** That line on which distances above or below (north or south) the reference line are marked, especially on a map, chart, or graph. { 'wī ˌak·səs }

y coordinate One of the coordinates of a point in a two- or three-dimensional coordinate system, equal to the directed distance of a point from the x axis in a two dimensional system, or from the plane of the x and z axes in a three-dimensional coordinate system, measured along a line parallel to the y axis. { 'wī kōˌȯrd·ən·ət }

Young's inequality An inequality that applies to a function $y = f(x)$ that is continuous and strictly increasing for $x \geqq 0$ and satisfies $f(0) = 0$, with inverse function $x = g(y)$; it states that, for any positive numbers a and b in the ranges of x and y, respectively, the product ab is equal to or less than the sum of the integral from 0 to a of $f(x)dx$ and the integral from 0 to b of $g(y)dy$. { 'yəŋz ˌin·ə'kwäl·əd·ē }

Z

z axis One of the three axes in a three-dimensional cartesian coordinate system; in a rectangular coordinate system it is perpendicular to the x and y axes. { 'zē ˌak·səs }

z coordinate One of the coordinates of a point in a three-dimensional coordinate system, equal to the directed distance of a point from the plane of the x and y axes, measured along a line parallel to the z axis. { 'zē kōˌȯrd·ən·ət }

Zeno's paradox An erroneous group of paradoxes dealing with motion; the most famous one concerns two objects, one chasing the other which has a given head start, where the chasing one moves faster yet seemingly never catches the other. { 'zē·nōz 'par·əˌdäks }

zero **1.** The additive identity element of an algebraic system. **2.** Any point where a given function assumes the value zero. { 'zir·ō }

zero divisor *See* divisor of zero. { ¦zir·ō di'vīz·ər }

zero-sum game A two-person game where the sum of the payoffs to the two players is zero for each move. { 'zir·ō ¦səm ˌgām }

zero vector The element 0 of a vector space such that, for any vector v in the space, the vector sum of 0 and v is v. { 'zir·ō ˌvek·tər }

zonal harmonics Spherical harmonics which do not depend on the azimuthal angle; they are proportional to Legendre polynomials of cos θ, where θ is the colatitude. { 'zōn·əl här'män·iks }

zone The portion of a sphere lying between two parallel planes that intersect the sphere. { zōn }

Zorn's lemma If every linearly ordered subset of a partially ordered set has a maximal element in the set, then the set has a maximal element. { 'zȯrnz 'lem·ə }

z-transform The z-transform of a sequence whose general term is f_n is the sum of a series whose general term is $f_n z^{-n}$, where z is a complex variable; n runs over the positive integers for a one-sided transform, over all the integers for a two-sided transform. { 'zē 'tranzˌfȯrm }

Appendix

Appendix

Equivalents of commonly used units for the U.S. Customary System and the metric system

1 inch = 2.5 centimeters (25 millimeters) 1 foot = 0.3 meter (30 centimeters) 1 yard = 0.9 meter 1 mile = 1.6 kilometers	1 centimeter = 0.4 inch 1 meter = 3.3 feet 1 meter = 1.1 yards 1 kilometer = 0.6 mile	1 inch = 0.08 foot 1 foot = 0.3 yard (12 inches) 1 yard = 3 feet (36 inches) 1 mile = 5280 feet (1760 yards)
1 acre = 0.4 hectare 1 acre = 4047 square meters	1 hectare = 2.47 acres 1 square meter = 0.0002 acre	
1 gallon = 3.8 liters 1 fluid ounce = 29.6 milliliters 32 fluid ounces = 946.4 milliliters	1 liter = 0.26 gallon 1 milliliter = 0.03 fluid ounce 1 liter = 1.1 quarts (0.3 gallon)	1 quart = 0.25 gallon (32 ounces; 2 pints) 1 pint = 0.125 gallon (16 ounces) 1 gallon = 4 quarts (8 pints)
1 quart = 0.9 liter 1 ounce = 28.4 grams 1 pound = 0.5 kilogram 1 ton = 907.18 kilograms	750 milliliters = 25.36 fluid ounces 1 gram = 0.04 ounce 1 kilogram = 2.2 pounds 1 kilogram = 1.1 × 10^{-3} ton	1 ounce = 0.6 pound 1 pound = 16 ounces 1 ton = 2000 pounds
°F = (1.8 × °C) + 32	°C = (°F − 32) ÷ 1.8	

Appendix

Conversion factors for the U.S. Customary System, metric system, and International System

A. UNITS OF LENGTH

Units	cm	m	in	ft	yd	mi
1 cm	= 1	0.01*	0.39	0.033	0.01	6.21×10^{-6}
1 m	= 100.	1	39.37	3.28	1.09	6.21×10^{-4}
1 in	= 2.54	0.03	1	0.08...	0.03...	1.58×10^{-5}
1 ft	= 30.48	0.30	12.	1	0.33...	$1.89... \times 10^{-4}$
1 yd	= 91.44	0.91	36.	3.	1	$5.68... \times 10^{-4}$
1 mile	= 1.61×10^{5}	1.61×10^{3}	6.34×10^{4}	5280.	1760.	1

B. UNITS OF AREA

Units	cm^2	m^2	in^2	ft^2	yd^2	mi^2
1 cm^2	= 1	10^{-4}	0.16	1.08×10^{-3}	1.20×10^{-4}	3.86×10^{-11}
1 m^2	= 10^{4}	1	1550.00	10.76	1.30	3.86×10^{-7}
1 in^2	= 6.45	6.45×10^{-4}	1	6.94×10^{-3}...	7.72×10^{-4}	2.49×10^{-10}
1 ft^2	= 929.03	0.09	1.44.	1	0.11...	3.59×10^{-8}
1 yd^2	= 8361.27	0.84	1296.	9.	1	3.23×10^{-7}
1 mi^2	= 2.59×10^{10}	2.59×10^{6}	4.01×10^{9}	2.79×10^{7}	3.10×10^{6}	1

C. UNITS OF VOLUME

Units	m^3	cm^3	liter	in^3	ft^3	qt	gal
1 m^3	= 1	10^6	10^3	6.10×10^4	35.31	1.057×10^3	264.17
1 cm^3	= 10^{-6}	1	10^{-3}	0.061	3.53×10^{-5}	1.057×10^{-3}	2.64×10^{-4}
1 liter	= 10^{-3}	1000.	1	61.02374	0.03531467	1.056688	0.26
1 in^3	= 1.64×10^{-5}	16.39	0.02	1	5.79×10^{-4}	0.02	4.33×10^{-3}
1 ft^3	= 2.83×10^{-2}	28316.85	28.32	1728.	1	2.99	7.48
1 qt	= 9.46×10^{-4}	946.35	0.95	57.75	0.03	1	0.25
1 gal (U.S.)	= 3.79×10^{-3}	3785.41	3.79	231.	0.13	4.	1

D. UNITS OF MASS

Units	g	kg	oz	lb	metric ton	ton
1 g	= 1	10^{-3}	0.04	2.20×10^{-3}	10^{-6}	1.10×10^{-6}
1 kg	= 1000.	1	35.27	2.20	10^{-3}	1.10×10^{-3}
1 oz (avdp)	= 28.35	0.028	1	0.06	2.83×10^{-5}	$5. \times 10^{-4}$
1 lb (avdp)	= 453.59	0.45	16.	1	4.54×10^{-4}	0.0005
1 metric ton	= 10^6	1000.	35273.96	2204.62	1	1.10
1 ton	= 907184.7	907.18	32000.	2000.	0.91	1

Conversion factors for the U.S. Customary System, metric system, and International System (cont.)

E. UNITS OF DENSITY

Units	$g \cdot cm^{-3}$	$g \cdot L^{-1}$, $kg \cdot m^{-3}$	$oz \cdot in^{-3}$	$lb \cdot in^{-3}$	$lb \cdot ft^{-3}$	$lb \cdot gal^{-1}$
1 $g \cdot cm^{-3}$	= 1	1000.	0.58	0.036	62.43	8.35
1 $g \cdot L^{-1}$, $kg \cdot m^{-3}$	= 10^{-3}	1	5.78×10^{-4}	3.61×10^{-5}	0.06	8.35×10^{-3}
1 $oz \cdot in^{-3}$	= 1.729994	1730	1	0.06	108.	14.44
1 $lb \cdot in^{-3}$	= 27.68	27679.91	16.	1	1728.	231.
1 $lb \cdot ft^{-3}$	= 0.02	16.02	9.26×10^{-3}	5.79×10^{-4}	1	0.13
1 $lb \cdot gal^{-1}$	= 0.12	119.83	4.75×10^{-3}	4.33×10^{-3}	7.48	1

F. UNITS OF PRESSURE

Units	Pa, $N \cdot m^{-2}$	$dyn \cdot cm^{-2}$	bar	atm	kg (wt) $\cdot cm^{-2}$	mmHg (torr)	in Hg	lb (wt) $\cdot in^{-2}$
1 Pa, 1 $N \cdot m^{-2}$	= 1	10	10^{-5}	9.87×10^{-6}	1.02×10^{-5}	7.50×10^{-3}	2.95×10^{-4}	1.45×10^{-4}
1 $dyn \cdot cm^{-2}$	= 0.1	1	10^{-6}	9.87×10^{-7}	1.02×10^{-6}	7.50×10^{-4}	2.95×10^{-5}	1.45×10^{-5}
1 bar	= 10^{5}	10^{6}	1	0.99	1.02	750.06	29.53	14.50
1 atm	= 101325.0	1013250.	1.01	1	1.03	760.	29.92	14.70
1 kg (wt) $\cdot cm^{-2}$	= 98066.5	980665.	0.98	0.97	1	735.56	28.96	14.22
1 mmHg (torr)	= 133.32	1333.22	1.33×10^{-3}	1.32×10^{-3}	1.36×10^{-3}	1	0.04	0.02
1 in Hg	= 3386.39	33863.88	0.03	0.03	0.03	25.4	1	0.49
1 lb (wt) $\cdot in^{-2}$	= 6894.76	68947.57	0.07	0.07	0.07	51.71	2.04	1

G. UNITS OF ENERGY

Units	g mass	J	int J	cal	cal_{IT}	Btu_{IT}	kWh	hp h	ft-lb (wt)	cu ft-lb (wt) in^2	liter-atm
1 g mass	= 1	8.99×10^{13}	8.99×10^{13}	2.15×10^{13}	2.15×10^{13}	8.52×10^{10}	2.50×10^{7}	3.35×10^{7}	6.63×10^{13}	4.60×10^{11}	8.87×10^{11}
1 J	$= 1.11 \times 10^{-14}$	1	1.00	0.24	0.24	9.48×10^{-4}	$2.78... \times 10^{-7}$	3.73	0.74	5.12×10^{-3}	9.87×10^{-3}
1 int J	$= 1.11 \times 10^{-14}$	1.00	1	0.24	0.24	9.48×10^{-4}	2.78×10^{-7}	3.73×10^{-7}	0.74	5.12×10^{-3}	9.87×10^{-3}
1 cal	$= 4.66 \times 10^{-14}$	4.18	4.18	1	1.00	3.97×10^{-3}	$1.16... \times 10^{-6}$	1.56×10^{-6}	3.09	2.14×10^{-2}	0.04
1 cal_{IT}	$= 4.66 \times 10^{-14}$	4.19	4.19	1.00	1	3.97×10^{-3}	1.16×10^{-6}	1.56×10^{-6}	3.09	2.14×10^{-2}	0.04
1 Btu_{IT}	$= 1.17 \times 10^{-11}$	1055.06	1054.88	252.16	252	1	2.93×10^{-4}	3.93×10^{-4}	778.17	5.40	10.41
1 kWh	$= 4.01 \times 10^{-8}$	3600000.	3599406.	860420.7	859845.2	3412.14	1	1.34	2655224.	18439.06	35529.24
1 hp h	$= 2.99 \times 10^{-8}$	2684519.	2684077.	641615.6	641186.5	2544.33	0.75	1	1980000.	13750.	26494.15
1 ft-lb (wt)	$= 1.51 \times 10^{-14}$	1.36	1.36	0.32	0.32	1.29×10^{-3}	3.77×10^{-7}	$5.05... \times 10^{-7}$	1	$6.94... \times 10^{-3}$	0.01
1 cu ft-lb (wt) in^2	$= 2.17 \times 10^{-12}$	195.24	195.21	46.66	46.63	0.19	5.42×10^{-5}	$7.27... \times 10^{-5}$	144.	1	1.93
1 liter-atm	$= 1.13 \times 10^{-12}$	101.33	101.31	24.22	24.20	0.10	2.81×10^{-5}	3.77×10^{-5}	74.73	0.52	1

Appendix

Mathematical notation, with definitions

Signs and symbols

$+$	Plus (sign of addition)
$+$	Positive
$-$	Minus (sign of subtraction)
$-$	Negative
$\pm$ ($\mp$)	Plus or minus (minus or plus)
$\times$	Times, by (multiplication sign)
$\cdot$	Multiplied by
$\div$	Sign of division
$/$	Divided by
$:$	Ratio sign, divided by, is to
$::$	Equals, as (proportion)
$<$	Less than
$>$	Greater than
$\ll$	Much less than
$\gg$	Much greater than
$=$	Equals
$\equiv$	Identical with
$\sim$	Similar to
$\approx$	Approximately equals
$\cong$	Approximately equals, congruent
$\leq$	Equal to or less than
$\geq$	Equal to or greater than
$\ne$ $\neq$	Not equal to
$\rightarrow$ $\doteq$	Approaches
$\propto$	Varies as
∞	Infinity
$\sqrt{\ }$	Square root of
$\sqrt[3]{\ }$	Cube root of
$\therefore$	Therefore
$\parallel$	Parallel to
( ) [ ] { }	Parentheses, brackets and braces; quantities enclosed by them to be taken together in multiplying, dividing, etc.
$\overline{AB}$	Length of line from A to B
π	(pi) = 3.14159+
$^\circ$	Degrees
$'$	Minutes

Mathematical notation, with definitions (cont.)

Signs and symbols (cont.)

$''$	Seconds
$\angle$	Angle
dx	Differential of x
Δ	(delta) difference
Δx	Increment of x
$\partial u/\partial x$	Partial derivative of u with respect to x
$\int$	Integral of
$\int_b^a$	Integral of, between limits a and b
$\oint$	Line integral around a closed path
Σ	(sigma) summation of
$f(x), F(x)$	Functions of x
∇	Del or nabla, vetor differential operator
∇^2	Laplacian operator
$\mathfrak{L}$	Laplace operational symbol
4!	Factorial $4 = 1 \times 2 \times 3 \times 4$
$\lvert x \rvert$	Absolute value of x
$\dot{x}$	First derivative of x with respect to time
$\ddot{x}$	Second derivative of x with respect to time
$\mathbf{A} \times \mathbf{B}$	Vector product; magnitude of **A** times magnitude of **B** times sine of the angle from **A** to **B**; $AB \sin \overline{AB}$
$\mathbf{A} \cdot \mathbf{B}$	Scalar product of **A** and **B**; magnitude of **A** times magnitude of **B** times cosine of the angle from **A** to **B**; $AB \cos \overline{AB}$

Mathematical logic

$p, q, P(x)$	Sentences, propositional functions, propositions
$-p, \sim p$, non p, Np	Negation, red "not p" ($\neq$: read "not equal")
$p \vee q, p + q, Apq$	Disjunction, read "p or q," "p, q," or both
$p \wedge q, p \cdot q, p\&q, Kpq$	Conjunction, read "p and q"
$p \rightarrow q, p \supset q, p \Rightarrow q$, Cpq	Implication, read "p implies q" or "if p then q"
$p \leftrightarrow q, p \equiv q, p \Leftrightarrow q$, Epq, p iff q	Equivalence, read "p is equivalent to q" or "p if and only if q"
n.a.s.c.	Read "necessary and sufficient condition"
( ), [ ], { }, . . ., · ·	Parentheses
$\forall$, $\forall$, Σ	Universal quantifier, read "for all" or "for every"
$\exists$, $\exists$, Π	Existential quantifier, read "there is a" or "there exists"
$\vdash$	Assertion sign ($p \vdash q$: read "q follows from p"; $\vdash p$: read "p is or follows from an axiom," or "p is a tautology"

Appendix

Mathematical notation, with definitions (cont.)

Mathematical logic (cont.)

Notation	Definition
0, 1	Truth, falsity (values)
$=$	Identity
$\overset{\text{Df}}{=}$, $\overset{\text{df}}{=}$, $=_{\text{df}}$, $=$	Definitional identity
■	"End of proof"; "QED"

Set theory, relations, functions

Notation	Definition
X, Y	Sets
$x \in X$	x is a member of the set X
$x \notin X$	x is not a member of X
$A \subset X$, $A \subseteq X$	Set A is contained in set X
$A \not\subset X$, $A \not\subseteq X$	A is not contained in X
$X \cup Y$, $X + Y$	Union of sets X and Y
$X \cap Y$, $X \cdot Y$	Intersecton of sets X and Y
$+$, $\dot{+}$, $\bigcirc$	Symmetric difference of sets
$\cup X_i$, ΣX_i	Union of all the sets X_i
$\cap X_i$, ΠX_i	Intersection of all the sets X_i
$\emptyset$, 0, Λ	Null set, empty set
X', **C**X, CX	Complement of the set X
$X - Y$, $X \backslash Y$	Difference of sets X and Y
$\hat{x}(P(x))$, $\{x \mid P(x)\}$, $\{x : P(x)\}$	The set of all x with the property P
(x,y,z), $\langle x,y,z \rangle$	Ordered set of elements x, y, and z; to be distinguished from (x,z,y), for example
$\{x,y,z\}$	Unordered set, the set whose elements are x, y, z, and no others
$\{a_1, a_2, \ldots, a_n\}$, $\{a_1\}_{i=1,2,\ldots,n}$, $\{a_1\}_{i=1}^{n}$	The set whose members are a_i, where i is any whole number from 1 to n
$\{a_1, a_2, \ldots\}$, $\{a_1\}_{i=1,2,\ldots}$, $\{a_1\}_{i=1}^{\infty}$	The set whose members are a_i, where i is any positive number
$X \times Y$	Cartesian product, set of all (x,y) such that $x \in X$, $y \in Y$
$\{a_i\}_{i \in I}$	The set whose elements are a_i, where $i \in I$
xRy, $R\{x,y\}$	Relation
$\equiv$, $\cong$, $\sim$, $\simeq$	Equivalence relations, for example, congruence
$\geqq$, $\geq$, $\succ$, $\geqslant$, $\leqq$, $\leq$, $<$	Transitive relations, for example, numerical order
$f: X \rightarrow Y$, $X \xrightarrow{f} Y$, $X \rightarrow Y$, $f \in Y^x$	Function, mapping, transformation
f^{-1}, $\overset{-1}{f}$, $X \xleftarrow{f-1} Y$	Inverse mapping
$g \circ f$	Composite functions: $(g \circ f)(x) = g(f(x))$

Mathematical notation, with definitions (cont.)

	Set theory, relations, functions (cont.)
$f(X)$	Image of X by f
$f^{-1}(X)$	Inverse-image set, counter image
1-1, one-one	Read "one-to-one correspondence"
$\begin{array}{ccc} X & \xrightarrow{f} & Y \\ \phi\downarrow & & \downarrow\psi \\ W & \xrightarrow{g} & Z \end{array}$	Diagram: the diagram is commutative in case $\psi \circ f = g \circ \phi$
$f \mid A$	Partial mapping, restriction of function f to set A
$\overline{\overline{X}}$, card X, $\lvert X \rvert$	Cardinal of the set A
$\aleph_0$, d	Denumerable infinity
$\mathfrak{c}$, c, $2^{\aleph_0}$	Power of continuum
ω	Order type of the set of positive integers
σ-	Read "countably"
	Number, numerical functions
1.4; 1,4; 1·4	Read "one and four-tenths"
1(1)20(10)100	Read "from 1 to 20 in intervals of 1, and from 20 to 100 in intervals of 10"
const	Constant
$A \geqq 0$	The number A is nonnegative, or, the matrix A is positive definite, or, the matrix A has nonnegative entries
$x \mid y$	Read "x divides y"
$x \equiv y \bmod p$	Read "x congruent to y modulo p"
$a_0 + \frac{1}{a_1 +} \frac{1}{a_2 +} \cdots,$ $a_0 + \left.\frac{1 \rfloor}{\lfloor a_1}\right. + \cdots$	Continued fractions
$[a,b]$	Closed interval
$[a,b)$, $[a,b[$	Half-open interval (open at the right)
(a,b), $]a,b[$	Open interval
$[a,\infty)$, $[a,\rightarrow[$	Interval closed at the left, infinite to the right
$(-\infty, \infty)$, $]\leftarrow,\rightarrow[$	Set of all real numbers
$\max_{x \in X} f(x)$, $\max \{f(x) \mid x \in X\}$	Maximum of $f(x)$ when x is in the set X
min	Minimum
sup, l.u.b.	Supremum, least upper bound
inf, g.l.b.	Infimum, greatest lower bound
$\lim_{x \to a} f(x) = b$, $\lim_{x=a} f(x) = b$, $f(x) \to b$ as $x \to a$	b is the limit of $f(x)$ as x approaches a

Appendix

Mathematical notation, with definitions (cont.)

	Number, numerical functions (cont.)
$\lim_{x\to a-} f(x)$, $\lim_{x=a-0} f(x)$, $f(a-)$	Limit of $f(x)$ as x approaches a from the left
lim sup, $\overline{\lim}$	Limit superior
lim inf, $\underline{\lim}$	Limit inferior
l.i.m.	Limit in the mean
$z = x + iy = re^{i\theta}$, $\zeta = \xi + i\eta$, $w = u + iv = \rho e^{i\phi}$	Complex variables
z^*	Complex conjugate
Re, $\Re$	Real part
Im, $\Im$	Imaginary part
arg	Argument
$\frac{\partial(u,v)}{\partial(x,y)}, \frac{D(u,v)}{D(x,y)}$	Jacobian, functional determinant
$\int_E f(x)\, d\mu(x)$	Integral (for example, Lebesgue integral) of function f over set E with respect to measure μ
$f(n) \sim \log n$ as $n \to \infty$	$f(n)/\log n$ approaches 1 as $n \to \infty$
$f(n) = O(\log n)$ as $n \to \infty$	$f(n)/\log n$ is bounded as $n \to \infty$
$f(n) = o(\log n)$	$f(n)/\log n$ approaches zero
$f(x) \nearrow b$, $f(x) \uparrow b$	$f(x)$ increases, approaching the limit b
$f(x) \downarrow b$, $f(x) \searrow b$	$f(x)$ decreases, approaching the limit b
a.e., p.p.	Almost everywhere
ess sup	Essential supremum
C^0, $C^0(X)$, $C(X)$	Space of continuous functions
C^k, $C^k[a,b]$	The class of functions having continuous kth derivative (on $[a,b]$)
C'	Same as C^1
Lip_α, Lip α	Lipschitz class of functions
L^p, L_p, $L^p[a,b]$	Space of functions having integrable absolute pth power (on $[a,b]$)
L'	Same as L^1
(C,α), (C,p)	Cesàro summability
	Special functions
$[x]$	The integral part of x
$\binom{n}{k}$, ${}^nC_k$, ${}_nC_k$	Binomial coefficient $n!/k!(n-k)!$
$\left(\frac{n}{p}\right)$	Legendre symbol
e^x, $\exp x$	Exponential function

Mathematical notation, with definitions (cont.)

	Special functions (cont.)
sinh x, cosh x, tanh x	Hyperbolic functions
sn x, cn x, dn x	Jacobi elliptic functions
$\wp(x)$	Weierstrass elliptic function
$\Gamma(x)$	Gamma function
$J_v(x)$	Bessel function
$\chi_X(x)$	Characteristic function of the set X: $\chi_X(x) = 1$ in case $x \in X$, otherwise $\chi_X(x) = 0$
sgn x	Signum: sgn $0 = 0$, while sgn $x = x/\lvert x \rvert$ for $x \neq 0$
$\delta(x)$	Dirac delta function
	Algebra, tensors, operators
$+, \cdot, \times, \circ, \mathsf{T}, \tau$	Laws of composition in albegraic systems
$e, 0$	Identity, unit, neutral element (of an additive system)
$e, 1, I$	Identity, unit neutral element (of a general albegraic system)
$e, \mathfrak{e}, E, P$	Idempotent
a^{-1}	Inverse of a
Hom(M,N)	Group of all homomorphisms of M into N
G/H	Factor group, group of cosets
$[K:k]$	Dimension of K over k
$\oplus, \dot{+}$	Direct sum
$\otimes$	Tensor product, Kronecker product
$\wedge$	Exterior product, Grassmann product
$\overrightarrow{x}$, $\mathbf{x}$, $\mathfrak{x}$, x	Vector
$\overrightarrow{x} \cdot \overrightarrow{y}$, $\mathbf{x} \cdot \mathbf{y}$, $(\mathfrak{x},\mathfrak{y})$	Inner product, scalar product, dot product
$\mathbf{x} \times \mathbf{y}$, $[\mathfrak{x},\mathfrak{y}]$, $\mathbf{x} \wedge \mathbf{y}$	Outer product, vector product, cross product
$\lvert x \rvert$, $\lvert x \rvert$, $\lVert x \rVert$, $\lVert x \rVert_p$	Norm of the vector x
Ax, xA	The image of x under the transformation A
δ_{ij}	Kronecker delta: $\delta_{ii} = 1$, while $\delta_{ij} = 0$ for $i \neq j$
A', tA, A^t, ${}^tA$	Transpose of the matrix A
A^*, $\tilde{A}$	Adjoint, Hermitian conjugate of A
tr A, Sp A	Trace of the matrix A
det A, $\lvert A \rvert$	Determinant of the matrix A
$\Delta^n f(x)$, $\Delta_h{}^n f$, $\underset{h}{\Delta^n} f(x)$	Finite differences
$[x_0,x_1]$, $[x_0,x_1,x_2]$, $\underset{x_1}{\Delta} u_{x_0}$, $[x_0,x_1]_f$	Divided differences
∇f, grad f	Read "gradient of f"
$\nabla \cdot \mathbf{v}$, div $\mathbf{v}$	Read "divergence of $\mathbf{v}$"
$\nabla \times \mathbf{v}$, curl $\mathbf{v}$, rot $\mathbf{v}$	Read "curl of $\mathbf{v}$"

Mathematical notation, with definitions (cont.)

	Algebra, tensors, operators (cont.)
∇^2, Δ, div grad	Laplacian
$[X,Y]$	Poisson bracket, or commutator, or Lie product
$GL(n,R)$	Full linear group of degree n over field R
$O(n,R)$	Full orthogonal group
$SO(n,R)$, $O^+(n,R)$	Special orthogonal group
	Topology
E^n	Euclidean n space
S^n	n sphere
$\rho(p,q)$, $d(p,q)$	Metric, distance (between points p and q)
$\bar{X}$, X^-, cl X, X^c	Closure of the set X
FrX. fr$X$, ∂X, bdry X	Frontier, boundary of X
int X, $\mathring{X}$	Interior of X
T_2 space	Hausdorff space
F_σ	Union of countably many closed sets
G_δ	Intersection of countably many open sets
dim X	Dimensionality, dimension of X
$\pi_1(X)$	Fundamental group of the space X
$\pi_n(X)$, $\pi_n(X,A)$	Homotopy groups
$H_n(X)$, $H_n(X,A;G)$, $H_*(X)$	Homology groups
$H^n(X)$, $H^n(X,A;G)$, $H^*(X)$	Cohomology groups
	Probability and statistics
X, Y	Random variables
$P(X \leqq 2)$, $\Pr\{X \leqq 2\}$	Probability that $X \leqq 2$
$P(X \leqq 2 \mid Y \geqq 1)$	Conditional probability
$E(X)$, $\mathrm{E}(X)$	Expectation of X
$E(X \mid Y \geqq 1)$	Conditional expectation
c.d.f.	Cumulative distribution function
p.d.f.	Probability density function
c.f.	Characteristic function
$\bar{x}$	Mean (especially, sample mean)
σ, s.d.	Standard deviation
σ^2, Var, var	Variance
$\mu_1, \mu_2, \mu_3, \mu_i, \mu_{ij}$	Moments of a distribution
ρ	Coefficient of correlation
$\rho_{12\cdot34}$	Partial correlation coefficient

Appendix

Symbols commonly used in geometry

Example	*Meaning*	*Comments*
$\overline{AB}$	Line segment having end points A and B	Could be named $\overline{BA}$
AB	Length of $\overline{AB}$	$AB = 5$ means $\overline{AB}$ is 5 units long
$\overleftrightarrow{XY}$	Line containing points X and Y	Could be named $\overleftrightarrow{YX}$
$\overrightarrow{PQ}$	Ray with end point P and containing Q	
$\angle V$	Angle with vertex V	Use only if no more than two rays have end point V
$\angle RST$	Angle formed by $\overrightarrow{SR}$ and $\overrightarrow{ST}$	Could be named $\angle TSR$
$\angle x$	Angle named x	
$\angle x = 30°$	Angle named x has measure 30°	
JK	Minor arc with end points J and K	
JLK	Major arc that contains point L	
$\odot C$	Circle with center C	
$\triangle XYZ$	Triangle with vertices X, Y, and Z	Could be named $\triangle YZX$, $\triangle XZY$, and so forth

Example	*Meaning*	*Comments*
$\square ABCD$	Rectangle with vertices A, B, C, and D	Adjacent letters must be adjacent vertices
$\square PQRS$	Square with vertices P, Q, R, and S	P and R are not adjacent vertices
$▱ABCD$	Parallelogram with vertices A, B, C, and D	
$ABCD \ldots X$	Polygon with vertices A, B, . . . , X	Adjacent letters must be adjacent vertices
$\overline{AB} \perp j$	A segment ($\overline{AB}$) is perpendicular to a line (j)	
$\overrightarrow{ZP} \parallel \overline{TV}$	A ray ($\overrightarrow{ZP}$) is parallel to a segment ($\overline{TV}$)	
$\overline{AB} \cong \overline{XY}$	Two segments ($\overline{AB}$ and $\overline{XY}$) are congruent	Congruent segments have equal lengths
$\triangle ABC \sim \triangle XYZ$	Two triangles ($\triangle ABC$ and $\triangle XYZ$) are similar	

Appendix

Common logarithm table, giving log (*a* + *b*)

a *b*:	.00	.01	.02	.03	.04	.05	.06	.07	.08	.09
1.0	.0000	.0043	.0086	.0128	.0170	.0212	.0253	.0294	.0334	.0374
1.1	.0414	.0453	.0492	.0531	.0569	.0607	.0645	.0682	.0719	.0755
1.2	.0792	.0828	.0864	.0899	.0934	.0969	.1004	.1038	.1072	.1106
1.3	.1139	.1173	.1206	.1239	.1271	.1303	.1335	.1367	.1399	.1430
1.4	.1461	.1492	.1523	.1553	.1584	.1614	.1644	.1673	.1703	.1732
1.5	.1761	.1790	.1818	.1847	.1875	.1903	.1931	.1959	.1987	.2014
1.6	.2041	.2068	.2095	.2122	.2148	.2175	.2201	.2227	.2253	.2279
1.7	.2304	.2330	.2355	.2380	.2405	.2430	.2455	.2480	.2504	.2529
1.8	.2553	.2577	.2601	.2625	.2648	.2672	.2695	.2718	.2742	.2765
1.9	.2788	.2810	.2833	.2856	.2878	.2900	.2923	.2945	.2967	.2989
2.0	.3010	.3032	.3054	.3075	.3096	.3118	.3139	.3160	.3181	.3201
2.1	.3222	.3243	.3263	.3284	.3304	.3324	.3345	.3365	.3385	.3404
2.2	.3424	.3444	.3464	.3483	.3502	.3522	.3541	.3560	.3579	.3598
2.3	.3617	.3636	.3655	.3674	.3692	.3711	.3729	.3747	.3766	.3784
2.4	.3802	.3820	.3838	.3856	.3874	.3892	.3909	.3927	.3945	.3962
2.5	.3979	.3997	.4014	.4031	.4048	.4065	.4082	.4099	.4116	.4133
2.6	.4150	.4166	.4183	.4200	.4216	.4232	.4249	.4265	.4281	.4298
2.7	.4314	.4330	.4346	.4362	.4378	.4393	.4409	.4425	.4440	.4456
2.8	.4472	.4487	.4502	.4518	.4533	.4548	.4564	.4579	.4594	.4609
2.9	.4624	.4639	.4654	.4669	.4683	.4698	.4713	.4728	.4742	.4757
3.0	.4771	.4786	.4800	.4814	.4829	.4843	.4857	.4871	.4886	.4900
3.1	.4914	.4928	.4942	.4955	.4969	.4983	.4997	.5011	.5024	.5038
3.2	.5052	.5065	.5079	.5092	.5105	.5119	.5132	.5145	.5159	.5172
3.3	.5185	.5198	.5211	.5224	.5237	.5250	.5263	.5276	.5289	.5302
3.4	.5315	.5328	.5340	.5253	.5366	.5378	.5391	.5403	.5416	.5428
3.5	.5441	.5453	.5465	.5478	.5490	.5502	.5515	.5527	.5539	.5551
3.6	.5563	.5575	.5587	.5599	.5611	.5623	.5635	.5647	.5658	.5670
3.7	.5682	.5694	.5705	.5717	.5729	.5740	.5752	.5763	.5775	.5786
3.8	.5798	.5809	.5821	.5832	.5843	.5855	.5866	.5877	.5888	.5899
3.9	.5911	.5922	.5933	.5944	.5955	.5966	.5977	.5988	.5999	.6010
4.0	.6021	.6031	.6042	.6053	.6064	.6075	.6085	.6096	.6107	.6117
4.1	.6128	.6138	.6149	.6160	.6170	.6180	.6191	.6201	.6212	.6222
4.2	.6232	.6243	.6253	.6263	.6274	.6284	.6294	.6304	.6314	.6325
4.3	.6335	.6345	.6355	.6365	.6375	.6385	.6395	.6405	.6415	.6425
4.4	.6435	.6444	.6454	.6464	.6474	.6484	.6493	.6503	.6513	.6522
4.5	.6532	.6542	.6551	.6561	.6571	.6580	.6590	.6599	.6609	.6618
4.6	.6628	.6637	.6646	.6656	.6665	.6675	.6684	.6693	.6702	.6712
4.7	.6721	.6730	.6739	.6749	.6758	.6767	.6776	.6785	.6794	.6803
4.8	.6812	.6821	.6830	.6839	.6848	.6857	.6866	.6875	.6884	.6893
4.9	.6902	.6911	.6920	.6928	.6937	.6946	.6955	.6964	.6972	.6981
5.0	.6990	.6998	.7007	.7016	.7024	.7033	.7042	.7050	.7059	.7067
5.1	.7076	.7084	.7093	.7101	.7110	.7118	.7126	.7135	.7143	.7152
5.2	.7160	.7168	.7177	.7185	.7193	.7202	.7210	.7218	.7226	.7235
5.3	.7243	.7251	.7259	.7267	.7275	.7284	.7292	.7300	.7308	.7316
5.4	.7324	.7332	.7340	.7348	.7356	.7364	.7372	.7380	.7388	.7396

Common logarithm table, giving log (*a* + *b*) (cont.)

a *b*:	.00	.01	.02	.03	.04	.05	.06	.07	.08	.09
5.5	.7404	.7412	.7419	.7427	.7435	.7443	.7451	.7459	.7466	.7474
5.6	.7482	.7490	.7497	.7505	.7513	.7520	.7528	.7536	.7543	.7551
5.7	.7559	.7566	.7574	.7582	.7589	.7597	.7604	.7612	.7619	.7627
5.8	.7634	.7642	.7649	.7657	.7664	.7672	.7679	.7686	.7694	.7701
5.9	.7709	.7716	.7723	.7731	.7738	.7745	.7752	.7760	.7767	.7774
6.0	.7782	.7789	.7796	.7803	.7810	.7818	.7825	.7832	.7839	.7846
6.1	.7853	.7860	.7868	.7875	.7882	.7889	.7896	.7903	.7910	.7917
6.2	.7924	.7931	.7938	.7945	.7952	.7959	.7966	.7973	.7980	.7987
6.3	.7993	.8000	.8007	.8014	.8021	.8028	.8035	.8041	.8048	.8055
6.4	.8062	.8069	.8075	.8082	.8089	.8096	.8102	.8109	.8116	.8122
6.5	.8129	.8136	.8142	.8149	.8156	.8162	.8169	.8176	.8182	.8189
6.6	.8195	.8202	.8209	.8215	.8222	.8228	.8235	.8241	.8248	.8254
6.7	.8261	.8267	.8274	.8280	.8287	.8293	.8299	.8306	.8312	.8319
6.8	.8325	.8331	.8338	.8344	.8351	.8357	.8363	.8370	.8376	.8382
6.9	.8388	.8395	.8401	.8407	.8414	.8420	.8426	.8432	.8439	.8445
7.0	.8451	.8457	.8463	.8470	.8476	.8482	.8488	.8494	.8500	.8506
7.1	.8513	.8519	.8525	.8531	.8537	.8543	.8549	.8555	.8561	.8567
7.2	.8573	.8579	.8585	.8591	.8597	.8603	.8609	.8615	.8621	.8627
7.3	.8633	.8639	.8645	.8651	.8657	.8663	.8669	.8675	.8681	.8686
7.4	.8692	.8698	.8704	.8710	.8716	.8722	.8727	.8733	.8739	.8745
7.5	.8751	.8756	.8762	.8768	.8774	.8779	.8785	.8791	.8797	.8802
7.6	.8808	.8814	.8820	.8825	.8831	.8837	.8842	.8848	.8854	.8859
7.7	.8865	.8871	.8876	.8882	.8887	.8893	.8899	.8904	.8910	.8915
7.8	.8921	.8927	.8932	.8938	.8943	.8949	.8954	.8960	.8965	.8971
7.9	.8976	.8982	.8987	.8993	.8998	.9004	.9009	.9015	.9020	.9025
8.0	.9031	.9036	.9042	.9047	.9053	.9058	.9063	.9069	.9074	.9079
8.1	.9085	.9090	.9096	.9101	.9106	.9112	.9117	.9122	.9128	.9133
8.2	.9138	.9143	.9149	.9154	.9159	.9165	.9170	.9175	.9180	.9186
8.3	.9191	.9196	.9201	.9206	.9212	.9217	.9222	.9227	.9232	.9238
8.4	.9243	.9248	.9253	.9258	.9263	.9269	.9274	.9279	.9284	.9289
8.5	.9294	.9299	.9304	.9309	.9315	.9320	.9325	.9330	.9335	.9340
8.6	.9345	.9350	.9355	.9360	.9365	.9370	.9375	.9380	.9385	.9390
8.7	.9395	.9400	.9405	.9410	.9415	.9420	.9425	.9430	.9435	.9440
8.8	.9445	.9450	.9455	.9460	.9465	.9469	.9474	.9479	.9484	.9489
8.9	.9494	.9499	.9504	.9509	.9513	.9518	.9523	.9528	.9533	.9538
9.0	.9542	.9547	.9552	.9557	.9562	.9566	.9571	.9576	.9581	.9586
9.1	.9590	.9595	.9600	.9605	.9609	.9614	.9619	.9624	.9628	.9633
9.2	.9638	.9643	.9647	.9652	.9657	.9661	.9666	.9671	.9675	.9680
9.3	.9685	.9689	.9694	.9699	.9703	.9708	.9713	.9717	.9722	.9727
9.4	.9731	.9736	.9741	.9745	.9750	.9754	.9759	.9764	.9768	.9773
9.5	.9777	.9782	.9786	.9791	.9795	.9800	.9805	.9809	.9814	.9818
9.6	.9823	.9827	.9832	.9836	.9841	.9845	.9850	.9854	.9859	.9863
9.7	.9868	.9872	.9877	.9881	.9886	.9890	.9894	.9899	.9903	.9908
9.8	.9912	.9917	.9921	.9926	.9930	.9934	.9939	.9943	.9948	.9952
9.9	.9956	.9961	.9965	.9969	.9974	.9978	.9983	.9987	.9991	.9996

Appendix

Values of trigonometric functions*

Degrees	*Radians*	*Sin*	*Csc*	*Tan*	*Cot*	*Sec*	*Cos*		
0° 0′	.0000	.0000	—	.0000	—	1.000	1.0000	1.5708	**90° 0′**
10′	029	029	343.8	029	343.8	000	000	679	50′
20′	058	058	171.9	058	171.9	000	000	650	40′
30′	.0087	.0087	114.6	.0087	114.5	1.000	1.0000	1.5621	**30′**
40′	116	116	85.95	116	85.94	000	.9999	592	20′
50′	145	145	68.76	145	68.75	000	999	563	10′
1° 0′	.0175	0.175	57.30	.0175	57.29	1.000	.9998	1.5533	**89° 0′**
10′	204	204	49.11	204	49.10	000	998	504	50′
20′	233	233	42.98	233	42.96	000	997	475	40′
30′	.0262	.0262	38.20	.0262	38.19	1.000	.9997	1.5446	**30′**
40′	291	291	34.38	291	34.37	000	996	417	20′
50′	320	320	31.26	320	31.24	001	995	388	10′
2° 0′	.0349	.0349	28.65	.0349	28.64	1.001	.9994	1.5359	**88° 0′**
10′	378	378	26.45	378	26.43	001	993	330	50′
20′	407	407	24.56	407	24.54	001	992	301	40′
30′	.0436	.0436	22.93	.0437	22.90	1.001	.9990	1.5272	**30′**
40′	465	465	21.49	466	21.47	001	989	243	20′
50′	495	495	20.23	495	20.21	001	988	213	10′
3° 0′	.0524	.0523	19.11	.0524	19.08	1.001	.9986	1.5184	**87° 0′**
10′	553	552	18.10	553	18.07	002	985	155	50′
20′	582	581	17.20	582	17.17	002	983	126	40′
30′	.0611	.0610	16.38	.0612	16.35	1.002	.9981	1.5097	**30′**
40′	640	640	15.64	641	15.60	002	980	068	20′
50′	669	669	14.96	670	14.92	002	978	039	10′
4° 0′	.0698	.0698	14.34	.0699	14.30	1.002	.9976	1.5010	**86° 0′**
10′	727	727	13.76	729	13.73	003	974	981	50′
20′	756	756	13.23	758	13.20	003	971	952	40′
30′	.0785	.0785	12.75	.0787	12.71	1.003	.9969	1.4923	**30′**
40′	814	814	12.29	816	12.25	003	967	893	20′
50′	844	843	11.87	846	11.83	004	964	864	10′

5° 0′	.0873	.0872	11.47	.0875	11.43	1.004	.9962	1.4835	**85° 0′**
10′	902	901	11.10	904	11.06	004	959	806	50′
20′	931	929	10.76	934	10.71	004	957	777	40′
30′	.0960	.0958	10.43	.0963	10.39	1.005	.9954	1.4748	**30′**
40′	989	987	10.13	992	10.08	005	951	719	20′
50′	.1018	.1016	9.839	.1022	9.788	005	948	690	10′
6° 0′	.1047	.1045	9.567	.1051	9.514	1.006	.9945	1.4661	**84° 0′**
10′	076	074	9.309	080	9.255	006	942	632	50′
20′	105	103	9.065	110	9.010	006	939	603	40′
30′	.1134	.1132	8.834	.1139	8.777	1.006	.9936	1.4573	**30′**
40′	164	161	8.614	169	8.556	007	932	544	20′
50′	193	190	8.405	198	8.345	007	929	515	10′
7° 0′	.1222	.1219	8.206	.1228	8.144	1.008	.9925	1.4486	**83° 0′**
10′	251	248	8.016	257	7.953	008	922	457	50′
20′	280	276	7.834	287	7.770	008	918	428	40′
30′	.1309	.1305	7.661	.1317	7.596	1.009	.9914	1.4399	**30′**
40′	338	334	7.496	346	7.429	009	911	370	20′
50′	367	363	7.337	376	7.269	009	907	341	10′
8° 0′	.1396	.1392	7.185	.1405	7.115	1.010	.9903	1.4312	**82° 0′**
10′	425	421	7.040	435	6.968	010	899	283	50′
20′	454	449	6.900	465	6.827	011	894	254	40′
30′	.1484	.1478	6.765	.1495	6.691	1.011	.9890	1.4224	**30′**
40′	513	507	6.636	524	6.561	012	886	195	20′
50′	542	536	6.512	554	6.435	012	881	166	10′
9° 0′	.1571	.1564	6.392	.1584	6.314	1.012	.9877	1.4137	**81° 0′**
		Cos	*Sec*	*Cot*	*Tan*	*Csc*	*Sin*	*Radians*	*Degrees*

*From G. E. F. Sherwood and A. E. Taylor, *Calculus*, 3d ed., Prentice-Hall, 1954.

Values of trigonometric functions (cont.)

Degrees	Radians	Sin	Csc	Tan	Cot	Sec	Cos		
9° 0′	.1571	.1564	6.392	.1584	6.314	1.012	.9877	1.4137	**81° 0′**
10′	600	593	277	614	197	013	872	108	50′
20′	629	622	166	644	084	013	868	079	40′
30′	.1658	.1650	6.059	.1673	5.976	1.014	.9863	1.4050	**30′**
40′	687	679	5.955	703	871	014	858	1.4021	20′
50′	716	708	855	733	769	015	853	992	10′
10° 0′	.1745	.1736	5.759	.1763	5.671	1.015	.9848	1.3963	**80° 0′**
10′	774	765	665	793	576	016	843	934	50′
20′	804	794	575	823	485	016	838	904	40′
30′	.1833	.1822	5.487	.1853	5.396	1.017	.9833	1.3875	**30′**
40′	862	851	403	883	309	018	827	846	20′
50′	891	880	320	914	226	018	822	817	10′
11° 0′	.1920	.1908	5.241	.1944	5.145	1.019	.9816	1.3788	**79° 0′**
10′	949	937	164	974	066	019	811	759	50′
20′	978	965	089	.2004	4.989	020	805	730	40′
30′	.2007	.1994	5.016	.2035	4.915	1.020	.9799	1.3701	**30′**
40′	036	.2022	4.945	065	843	021	793	672	20′
50′	065	051	876	095	773	022	787	643	10′
12° 0′	.2094	.2079	4.810	.2126	4.705	1.022	.9781	1.3614	**78° 0′**
10′	123	108	745	156	638	023	775	584	50′
20′	153	136	682	186	574	024	769	555	40′
30′	.2182	.2164	4.620	.2217	4.511	1.024	.9763	1.3526	**30′**
40′	211	193	560	247	449	025	757	497	20′
50′	240	221	502	278	390	026	750	468	10′
13° 0′	.2269	.2250	4.445	.2309	4.331	1.026	.9744	1.34329	**77° 0′**
10′	298	278	390	339	275	027	737	410	50′
20′	327	306	336	370	219	028	730	381	40′
30′	.2356	.2334	4.284	.2401	4.165	1.028	.9724	1.3352	**30′**
40′	385	363	232	432	113	029	717	323	20′
50′	414	391	182	462	061	030	710	294	10′

14° 0′	.2443	.2419	4.134	.2493	4.011	1.031	.9703	1.3265	**76° 0′**
10′	473	447	086	524	3.962	031	696	235	50′
20′	502	476	039	555	914	032	689	206	40′
30′	.2531	.2504	3.994	.2586	3.867	1.033	.9681	1.3177	**30′**
40′	560	532	950	617	821	034	674	148	20′
50′	589	560	906	648	776	034	667	119	10′
15° 0′	.2618	.2588	3.864	.2679	3.732	1.035	.9659	1.3090	**75° 0′**
10′	647	616	822	711	689	036	652	061	50′
20′	676	644	782	742	647	037	644	032	40′
30′	.2705	.2672	3.742	.2773	3.606	1.038	.9636	1.3003	**30′**
40′	734	700	703	805	566	039	628	974	20′
50′	763	728	665	836	526	039	621	945	10′
16° 0′	.2793	.2756	3.628	.2867	3.487	1.040	.9613	1.2915	**74° 0′**
10′	822	784	592	899	450	041	605	886	50′
20′	851	812	556	931	412	042	596	857	40′
30′	.2880	.2840	3.521	.2962	3.376	1.043	.9588	1.2828	**30′**
40′	909	868	487	994	340	044	580	799	20′
50′	938	896	453	.3026	305	045	572	770	10′
17° 0′	.2967	.2924	3.420	.3057	3.271	1.046	.9563	1.2741	**73° 0′**
10′	996	952	388	089	237	047	555	712	50′
20′	.3025	979	357	121	204	048	546	683	40′
30′	.3054	.3007	3.326	.3153	3.172	1.048	.9537	1.2654	**30′**
40′	083	035	295	185	140	049	528	625	20′
50′	113	062	265	217	108	050	520	595	10′
18° 0′	.3142	.3090	3.236	.3249	3.078	1.051	.9511	1.2566	**72° 0′**
		Cos	*Sec*	*Cot*	*Tan*	*Csc*	*Sin*	*Radians*	*Degrees*

Values of trigonometric functions (cont.)

Degrees	*Radians*	*Sin*	*Csc*	*Tan*	*Cot*	*Sec*	*Cos*		
18° 0′	.3142	.3090	3.236	.3249	3.078	1.051	.9511	1.2566	**72° 0′**
10′	171	118	207	281	047	052	502	537	50′
20′	200	145	179	314	018	053	492	508	40′
30′	.3229	.3173	3.152	.3346	2.989	1.054	.9483	1.2479	**30′**
40′	258	201	124	378	960	056	474	450	20′
50′	287	228	098	411	932	057	465	421	10′
19° 0′	.3316	.3256	3.072	.3443	2.904	1.058	.9455	1.2392	**71° 0′**
10′	345	283	046	476	877	059	446	363	50′
20′	374	311	021	508	850	060	436	334	40′
30′	.3403	.3338	2.996	.3541	2.824	1.061	.9426	1.2305	**30′**
40′	432	365	971	574	798	062	417	275	20′
50′	462	393	947	607	773	063	407	246	10′
20° 0′	.3491	.3420	2.924	.3640	2.747	1.064	.9397	1.2217	**70° 0′**
10′	520	448	901	673	723	065	387	188	50′
20′	549	475	878	706	699	066	377	159	40′
30′	.3578	.3502	2.855	.3739	2.675	1.068	.9367	1.2130	**30′**
40′	607	529	833	772	651	069	356	101	20′
50′	636	557	812	805	628	070	346	072	10′
21° 0′	.3665	.3584	2.790	.3839	2.605	1.071	.9336	1.2043	**69° 0′**
10′	694	611	769	872	583	072	325	1.2014	50′
20′	723	638	749	906	560	074	315	985	40′
30′	.3752	.3665	2.729	.3939	2.539	1.075	.9304	1.1956	**30′**
40′	782	692	709	973	517	076	293	926	20′
50′	811	719	689	.4006	496	077	283	6897	10′
22° 0′	.3840	.3746	2.669	.4040	2.475	1.079	.9272	1.1868	**68° 0′**
10′	869	773	650	074	455	080	261	839	50′
20′	898	800	632	108	434	081	250	810	40′
30′	.3927	.3827	2.613	.4142	2.414	1.082	.9239	1.1781	**30′**
40′	956	854	595	176	394	084	228	752	20′
50′	985	881	577	210	375	085	216	723	10′

23° 0′	.4014	.3907	2.559	.4245	2.356	1.086	.9205	1.1694	**67° 0′**
10′	043	934	542	279	337	088	194	665	50′
20′	072	961	525	314	318	089	182	636	40′
30′	.4102	.3987	2.508	.4348	2.300	1.090	.9171	1.1606	**30′**
40′	131	.4014	491	383	282	092	159	577	20′
50′	160	041	475	417	264	093	147	548	10′
24° 0′	.4189	.4067	2.459	.4452	2.246	1.095	.9135	1.1519	**66° 0′**
10′	218	094	443	487	229	096	124	490	50′
20′	247	120	427	522	211	097	112	461	40′
30′	.4276	.4147	2.411	.4557	2.194	1.099	.9100	1.1432	**30′**
40′	305	173	396	592	177	100	088	403	20′
50′	334	200	381	628	161	102	075	374	10′
25° 0′	.4363	.4226	2.366	.4663	2.145	1.103	.9063	1.1345	**65° 0′**
10′	392	253	352	699	128	105	051	316	50′
20′	422	279	337	734	112	106	038	286	40′
30′	.4451	.4305	2.323	.4770	2.097	1.108	.9026	1.1257	**30′**
40′	480	331	309	806	081	109	013	228	20′
50′	509	358	295	841	066	111	001	199	10′
26° 0′	.4538	.4384	2.281	.4877	2.050	1.113	.8988	1.1170	**64° 0′**
10′	567	410	268	913	035	114	975	141	50′
20′	596	436	254	950	020	116	962	112	40′
30′	.4625	.4462	2.241	.4986	2.006	1.117	.8949	1.1083	**30′**
40′	654	488	228	.5022	1.991	119	936	054	20′
50′	683	514	215	059	977	121	923	1.1025	10′
27° 0′	.4712	.4540	2.203	.5095	1.963	1.122	.8910	1.0996	**63° 0′**
		Cos	*Sec*	*Cot*	*Tan*	*Csc*	*Sin*	*Radians*	*Degrees*

Values of trigonometric functions (cont.)

Degrees	*Radians*	*Sin*	*Csc*	*Tan*	*Cot*	*Sec*	*Cos*		
27° 0′	.4712	.4540	2.203	.5095	1.963	1.122	.8910	1.0996	**63° 0′**
10′	741	566	190	132	949	124	897	966	50′
20′	771	592	178	169	935	126	884	937	40′
30′	.4800	.4617	2.166	.5206	1.921	1.127	.8870	1.0908	**30′**
40′	829	643	154	243	907	129	857	879	20′
50′	858	669	142	280	894	131	843	850	10′
28° 0′	.4887	.4695	2.130	.5317	1.881	1.133	.8829	1.0821	**62° 0′**
10′	916	720	118	354	868	134	816	792	50′
20′	945	746	107	392	855	136	802	763	40′
30′	.4974	.4772	2.096	.5430	1.842	1.138	.8788	1.0734	**30′**
40′	.5003	797	085	467	829	140	774	705	20′
50′	032	823	074	505	816	142	760	676	10′
29° 0′	.5061	.4848	2.063	.5543	1.804	1.143	.8746	1.0647	**61° 0′**
10′	091	874	052	581	792	145	732	617	50′
20′	120	899	041	619	780	147	718	588	40′
30′	.5149	.4924	2.031	.5658	1.767	1.149	.8704	1.0559	**30′**
40′	178	950	020	696	756	151	689	530	20′
50′	207	975	010	735	744	153	675	501	10′
30° 0′	.5236	.5000	2.000	.5744	1.732	1.155	.8660	1.0472	**60° 0′**
10′	265	025	1.990	812	720	157	646	443	50′
20′	294	050	980	851	709	159	631	414	40′
30′	.5323	.5075	1.970	.5890	1.698	1.161	.8616	1.0385	**30′**
40′	352	100	961	930	686	163	601	356	20′
50′	381	125	951	969	675	165	587	327	10′
31° 0′	.5411	.5150	1.942	.6009	1.664	1.167	.8572	1.0297	**59° 0′**
10′	440	175	932	048	653	169	557	268	50′
20′	469	200	923	088	643	171	542	239	40′
30′	.5498	.5225	1.914	.6128	1.632	1.173	.8526	1.0210	**30′**
40′	527	250	905	168	621	175	511	181	20′
50′	556	275	896	208	611	177	496	152	10′

32° 0′	.5585	.5299	1.887	.6249	1.600	1.179	.8480	1.0123	**58° 0′**
10′	614	324	878	289	590	181	465	094	50′
20′	643	348	870	330	580	184	450	065	40′
30′	.5672	.5373	1.861	.6371	1.570	1.186	.8434	1.0036	**30′**
40′	701	398	853	412	560	188	418	1.0007	20′
50′	730	422	844	453	550	190	403	977	10′
33° 0′	.5760	.5446	1.836	.6494	1.540	1.192	.8387	.9948	**57° 0′**
10′	789	471	828	536	530	195	371	919	50′
20′	818	495	820	577	520	197	355	890	40′
30′	.5847	.5519	1.812	.6619	1.511	1.199	.8339	.9861	**30′**
40′	876	544	804	661	501	202	323	832	20′
50′	905	568	796	703	1.492	204	307	803	10′
34° 0′	.5934	.5592	1.788	.6745	1.483	1.206	.8290	.9774	**56° 0′**
10′	963	616	781	787	473	209	274	745	50′
20′	992	640	773	830	464	211	258	716	40′
30′	.6021	.5664	1.766	.6873	1.455	1.213	.8241	.9687	**30′**
40′	050	688	758	916	446	216	225	657	20′
50′	080	712	751	959	437	218	208	628	10′
35° 0′	.6109	.5736	1.743	.7002	1.428	1.221	.8192	.9599	**55° 0′**
10′	138	760	736	046	419	223	175	570	50′
20′	167	783	729	089	411	226	158	541	40′
30′	.6196	.5807	1.722	.7133	1.402	1.228	.8141	.9512	**30′**
40′	225	831	715	177	393	231	124	483	20′
50′	254	854	708	221	385	233	107	454	10′
36° 0′	.6283	.5878	1.701	.7265	1.376	1.236	.8090	.9425	**54° 0′**
		Cos	*Sec*	*Cot*	*Tan*	*Csc*	*Sin*	*Radians*	*Degrees*

Values of trigonometric functions (cont.)

Degrees	*Radians*	*Sin*	*Csc*	*Tan*	*Cot*	*Sec*	*Cos*		
36° 0′	.6283	.5878	1.701	.7265	1.376	1.236	.8090	.9425	**54° 0′**
10′	312	901	695	310	368	239	073	396	50′
20′	341	925	688	355	360	241	056	367	40′
30′	.6370	.5948	1.681	.7400	1.351	1.244	.8039	.9338	**30′**
40′	400	972	675	445	343	247	021	308	20′
50′	429	995	668	490	335	249	004	279	10′
37° 0′	.6458	.6018	1.662	.7536	1.327	1.252	.7986	.9250	**53° 0′**
10′	487	041	655	581	319	255	969	221	50′
20′	516	065	649	627	311	258	951	192	40′
30′	.6545	.6088	1.643	.7673	1.303	1.260	.7934	.9153	**30′**
40′	574	111	636	720	295	263	916	134	20′
50′	603	134	630	766	288	266	898	105	10′
38° 0′	.6632	.6157	1.624	.7813	1.280	1.269	.7880	.9076	**52° 0′**
10′	661	180	618	860	272	272	862	047	50′
20′	690	202	612	907	265	275	844	.9018	40′
30′	.6720	.6225	1.606	.7954	1.257	1.278	.7826	.8988	**30′**
40′	749	248	601	.8002	250	281	808	959	20′
50′	778	271	595	050	242	284	790	930	10′
39° 0′	.6807	.6293	1.589	.8098	1.235	1.287	.7771	.8901	**51° 0′**
10′	836	316	583	146	228	290	753	872	50′
20′	865	338	578	195	220	293	735	843	40′
30′	.6894	.6361	1.572	.8243	1.213	1.296	.7716	.8814	**30′**
40′	923	383	567	292	206	299	698	785	20′
50′	952	406	561	342	199	302	679	756	10′
40° 0′	.6981	.6428	1.556	.8391	1.192	1.305	.7660	.8727	**50° 0′**
10′	.7010	450	550	441	185	309	642	698	50′
20′	039	472	545	491	178	312	623	668	40′
30′	.7069	.6494	1.540	.8541	1.171	1.315	.7604	.8639	**30′**
40′	098	517	535	591	164	318	585	610	20′
50′	127	539	529	642	157	322	566	581	10′

41° 0′	.7156	.6561	1.524	.8693	1.150	1.325	.7547	.8552	**49° 0′**
10′	185	583	519	744	144	328	528	523	50′
20′	214	604	514	796	137	332	509	494	40′
30′	.7243	.6626	1.509	.8847	1.130	1.335	.7490	.8465	**30′**
40′	272	648	504	899	124	339	470	436	20′
50′	301	670	499	952	117	342	451	407	10′
42° 0′	.7330	.6691	1.494	.9004	1.111	1.346	.7431	.8378	**48° 0′**
10′	359	713	490	057	104	349	412	348	50′
20′	389	734	485	110	098	353	392	319	40′
30′	.7418	.6756	1.480	.9163	1.091	1.356	.7373	.8290	**30′**
40′	447	777	476	217	085	360	353	261	20′
50′	476	799	471	271	079	364	333	232	10′
43° 0′	.7505	.6820	1.466	.9325	1.072	1.367	.7314	.8203	**47° 0′**
10′	534	841	462	380	066	371	294	174	50′
20′	563	862	457	435	060	375	274	145	40′
30′	.7592	.6884	1.453	.9490	1.054	1.379	.7254	.8116	**30′**
40′	621	905	448	545	048	382	234	087	20′
50′	650	926	444	601	042	386	214	058	10′
44° 0′	.7679	.6947	1.440	.9657	1.036	1.390	.7193	.8029	**46° 0′**
10′	709	967	435	713	030	394	173	999	50′
20′	738	988	431	770	024	398	153	970	40′
30′	.7767	.7009	1.427	.9827	1.018	1.402	.7133	.7941	**30′**
40′	796	030	423	884	012	406	112	912	20′
50′	825	050	418	942	006	410	092	883	10′
45° 0′	.7854	.7071	1.414	1.000	1.000	1.414	.7071	.7854	**45° 0′**
		Cos	*Sec*	*Cot*	*Tan*	*Csc*	*Sin*	*Radians*	*Degrees*

Appendix

Compound amount: $(1 + r)n$*

n \ r	1%	$1\frac{1}{4}$%	$1\frac{1}{2}$%	2%	$2\frac{1}{2}$%	3%	4%	5%	6%
1	1.0100	1.0125	1.0150	1.0200	1.0250	1.0300	1.0400	1.0500	1.0600
2	1.0201	1.0252	1.0302	1.0404	1.0506	1.0609	1.0816	1.1025	1.1236
3	1.0303	1.0380	1.0457	1.0612	1.0769	1.0927	1.1249	1.1576	1.1910
4	1.0406	1.0509	1.0614	1.0824	1.1038	1.1255	1.1699	1.2155	1.2625
5	1.0510	1.0641	1.0773	1.1041	1.1314	1.1593	1.2167	1.2763	1.3382
6	1.0615	1.0774	1.0934	1.1262	1.1597	1.1941	1.2653	1.3401	1.4185
7	1.0721	1.0909	1.1098	1.1487	1.1887	1.2299	1.3159	1.4071	1.5036
8	1.0829	1.1045	1.1265	1.1717	1.2184	1.2668	1.3688	1.4775	1.5938
9	1.0937	1.1183	1.1434	1.1951	1.2489	1.3048	1.4233	1.5513	1.6895
10	1.1046	1.1323	1.1605	1.2190	1.2801	1.3439	1.4802	1.6289	1.7908
11	1.1157	1.1464	1.1779	1.2434	.13121	1.3842	1.5395	1.7103	1.8983
12	1.1268	1.1608	1.1956	1.2682	1.3449	1.4258	1.6010	1.7959	2.0122
13	1.1381	1.1753	1.2136	1.2936	1.3785	1.4685	1.6651	1.8856	2.1329
14	1.1495	1.1900	1.2318	1.3195	1.4130	1.5126	1.7317	1.9799	2.2609
15	1.1610	1.2048	1.2502	1.3459	1.4483	1.5580	1.8009	2.0789	2.3966
16	1.1726	1.2199	1.2690	1.3728	1.4845	1.6047	1.8730	2.1829	2.5404
17	1.1843	1.2351	1.2880	1.4002	1.5216	1.6528	1.9479	2.2920	2.6928
18	1.1961	1.2506	1.3073	1.4282	1.5597	1.7024	2.0258	2.4066	2.8543
19	1.2081	1.2662	1.3270	1.4568	1.5987	1.7535	2.1068	2.5270	3.0256
20	1.2202	1.2820	1.3469	1.4859	1.6386	1.8061	2.1911	2.6533	3.2071
21	1.2324	1.2981	1.3671	1.5157	1.6796	1.8603	2.2788	2.7860	3.3996
22	1.2447	1.3143	1.3876	1.5460	1.7216	1.9161	2.3699	2.9253	3.6035
23	1.2572	1.3307	1.4084	1.5769	1.7646	1.9736	2.4647	3.0715	3.8197
24	1.2697	1.3474	1.4295	1.6084	1.8087	2.0328	2.5633	3.2251	4.0489
25	1.2824	1.3642	1.4509	1.6406	1.8539	2.0938	2.6658	3.3864	4.2919
26	1.2953	1.3812	1.4727	1.6734	1.9003	2.1566	2.7725	3.5557	4.5494
27	1.3082	1.3985	1.4948	1.7069	1.9478	2.2213	2.8834	3.7335	4.8223
28	1.3213	1.4160	1.5172	1.7410	1.9965	2.2879	2.9987	3.9201	5.1117
29	1.3345	1.4337	1.5400	1.7758	2.0464	2.3566	3.1187	4.1161	5.4184
30	1.3478	1.4516	1.5631	1.8114	2.0976	2.4273	3.2434	4.3219	5.7435

n									
31	1.3613	1.4698	1.5865	1.8476	2.1500	2.5001	3.3731	4.5380	6.0881
32	1.3749	1.4881	1.6103	1.8845	2.2038	2.5751	3.5081	4.7649	6.4534
33	1.3887	1.5067	1.6345	1.9222	2.2589	2.6523	3.6484	5.0032	6.8406
34	1.4026	1.5256	1.6590	1.9607	2.3153	2.7319	3.7943	5.2533	7.2510
35	1.4166	1.5446	1.6839	1.9999	2.3732	2.8139	3.9461	5.5160	7.6861
36	1.4308	1.5639	1.7091	2.0399	2.4325	2.8983	4.1039	5.7918	8.1473
37	1.4451	1.5835	1.7348	2.0807	2.4933	2.9852	4.2681	6.0814	8.6361
38	1.4595	1.6033	1.7608	2.1223	2.5557	3.0748	4.4388	6.3855	9.1543
39	1.4741	1.6233	1.7872	2.1647	2.6196	3.1670	4.6164	6.7048	9.7035
40	1.4889	1.6436	1.8140	2.2080	2.6851	3.2620	4.8010	7.0400	10.2857
41	1.5038	1.6642	1.8412	2.2522	2.7522	3.3599	4.9931	7.3920	10.9029
42	1.5188	1.6850	1.8688	2.2972	2.8210	3.4607	5.1928	7.7616	11.5570
43	1.5340	1.7060	1.8969	2.3432	2.8915	3.5645	5.4005	8.1497	12.2505
44	1.5493	1.7274	1.9253	2.3901	2.9638	3.6715	5.6165	8.5572	12.9855
45	1.5648	1.7489	1.9542	2.4379	3.0379	3.7816	5.8412	8.9850	13.7646
46	1.5805	1.7708	1.9835	2.4866	3.1139	3.8950	6.0748	9.4343	14.5905
47	1.5963	1.7929	2.0133	2.5363	3.1917	4.0119	6.3178	9.9060	15.4659
48	1.6122	1.8154	2.0435	2.5871	3.2715	4.1323	6.5705	10.4013	16.3939
49	1.6283	1.8380	2.0741	2.6388	3.3533	4.2562	6.8333	10.9213	17.3775
50	1.6446	1.8610	2.1052	2.6916	3.4371	4.3839	7.1067	11.4674	18.4202

*If a principal P is deposited at interest rate r (in decimals) compounded annually, then at the end of n years the accumulated amount $A = P(1 + r)^n$.

SOURCE: Murray R. Spiegel, *Mathematical Handbook of Formulas and Tables*, Schaum's Outline Series, McGraw-Hill, 1968.

Appendix

Regular polytopes in n dimensions

Polytope	*Schläfli symbol*	*Vertices*	*Edges*	*Faces*	*Solid cells*	*Hypersolid cells*
$n = 2$						
p-gon	$\{p\}$	p	p			
$n = 3$						
Tetrahedron	{3,3}	4	6	4		
Cube	{4,3}	8	12	6		
Octahedron	{3,4}	6	12	8		
Dodecahedron	{5,3}	20	30	12		
Icosahedron	{3,5}	12	30	20		
$n = 4$						
5-cell	{3,3,3}	5	10	10	5	
8-cell	{4,3,3}	16	32	24	8	
16-cell	{3,3,4}	8	24	32	16	
24-cell	{3,4,3}	24	96	96	24	
120-cell	{5,3,3}	600	1200	720	120	
600-cell	{3,3,5}	120	720	1200	600	
$n > 4$						
Simplex	{3,3,...,3}	$n + 1$	$\frac{1}{2}n(n + 1)$	. . . .		$n + 1$
Hypercube	{4,3,...,3}	2^n	$2^{n-1}n$	. . . .		$2n$
Cross polytope	{3,...,3,4}	$2n$	$2n(n - 1)$	. . . .		2^n